Security of e-Systems and Computer Networks

e-Based systems and computer networks are ubiquitous in the modern world, with applications spanning e-commerce, WLANs, healthcare and governmental organizations, among others. The secure transfer of information has therefore become a critical area of research, development, and investment. This book presents the fundamental concepts, tools, and protocols of e-based system and computer network security and its wide range of applications.

The core areas of e-system and computer network security, such as authentication of users; system integrity; confidentiality of communication; availability of business service; and non-repudiation of transactions, are covered in detail. Throughout the book, the major trends, challenges, and applications of e-security are presented, with emphasis on public key infrastructure (PKI) systems, biometric-based security systems, trust management systems, and the e-service paradigm. Intrusion detection technologies, virtual private networks (VPNs), malware, WLANs security, and risk management are also discussed. Moreover, applications such as e-commerce, e-government, and e-service are all dealt with.

Technically oriented with many practical examples, this book is an invaluable reference for practitioners in network and information security. It will also be of great interest to researchers and graduate students in electrical and computer engineering, telecommunication engineering, computer science, informatics, and system and software engineering.

Moreover, the book can be used as a text for a graduate or senior undergraduate course in security of e-systems and computer networks, security of information systems, security of communication networks, or security of e-systems.

MOHAMMAD S. OBAIDAT, recognized around the world for his pioneering and lasting contributions to several areas including networks and information security, is a Professor of Computer Science at Monmouth University, New Jersey. He is the author of several books and numerous publications. He obtained his Ph.D. from the Ohio State University and has received numerous awards. He is a Fellow of the SCS and a Fellow of the IEEE.

NOUREDDINE A. BOUDRIGA received his Ph.D. in Mathematics from the University Paris XI, France and in Computer Science from the University Tunis II, Tunisia. He is currently based at the University of 7th November at Carthage in Tunisia, where he is a Professor of Telecommunications at the School of Communication Engineering, and Director of the Research Laboratory on Networks and Security.

Security of e-Systems and Computer Networks

MOHAMMAD S. OBAIDAT
Monmouth University, New Jersey

NOUREDDINE A. BOUDRIGA
University of 7th November at Carthage, Tunisia

CAMBRIDGE UNIVERSITY PRESS
Cambridge, New York, Melbourne, Madrid, Cape Town, Singapore, São Paulo

Cambridge University Press
The Edinburgh Building, Cambridge CB2 8RU, UK

Published in the United States of America by Cambridge University Press, New York

www.cambridge.org
Information on this title: www.cambridge.org/9780521837644

© Cambridge University Press 2007

First published 2007

Printed in the United Kingdom at the University Press, Cambridge

A catalog record for this publication is available from the British Library

ISBN 978-0-521-83764-4 hardback

To our families

Contents

Preface

Security of e-based systems and computer networks has become an important issue recently due to the increased dependence of organizations and people on such systems. The risk of accessing an e-commerce, or e-government system or Web site ranges from invasion of privacy and loss of money to exposing national security information and catastrophe. E-security solutions aim to provide five important services: authentication of users and actors, integrity, confidentiality of communication, availability of business services and non-repudiation of transactions. Most e-security solutions that are provided by the literature use two main cryptographic techniques: public key cryptosystems and digital signatures. Efficient solutions also should be compliant with the national legal framework.

There are multibillion dollars being invested in computer networks and e-systems; therefore, securing them is vital to their proper operation as well as to the future of the organizations and companies and national security. Due to the difficulties in securing the different platforms of e-systems, and the increasing demand for better security and cost-effective systems, the area of e-system and network security is an extremely rich field for research, development and investment. Security of e-systems provides in-depth coverage of the wide range of e-system security aspects including techniques, applications, trends, challenges, etc.

This book is the first book that is dedicated entirely to security of e-systems and networks. It consists of four main parts with a total of 14 chapters.

Chapter 1 describes the importance of system security and presents some relevant concepts in network security and subscribers' protection. It also introduces some basic terminology that is used throughout the book to define service, information, computer security and network security. Moreover, the chapter covers important related topics such as security costs, services, threats and vulnerabilities.

Chapter 2 discusses encryption and its practical applications. It focuses on the techniques used in public key cryptosystems. It also details various types of ciphers and their applications to provide the basic e-service solutions. It provides the reader with simple examples that explain how the main concepts and procedures work. Topics such as public key cryptosystems with emphasis on symmetric encryption, RSA and ElGamel algorithms, management of public key, life cycle, key distribution, and attacks against public key cryptosystems are all discussed in this chapter.

Chapter 3 covers the authentication of users and messages. It details the main schemes of digital signature and their applications. It also addresses the notions of hash function and key establishment. These notions are important because they constitute the hidden part of any protection process that uses public key-based systems. Topics such as weak and strong authentication schemes, attacks on authentication digital signature frameworks, hash functions and authentication applications and services are discussed.

Chapter 4 provides details on the public key infrastructure (PKI) systems covering aspects such as the PKI architecture model, management functions, public key certificates, trust hierarchical models, certification path processing and deployment of PKI. A particular emphasis is given to the definition of certificate generation, certificate verification and certificate revocation. Several other related issues are discussed including cross-certification, PKI operation, PKI assessment and PKI protection.

Chapter 5 introduces biometrics schemes as a way to secure systems. The various techniques of biometrics are reviewed and elaborated. Accuracy of biometrics schemes is analyzed and compared with each other. We also shed some light on the different issues and challenges of biometric systems.

Chapter 6 discusses trust management in communication networks. It covers topics such as trust definition as related to security, digital credentials including active credentials and SPKI. It also sheds some light on the authorization and access control systems, trust policies, and trust management applications such as clinical information systems, e-payment systems, and distributed firewalls.

The purpose of Chapter 7 is to examine the e-service paradigm, discuss the technical features it depicts and study the security challenges it brings forward. It also describes well established e-services and shows how they are composed and delivered. Other topics covered include the UDDI/SOAP/WSDL and ebXML initiatives, message protection mechanisms, and securing registry security.

Chapter 8 provides key support to service providers wishing to provide e-government services in a trusted manner. It lays the foundations for enabling secure services that will really transform the way citizens and businesses interact with government. The chapter covers topics such as e-government concepts and practices, authentication and privacy in e-government, e-voting security, engineering secured e-government, monitoring e-government security along with advanced issues such as response support system.

Chapter 9 discusses the e-commerce requirements and defines the major techniques used to provide and protect e-commerce. A special interest is given to the SSL, TLS and SET protocols. Electronic payment, m-commerce and transaction security with SET process are also addressed.

Chapter 10 reviews and investigates the security of wireless local area networks (WLANs). The major techniques and their advantages and drawbacks are presented. Moreover, the chief issues related to WLANs security are discussed. Attacks on WLANs, security services, Wired Equivalent Privacy (WEP) protocol and its features and drawbacks, Wi-Fi Protected Access (WPA) protocol and its advantages, mobile IP and Virtual Private Networks (VPNs) are all discussed in this chapter.

In Chapter 11, a global view is proposed to the reader through a presentation of the intrusion classification. Several approaches for the detection of malicious traffic and abnormal activities are addressed including pattern matching, signature-based, traffic-anomaly-based, heuristic-based, and protocol-anomaly-based analysis. A model is proposed to describe events, alerts, and correlation. It defines the fundamentals of most intrusion detection methodologies currently used by enterprises. A survey of the main concepts involved in the model is presented. The chapter also discusses the definition and role of the correlation function, detection techniques and advanced issues in intrusion detection systems.

Chapter 12 presents the basics and techniques of virtual private networks (VPNs). It also reviews VPN services that include Intranet, Extranet and Remote Access VPNs. Security concerns that arise when transmitting data over shared networks using VPNs technology are also addressed in detail. The protocols used in VPNs such as PPTP and L2TP as well as security aspects are discussed. The quality of service provision in VPNs is also reviewed.

Chapter 13 discusses malware definition and classification. It describes the ways that major classes of malware, such as viruses, worms, and Trojans, are built and propagated. It also describes the major protection measures that an enterprise needs to develop and presents a non-exhaustive set of guidelines to be followed to make the protection better. Other topics discussed in this chapter include firewall-based protection and invasion protection schemes, protection guidelines and polymorphism challenges.

Finally, Chapter 14 investigates the characteristics that a risk management framework should possess. It discusses the typical risk management approaches that have been proposed. The chapter highlights some of the structured methodologies that are developed based on a set of essential concepts including vulnerability analysis, threat analysis, risk analysis and control implementation. The chapter also stresses the limits and use of these approaches as well as the role of risk analysis and risk assessment techniques. Other topics covered in this chapter include management risk libraries, risk assessment, and schemes of monitoring the system state such as the pattern-based monitoring and the behavior-based monitoring.

The book will be an ideal reference to practitioners and researchers in information and e-security systems as well as a good textbook for graduate and senior undergraduate courses in information security, e-security, network security, information systems security and e-commerce and e-government security.

We would like to thank the reviewers of the original book proposal for their constructive suggestions. Also, we thank our students for some of the feedback that we received while trying the manuscript in class. Many thanks go to the editors and editorial assistants of Cambridge University Press for their cooperation and fine work.

Part I

E-security

Introduction to Part I

In enterprise systems, a security exposure is a form of possible damage in the organization's information and communication systems. Examples of exposures include unauthorized disclosure of information, modification of business or employees' data, and denial of legal access to the information system. A vulnerability is a weakness in the system that might be exploited by an adversary to cause loss or damage. An intruder is an adversary who exploits vulnerabilities, and commits security attacks on the information/production system.

Electronic security (e-security) is an important issue to businesses and governments today. E-security addresses the security of a company, locates its vulnerabilities, and supervises the mechanisms implemented to protect the on-line services provided by the company, in order to keep adversaries (hackers, malicious users, and intruders) from getting into the company's networks, computers, and services. E-service is a very closely related concept to e-privacy and it is sometimes hard to differentiate them from each other. E-privacy issues help tracking users or businesses and what they do on-line to access the enterprise's web sites.

Keeping the company's business secure should be a major priority in any company no matter how small or large is the business of the company, and no matter how open or closed the company network is. For this intent, a security policy should be set up within the company to include issues such as password usage rules, access control, data security mechanisms and business transaction protection. A set of good practices that should be followed by any company includes: (a) keep virus scanning software and reactive tools updated; (b) consider the use of stand-alone computers (i.e., computers that are not connected to the network) for sensitive data; (c) define appropriate trusted domains based on the organization of the business activity; (d) install malicious activity detection systems in conformance with the security policy; and (e) define a strong practice of e-mails management (particularly when they are issued from unknown sources and receiving files with known extensions).

E-security solutions aim to provide five important services. These services are: authentication of users and actors, integrity, confidentiality of communication, availability of business services, and non-repudiation of transactions. Most e-security solutions that are

1

provided by the literature use two main cryptographic techniques: public key cryptosystems and digital signatures. Efficient solutions also should be compliant with the national legal framework, if any.

The first part of the book aims at defining the main concepts used in e-security. It also describes the major techniques and challenging issues in a company's security system. It classifies security attacks and security services and develops the main issues. This part contains three chapters.

Chapter 1 describes the importance of e-security and presents some relevant concepts in network security and subscribers' protection. It also introduces some basic terminology that is used throughout the book to define service, information, computer security, and network security. This chapter aims to provide self-contained features for this book.

Chapter 2 discusses encryption and its practical applications. It focuses on the techniques used in public key cryptosystems. It also details various types of ciphers and their applications to provide the basic e-service solutions. It provides the reader with simple examples that explain how the main concepts and procedures work.

Chapter 3 covers the authentication of users and messages. It details the main schemes of digital signature and their applications. It also addresses the notions of hash function and key establishment. These notions are important because they constitute the hidden part of any protection process that uses public key-based systems.

1 Introduction to e-security

This chapter discusses the importance and role of e-security in business environments and networked systems. It presents some relevant concepts in network security and subscribers protection. It also introduces some basic terminology that is used throughout the book to define service, information, computer security, and network security. This chapter aims at providing self contained features to this book.

1.1 Introduction

Every organization, using networked computers and deploying an information system to perform its activity, faces the threat of hacking from individuals within the organization and from its outside. Employees (and former employees) with malicious intent can represent a threat to the organization's information system, its production system, and its communication networks. At the same time, reported attacks start to illustrate how pervasive the threats from outside hackers have become. Without proper and efficient protection, any part of any network can be prone to attacks or unauthorized activity. Routers, switches, and hosts can all be violated by professional hackers, company's competitors, or even internal employees. In fact, according to various studies, more than half of all network attacks are committed internally.

One may consider that the most reliable solution to ensure the protection of organizations' information systems is to refrain from connecting them to communication networks and keep them in secured locations. Such a solution could be an appropriate measure for highly sensitive systems. But it does not seem to be a very practical solution, since information systems are really useful for the organization's activity when they are connected to the network and legitimately accessed. Moreover, in today's competitive world, it is essential to do business electronically and be interconnected with the Internet. Being present on the Internet is a basic requirement for success.

Organizations face three types of economic impact as possible results of malicious attacks targeting them: the immediate, short-term, and midterm economic impacts. The immediate economic impact is the cost of repairing, modifying, or replacing systems (when needed) and the immediate losses due to disruption of business operations, transactions, and cash flows. Short-term economic impact is the cost on an organization, which includes the loss of contractual relationships or existing customers because of the inability to deliver products or services as well as the negative impact on the reputation of the organization. Long-term economic impact is induced by the decline in an organization's market appraisal.

During the last decade, enterprises, administrations, and other business structures have spent billions of dollars on expenditures related to network security, information protection, and loss of assets due to hackers' attacks. The rate at which these organizations are expending funds seems to be impressively increasing. This requires the business structures to build and plan efficient strategies to address these issues in a cost-effective manner. They also need to spend large amounts of money for security awareness and employees' training (Obaidat, 1993a; Obaidat, 1993b; Obaidat, 1994).

1.2   Security costs

Network attacks may cause organizations hours and days of system downtime and serious violations in data confidentiality, resource integrity, and client/employee privacy. Depending on the level of the attack and the type of information that has been compromised, the consequences of network attacks vary in degree from simple annoyance and inconvenience to complete devastation. The cost of recovery from attacks can range from hundreds to millions of dollars. Various studies including a long-running annual survey conducted by the Federal Bureau of Investigation (FBI, Los Angeles), and the American Computer Security Institute (CSI) have highlighted some interesting numbers related to these costs. The Australian computer crime and security survey has found similar findings (Gordon *et al.*, 2004; Aust, 2004). The surveys have mainly determined the expenditures from a large number of responses collected from individuals operating in the computer and network security of business organizations. The findings of the surveys are described in the following subsections to highlight the importance of security in business structures.

1.2.1   The CSI/FBI computer crime and security survey

Based on responses collected from about 500 information security practitioners in US enterprises, financial institutions, governmental agencies, university centres, and medical institutions, the conclusions of the *2005 Computer Crime and Security Survey* confirmed that the threat from computer hacking and other information security breaches continues to damage the information systems and resources in the surveyed organizations. It also confirmed that the financial cost of the privilege of using the information technologies is increasing for the tenth year. It reports also that, except for the abuse of wireless networks, all the categories of attacks of information systems have been slowly decreasing over many years. Major highlights of the *Survey* include:

- Virus attacks and denial of service attacks continue to be the major source of financial losses, while unauthorized accesses show an important cost increase.
- Over 87% of the surveyed organizations have conducted security audits during 2005 to assess the efficiency of their security solutions. Only 82% had conducted security audits in 2004.

- The majority of the organizations did not outsource system security activities.
- The average reported computer security operating system and investment per employee was high for firms with low annual sales and decreased for companies with very high annual sales.
- The large majority of the organizations have considered security awareness and training as an important task, although (on average) respondents from all sectors have declared that they do not believe that their organization invests enough in this area.

The survey has identified four areas of interest to measure the importance of security issues in conducting business and competing with other organizations. These areas are: (a) budgeting, (b) nature and cost of cyber-security breaches, (c) security audits and security awareness, and (d) information sharing. The overall findings of the survey can be summarized by the following issues:

Budgeting issues

These issues consider the costs associated with security breaches, financial aspects of information security management, and solutions implementation. Two major indicators have been considered. They are: (a) the percentage of the information security budget relative to the organization's overall IT budget and (b) the average computer security operating expense and investment per employee.

The 2004 survey has reported that 5.46% of the respondents have indicated that their organizations allocated between 1% and 5% of the total IT budget to security, while only 16% of the respondents have indicated that security received less than 1% of the IT budget. Moreover, 23% of respondents indicated that security received more than 5% of the budget, however, 14% of respondents indicated that the portion was unknown to them. The results of the survey also demonstrate that as an organization grows, computer security operating and capital expenditures grow less rapidly. This highlights the fact that there is economy of scale when it comes to information security.

On the other hand, the 2005 survey has shown another tendency. Firms with annual sales under $10 million spent an average of $643 per employee on operating expenses and investment in computer security. The largest firms have only spent an average of $325 per employee.

Spending per employee on information security, broken down by sector, shows a slightly different scenario compared to the results provided in the 2004 survey. The highest average computer security spending was reported by state governments. The next highest sectors are utilities, transportation, and telecommunications. The highest sectors reported in the 2004 survey were transportation, Federal government, and telecommunications. Two observations should be mentioned, however:

1. Securing state governments is a very hot issue nowadays.
2. Managers responsible for computer security are increasingly asked to justify their budget requests in economic terms (ROI, for example).

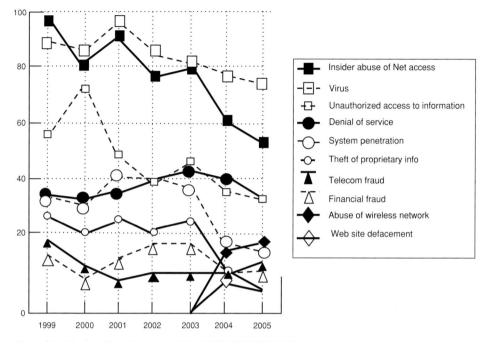

Figure 1.1   Types of attacks reported for 2005 (BFI/CSI 2005 survey).

Nature of attacks

Figure 1.1 depicts the percentage of respondents detecting attacks per type of attack. It shows that detected attacks and misuses have been slowly decreasing over the last years. Two categories, however, have shown an important increase: Web site defacement and the abuse of wireless networks.

The 2004 survey demonstrates that the denial of service category of attacks has emerged for the first time as the incident type generating the largest total losses (replacing theft of proprietary information, which had been the most expensive category of loss for the five previous years). It has shown also that the respondents have reported abuse of wireless networks (15% of the respondents), Web site defacement (7%) and misuse of public Web applications (10%).

The 2005 survey reports that the total losses were dramatically decreasing. But, beyond the overall decline, viruses, unauthorized accesses, and denial of services are generating 61% of financial losses. In addition, two areas of increase can be noticed, unauthorized access to information, where the average loss per respondent moved up from $51 000 to $300 000, and theft, where the average loss moved from $168 000 to $355 000.

Security audits and awareness

The 2004 survey found that 82% of respondents indicated that their organizations conducted security audits. This percentage has increased to 87% in the 2005 survey. In addition to security audits the surveys demonstrate that investing in security learning did not reach an acceptable level since on average, respondents from all sectors reported in the 2004 and 2005 surveys do not believe their organizations invest enough in security awareness. Respondents also share the following thoughts:

1. Security awareness training was perceived most valuable in the areas of security policy and security management (70%), followed by access control systems (64%), network security (63%), and cryptography (51%).
2. The two areas in which security awareness was perceived to be the least valuable were: security systems architecture, and investigations and legal issues.

Information sharing

The 2004 survey shows that only half of all respondents indicated that their organizations share information about security breaches and that more than 90% of respondents indicated that their organization responds by patching security holes. However, 57% of the respondents indicated that their organization does not belong to any incident/response information-sharing organization. Over 50% of respondents (among those indicating that their organization would not report an intrusion to law enforcement agencies) declared as very important the perception that the negative publicity would hurt their organization's stock and/or image. The findings of the 2005 survey are similar. This latter survey has shown that 46% of the respondents indicated that their organization does not belong to any information-sharing group.

1.2.2 *The Australian computer crime and security survey*

The Australian *Computer Crime and Security Survey* provides a unique insight into the information security operations of Australian organizations ranging from single person enterprises to large corporations. The results of the 2005 survey presents some similarities with the results reported by the *2004 Computer Crime and Security Survey* (Gordon *et al.*, 2004) and show the following key findings:

1. More respondent organizations have experienced electronic attacks that harmed the confidentiality, integrity, or availability of network data or systems compared to the previous year.
2. Infections from viruses, worms or Trojans were the most common form of electronic attacks reported by respondents for the consecutive year. They were the greatest cause of financial losses and accounted for 45% of total losses for 2004. However, denial of service attacks that have been reported by the 2005 survey were the greatest cause of financial losses.
3. The readiness of organizations to protect their IT systems has improved in three major areas: (a) the use of information security policies, practices, and procedures; (b) the use of information security standards; and (c) the number of organizations with experienced, trained, qualified or certified staff.

4. Unprotected software vulnerabilities and inadequate staff training and education in security practices were identified as the two most common factors which contributed to harmful electronic attacks.
5. The most common challenges and difficulties that respondent organizations faced were changing user attitudes and behavior and keeping up to date with information about the latest computer threats and vulnerabilities.

Therefore, the effort being made by responding organizations to improve their readiness to protect their systems appeared to be insufficient to cope with the changing nature of the threats and vulnerabilities. This includes the increased number and severity of system vulnerabilities as well as the growing number and rapid propagation of Internet worms and viruses.

Nature and impact of electronic attacks

The survey shows that 95% of respondents have experienced one or more security attacks, computer crime, computer access misuse, and abuse in the last 12 months. The most common incidents were virus, worm and Trojan infections (88% in 2004, compared to 80% in 2003 and 76% in 2002); insider abuse of Internet access, email or computer system resources (69% in 2004, compared to 62% in 2003 and 80% in 2002); and laptop theft (reported 58% in 2004, compared to 53% in 2003 and 74% in 2002).

Impact can be measured in direct and indirect costs, time to recover, and intangible impacts such as damage to an organization's credibility, trustworthiness, or reputation. The impact of electronic attacks, computer crime and computer access misuse ranges from negligible to grave, in both cost and time. Overall, the losses experienced by respondent organizations as a whole have got worse (20% higher than in 2003) with average of $116 212 for each organization that quantified its losses. By comparison, in 2003 the average loss was $93 657 and in 2002 it was $77 084.

The cost of computer crime

The survey ranges the cost based on a set of sixteen causes of loss that were incurred including: (a) virus, worm, and Trojan infections (54% of the total losses); (b) computer facilitated financial fraud (15% of the total losses); (c) degradation of network performance (11% of the total loss); (d) laptop theft; and (e) theft/violation of proprietary or confidential information. The survey, however, demonstrates that sabotage of data or communication networks, telecommunications fraud, denial of service attacks, system penetration by outsiders, and unauthorized access to information by insiders do not exceed 9% of the total annual losses.

For the vast majority of electronic attacks, computer crimes, computer access misuse incidents, recovery time was between one to seven days or less than a day. For respondents that estimated the time it took to recover from the most serious incident they had in each of the sixteen listed categories, 60% estimated that they recovered in less than a day; 74% estimated that recovery took between one to seven days; 28% estimated that recovery took between eight days to four weeks; 13% estimated that recovery took more than one month; and 5% experienced incidents which they assessed they may never recover from.

1.3   Security services

Information and network security risks are increasing tremendously with the growth of the number of threats and the sophistication of attacks. To cope with this growth, incidents of viruses, hackers, theft, and sabotage are being publicized more frequently and the enterprise management has started keeping an interest on the archives developed. A non-exhaustive list of simple security incidents contains, but is not limited to, the following examples of security breaches (Stallings, 2001):

- A user, named Us_A, transmits a file to another user named Us_B. The file containing sensitive information (e.g., financial transaction and private data) should be protected from disclosure. User Us_H, who is not authorized to access the file content, is able to capture a copy of the file during its transmission on the network, if the file is not well protected.

- A network manager, named Man_D, transmits a message to a networked computer, called Com_E, to update an authorization file and include the identities of new users who are to be granted access to Com_E. User Us_H, who is an adversary, intercepts the message, alters its content, and forwards it to Com_E. Computer Com_E accepts the message as coming from Man_D and updates its authorizations accordingly.

- A message is sent from a customer Us_A to a stockbroker Us_B with instructions to execute various transactions. User Us_H may intercept the message and get a copy of it. Subsequently, Us_H sends several copies of the message inducing a multiple execution of the initial transaction, generating by this way financial losses to Us_A.

- An employee is discharged without warning. The personnel manager sends a message to a server system to invalidate the employee's account. When the invalidation is accomplished, the server posts a notice to the employee's file as confirmation of the action. The employee is able to intercept the message and delay it long enough to make a final access to the server to retrieve sensitive information.

As information systems become essential to conduct business, electronic information takes care of many of the roles traditionally performed by paper documents. According to this consideration, functions that are traditionally associated with paper documents must be performed on documents that exist in electronic form. To assess effectively the security needs of an organization and evaluate or select security products and policies, the manager responsible for the organization's security may need a systematic way of defining the requirements for security and characterizing the approaches to satisfy those requirements. For this, three aspects of information security need to be considered:

- **Security attacks** A security attack is defined to be any action that compromises (or attempts to compromise) the security of the information system (or information resources) owned by an organization, an employee, or a customer.

- **Security mechanisms** These are mechanisms (procedures, applications, or devices) that are designed to detect, prevent, or recover from security attacks.
- **Security services** These are services that enhance the security of the data processing and the information transfers of an organization. The services are intended to counter security attacks and assumed to make use of one or more security mechanisms.

1.3.1   Security services

Several aspects of electronic documents make the provision of security functions or services challenging (Stallings, 2001). These services include but are not limited to the following:

Authentication

The authentication service aims at assuring that a communication is authentic (genuine). This requires that the origin of a message must be correctly identified, with assurance that the identity is not false. In the case of a single message, the authentication service assures the recipient that the message is issued from the source that it claims to be from. In the case of an ongoing interaction, two aspects are involved. First, at the time of connection establishment, the service assures that the two communicating entities are authentic. That is, each entity has the identity that it claims to have. Second, the service assures that the connection is not interfered with by another connection in such a way that a third party can masquerade as one of the two legitimate parties for the purposes of unauthorized transmission or reception of sensitive information.

Confidentiality

This service requires that the information in a computer system, as well as the transmitted information, be accessible only for reading by authorized parties. Therefore, confidentiality is the protection of static and flowing data from attacks. It is also related to the protection of information from unauthorized access, regardless of where the information is located, how it is stored, or in which form it is transmitted. Several levels of protection can be identified and implemented to guarantee such service. The broadest service protects all user data transmitted between two users over a period of time. Limited forms of the confidentiality service can address the protection of a single message or even specific fields within the message.

Integrity

This service ensures that computer systems resources and flowing information can only be modified by authorized parties. Integrity is the protection of information, applications, systems, and networks from intentional, unauthorized, or accidental changes. It can apply to a stream of messages, a single message, or specific parts of a message. Two classes of integrity services can be considered: connection-oriented integrity and connectionless integrity. A connection-oriented integrity service guarantees that messages in a given connection are received as sent, with no duplication, insertion, modification, reordering, or replay. The delete/destruction of data is also controlled by this service. On the other hand, a connectionless integrity service deals with individual messages and typically provides protection against message modification only.

Non-repudiation

This service guarantees that neither the sender of a message can deny the transmission of the message nor the receiver of a message is able to deny the reception of the message. Therefore, when a message is sent, the receiver can prove that the message was in fact sent by the alleged sender. Similarly, when a message is received, the sender can prove that the message was in fact received by the legitimate receiver. The technical control used to protect against non-repudiation can be provided by the so-called digital signature.

Access control

This service guarantees that access to information be controlled by (or on behalf of) the target system. In the context of network security, access control is the ability to limit and control the access to host systems and applications via communication links. To achieve this, each entity trying to gain access must first be identified, or authenticated, so that access rights can be tailored to the individuals. Access control protects against unauthorized use or manipulation of resources through authentication and authorization capabilities. Authorization services ensure that access to information and processing resources is restricted based on the management of a security policy.

Availability function

This is the property of a system, network, or a resource being accessible and usable any time upon demand by an authorized system entity, according to performance specifications for the system. This means that a system is available if it provides services according to the system design whenever users require them. The X.800 standard considers availability as a property to be associated with various security services such as services protecting from denial of service.

1.3.2   Security attacks

Security attacks include interruption (i.e., an asset is destroyed or becomes unavailable), interception (this happens when an unauthorized party gains access to an asset), modification, and fabrication (particularly visible when an unauthorized party inserts counterfeit objects such as messages or files). A primary way to classify security attacks considers that attacks can be either passive or active. Passive attacks aim to gain information that is being transmitted. Passive attacks are very difficult to detect because they do not involve any alteration of the observed data or the network resources. However, it is feasible to prevent the success of these attacks, usually by means of encryption for example. A particular example of passive attacks of some importance is the traffic analysis attack, which permits the intruder to observe communications and learn about transmission time, size, and structure of transmitted messages. They allow getting copy of the exchanged data and using it to extract confidential information. They also determine the location and the identity of communicating hosts (e.g., service location, used ports, and IP addresses).

Active attacks involve some modification of the data flow, to be attacked, or the creation of a false data stream. They can be subdivided into different categories: impersonate attacks, replay, modification, and denial of service. An impersonate attack takes place when an entity pretends to be a different entity in order to obtain extra privileges. It usually includes one of the forms of active attacks. A valid authentication message can be captured and resent to gain specific interest. A replay attack involves the passive capture of a data unit and its subsequent retransmission to produce an unauthorized effect. A modification means that some portion of a legitimate message (or a file) is altered, delayed or reordered to produce unauthorized effects. Examples of modifications include modifying routing tables within a network router and modifying the content of transactions.

A denial of service prevents or reduces the normal use, operation, or management of communication facilities. Denial of service attacks attempt to block or disable access to resources at the victim system. These resources can be network bandwidth, computing power, buffers, or system data structures. The different categories of denial of service attacks can be mainly classified into software exploits and flooding attacks. In software exploits, the attacker sends a few packets to exercise specific software bugs within the target service or application, disabling the victim. In flooding attacks, one or more attackers can send continuous streams of packets for the purpose of overwhelming a communication protocol or computing resources at the target system.

Based on the location or observation points, attacks can be classified as single-source when a single zombie (i.e., procedure that acts on behalf of the attacker) is observed performing the attack to the targeted victim, and as direct multi-source when multiple distributed zombies are attacking the victim system. Often, zombies are typically injected procedures or insecure objects that have been compromised by a malicious user.

1.4 Threats and vulnerabilities

In this section, we consider the notion of vulnerabilities noticeable on each asset within an enterprise information and communication system. We will also consider threats to hardware, applications, and information. Since physical devices are so visible, they are rather simple targets to attack. Serious physical attacks include machine-slaughter, in which an adversary actually intends to harm an information system. Fortunately, however, reasonable safeguards can be put down to protect physical assets (West-Brown *et al.*, 1998; Allen, 2001).

A secure processing environment within an organization includes making sure that only those who are authorized are allowed to have access to sensitive information, that the information is processed correctly and securely, and that the processing is available when necessary. To apply appropriate security controls to an operating environment, it is necessary to understand who or what poses a threat to the operating environment; and therefore to understand the risk or danger that the threat can cause. When the risk is assessed, management procedures must decide how to mitigate the risk to an acceptable level. Another approach for managing security controls resides in providing some degree of compensation for losses/damages by contracting insurance.

Vulnerabilities, threats, and risks are defined in what follows. They will be studied to a large extent in Chapter 14 where the risk management is completely addressed.

Vulnerabilities

A security vulnerability is a weakness (e.g., a flaw or a hole) in an application, product, or asset that makes it infeasible to prevent an attacker from gaining privileges on the organization's system, compromising data on it, modifying its operation, or assuming non-granted trust. The following simple examples would constitute security violations:

- A flaw in a web server that enables a visitor to read a file that he/she is authorized to read.
- A flaw that allows an unauthorized user to read another user's files, regardless of the permissions on the files.
- The fact that an attacker could send a large number of legitimate requests to a server and cause it to fail.
- A flaw in a payment gateway allowing price manipulation to be transmitted unnoticed.

Threats

An organization's information can be accessed, copied, modified, compromised, or destroyed by three types of threats: intentional, unintentional, or natural threats. Intentional threats are unauthorized users who inappropriately access data and information that they are not granted privilege to read, write, or use. These users can be external or internal to the organization and can be classified as malicious. The malicious users have intent to steal, compromise, or destroy assets. Intrusion by malicious unauthorized users first attempts to get results in a breach of confidentiality and second by causing breaches in integrity and availability.

Unintentional threats are typically caused by careless or untrained employees who have not taken the steps to ensure that their privileges are protected, their computers are properly secured, or that virus scanning software is frequently updated. Unintentional threats also include programmers or data processing personnel who do not follow established policies or procedures designed to build controls into the operating environment. Unintentional threats can affect the integrity of the entire application and any integrated applications using common information.

Natural threats include, but are not restricted to, equipment failures or natural disasters such as fire, floods, and earthquakes that can result in the loss of equipment and data. Natural threats usually affect the availability of processing resources and information.

Risks

There are many events that can be generated if a breach of confidentiality, integrity, or unavailability occurs. From the business point of view, there is always financial loss involved. The business risks include contractual liability, financial misstatement, increased costs, loss of assets, lost business, and public embarrassment. Increased costs occur from many types of security incidences. A lack of integrity in systems and data will result in significant cost to find the problem and fix it. This could be caused by the modification, testing, and installation of new programs and rebuilding systems.

Controls

Controls fall into three major categories: physical, administrative, and technical. Physical controls constrain direct physical access to equipment. They consist of cipher locks, keyboard locks, security guards, alarms, and environmental systems for water, fire, and smoke detection. Physical controls also consist of system backups and backup power such as batteries and uninterruptible power supply.

Administrative controls include the enforcement of security policies and procedures. Policies inform users on what they can and cannot do while they use the organization's computing resources. Procedures help the management review the changes to code, authorization for access to resources, and periodic analysis. They tell how to check that an access is still appropriate, perform separation of tasks, realize security awareness, and plan for technical training. Administrative controls can be continuous, periodic, or infrequent. They contain documents on IT management and supervision, disaster recovery, and contingency plans. Administrative controls also include security review and audit reporting that are used to identify if users are adhering to security policies and procedures. Technical controls are implemented through hardware or software, which typically work without human intervention. They are usually referred to as logical controls. Specific software controls include anti-virus, digital signatures, encryption, dial-up access control, callback systems, audit trails, and intrusion detection systems.

Administrative, technical, and physical controls can be further subdivided into two categories: preventive or detective. Preventive controls, such as technical training, attempt to avoid the occurrence of unwanted events, whereas detective controls, such as security reviews and audits, attempt to identify unwanted events after they happen. When an event does occur, controls involving the detection and analysis of related objects are designed to catch the problem quickly enough to minimize the damage and be able to accurately assess the magnitude of the damage caused.

1.5   Basics of protection

Security solutions attempt to establish perimeters of protection (for security domains) to filter the entering and the available data. A security solution depends on the correctness and the reliability of three related components: security policy, implementation mechanisms, and assurance measures. In the following, we attempt to characterize the notions of security domain and security policy in an enterprise, and show how they are related to each other.

1.5.1 Security management

Security policies set the guidelines for operational procedures and security techniques that reduce security risks with controls and protective measures. A security policy has a direct impact on the rules and policing actions that ensure proper operation of the implemented mechanisms. Policy has an indirect influence on users; they see security applications and access services, not policies. The security policy of an organization should also determine the balance between users' ease of use and the level of responsibility. The amounts of controls and countermeasures should also be considered.

The goal of a security manager is to apply and enforce consistent security policies across system boundaries and throughout the organization. The challenges in achieving a functional security system include two constraints. First, a consistent and complete specification of the desired security policy needs to be defined independently of the implementing technologies. Second, satisfying the need for a unified scheme to enforce the applicable security policies using and reusing available tools, procedures, and mechanisms. The difficult task in achieving an acceptable *state of security* is not obtaining the necessary tools, but choosing and integrating the right tools to provide a comprehensive and trustworthy environment.

Security management can be defined as the real-time monitoring and control of active security applications that implement one or more security services in order to keep the organization's system at an acceptable state of security. The purpose of security management is to ensure that the security measures are operational and compliant with the security policy established by the organization. Not only must the security services function correctly, but they must counteract existing threats to generate justifiable confidence in the system trustworthiness.

Network security management is by nature a distributed function. Applications that may be under the coverage of security management include firewalls, intrusion detection systems, and security control applications. Security management faces the same security threats that other distributed applications have to face. Coordinated management of security is not feasible without a secure management infrastructure that protects the exchanged messages from blocking, modification, spoofing, and replay. Although the discussion of security management architecture is beyond the scope of this section, it is obvious that management, access control, and reliable implementation of management software are also critical.

In its simplest form, security management could require the presence of skilled employees at each security device. It also requires manual evaluation of all significant events. On the other hand, security experts do believe that remote monitoring with computer-assisted correlation of alerts and management of system events is just as efficient for security management as it is for network management. In fact, it may be argued that detection of sophisticated attacks needs the help of sophisticated decision-support systems. In its more advanced form, security management should integrate what is called security assurance and rely on the notion of trust domain.

Assurance is the conventional term for methods that are applied to assess and ensure a security system enforces and complies with intended security policies. One may use assurance tools before, during, or after security mechanism operations. A trusted domain comprises systems (and network components) or parts of systems (e.g. security modules, computers, and routers), with the assumption that no attackers are assumed within a trusted domain. This restriction induces the fact that a trusted domain is always related to a single user or group of users accessing specific applications.

1.5.2   Security policies

Defining a Security Policy (SP) has been the subject of a big debate in the security community. Indeed, although this concept has been addressed by many specialists, it turns out to be hard to find a definition that relates a uniform view. In Hare (2000), SP has been presented as a "*high level statement that reflects organization's belief related to information security.*" We believe that an information security policy is designed to inform all employees operating within an organization about how they should behave related to a specific topic, how the executive management of the organization should behave about that topic, and what specific actions the organization is prepared to take. Security policy should specify the policy scope, responsibilities, and management's intention for implementing it. It addresses security activities that include designing controls into applications, conducting investigation of computer crimes, and disciplining employees for security violation. Security policies also include standards, procedures, and documents.

Standards constitute a specific set of rules, procedures or conventions that are agreed upon between parties in order to operate more uniformly, effectively, and regardless of technologies. Standards have a large impact on the implementation of security. When they are taken into consideration, they can impact the amount of support required to meet the enterprise security objectives. This will have a definite impact on both cost and the level of security risk. Procedures are plans, processes, or operations that address the conditions of how to go about a particular action.

The constituency of the SP is also a fundamental issue that is tightly related to both objectives and requirements. The major components of a good SP should include (IETF, 1997):

1. An *Access Policy* which defines privileges that are granted to system users in order to protect assets from loss or misuse. It should specify guidelines for external connections and adding new devices or software components to the information system.
2. An *Accountability Policy* that defines the responsibilities of users, operations staff, and management. It should specify the audit coverage and operations and the incident handling guidelines.
3. An *Authentication Policy* which addresses different authentication issues such as the use of Operating System (OS) passwords, authentication devices or digital certificates. It should provide guidelines for use of remote authentication and authentication devices.

4. An *Availability Policy*, which defines a set of user's expectations for the availability of resources. It should describe recovery issues and redundancy of operations during down time periods.
5. A *Maintenance Policy*, which describes how the maintenance people are allowed to handle the information system and network. It should specify how remote maintenance can be performed, if any.
6. A *Violations Reporting Policy*, which describes all types of violations that must be reported and how reports are handled.
7. A *Privacy Policy*, which defines the barrier that separates the security objectives from the privacy requirements. Ideally, this barrier must not be crossed in both directions.
8. A set of *supporting information*, which provides systems users and managers with useful information for each type of policy violation.

Generally, the most important goals that might be achieved by a security policy are described through the three following points:

1. The measures of the SP should maintain the security of critical components at an acceptable level. In other terms, the security policy helps enterprises in reducing the likelihood of harmful adverse events.
2. The security policy must include some response schemes that make the system recover if an incident (e.g., security attack, or natural disaster) occurs.
3. The security policy must ensure the continuity of the critical processes conducted by an enterprise whenever an incident occurs.

Achieving these objectives requires a solid interaction between the security policy and the remaining strategic documents of the organization such as the Disaster Recovery Plan (DRP), the Incident Response Plan (IRP), and the Business Continuity Plan (BCP).

To develop security policies and procedures, one must undertake the following activities: (a) look at what the organization desires to protect, by taking an inventory of all applications, operating systems, databases, and communication networks that compose the organization's information system; (b) determine who is responsible for these components; (c) look with the resource owners at the resource from a confidentiality, integrity, and availability point of view; and (d) determine how to protect the resource in a cost-effective manner based on its value to the enterprise.

Threat and realistic risk estimation are necessary components of this activity. Security baseline of risk estimation is typically an important step to be performed because of security audit requirements and the need to classify the state of security of a system. Policies, standards, and procedures should be reviewed continuously and improved every time a weakness is found. Risk estimation should also be a continuous process because of two facts: (a) the distributed organization's computing environment is always changing and (b) threats and attacks are increasing in number, complexity, and ability to create damage.

Particular policies that relate to protection are the e-mail policy, Internet policy, and e-commerce policy. The e-mail policy is designed to protect business partners' interest, client information, the employee, and the enterprise from liability. The e-mail security policy covers several subjects including the confidentiality and privacy of mails, passwords, and mail servers. It also discusses how to protect sensitive information through the use of encryption, the limits of personal use of the e-mail system, procedure for communicating with individuals outside the enterprise, and restriction related to communications or information provided to corporate entities, business partners, or customers.

In addition to e-mail, and other Internet services such as the transfer of files (FTP), the ability to log on remotely to a host at another location (Telnet), web service discovery and brokering are used. Such services show the necessity to establish an Internet security policy since there are significant risks related to providing a network connection from the internal network of the enterprise to the Internet. The major risks include the destruction, downtime, and maintenance costs dues to viruses, potential access to internal information, and vulnerabilities from competent, and technically proficient yet destructive hackers.

1.6   Protections of users and networks

1.6.1   Protection of employees

Protection of users includes the following major tools: the shared key encryption schemes, crypto-hash functions, the public key encryption schemes and the digital signature schemes.

Shared key encryption schemes

In its simplest terms, encryption is the process of making information and messages flowing through the network unreadable by unauthorized individuals and entities. The process may be manual, mechanical, or electronic. We, however, consider that the encryption systems in this book are only electronic and are achieved through well-established standards. The conceptual model of an encryption model consists of a sender and a receiver, a message (called the *plain text*), the encrypted message (called the *cipher text*), and an item called secret key. The encryption process (or cryptographic algorithm) transforms the plain text into the cipher text using the key. The key is used to select a specific instance of the encryption process embodied in the transformation.

Examples of encryption algorithms include DES, 3DES, IDEA, AES, (Stallings, 2001). These schemes can handle data messages of several Gigabyte/s and use keys of varying size in the range of 56–128–192–256 bits. Although encryption is an effective solution, its effectiveness is limited by the difficulty in managing encryption keys (i.e., of assigning keys to users and recovering keys if they are lost or forgotten).

Hash functions

A Hash Function H generates a value $H(m)$, for every variable-length message $H(m)$ called the message digest of m. A hash function assumes first that it is easy to compute any message digest, but assumes that it is virtually impossible to generate the message code knowing only its digest. Second, collision resistance is provided; i.e., an alternative message hashing to the same value cannot be found. This prevents forgery when encrypted hash code is used and guarantees message authentication of the message given its digest.

Digital signature schemes

Digital signature is an electronic process that is analogous to the hand written signature. It is the process of binding some information (e.g., a document or a message) to its originator (e.g., the signer). The essential characteristic of a digital signature is that the signed data unit cannot be created without using a key private to the signer, called *private key*. This means three things:

1. The signed data message cannot be created by any individual other than the holder of the private key used for the signing.
2. The recipient cannot create the signed data unit, but can verify it using the related publicly available information; the public key.
3. The sender cannot deny signing and sending the signed data unit.

Therefore, using only publicly available information, it is possible to identify the signer of a data unit as the possessor of the related private key. It is also possible to prove the identity of the signer of the data unit to a reliable third party in case a dispute arises. A digital signature certifies the integrity of the message content, as well as the identity of the signer. The smallest change in a digitally signed document will cause the digital signature verification process to fail. However, if the signature verification fails, it is in general difficult to determine whether there was an attempted forgery or simply a transmission error.

1.6.2 Protection of networks

Various mechanisms are typically used to protect the components of a communication network. They include, but are not limited to, the following mechanisms:

Firewalls

A firewall is a security barrier between two networks that analyzes traffic coming from one network to the other to accept or reject connections and service requests according to a set of rules, which is often part of the overall security policy or separated for updating these rules to cope with the attack growth. If configured properly, a firewall addresses a large number of threats that originate from outside a network without introducing any significant security danger. Firewalls can protect against attacks on network nodes, hosts, and applications. They also provide a central method for administering security on a network and logging incoming and outgoing traffic to allow for accountability of users' actions and for triggering incident response activity if unauthorized activities occur.

Firewalls are typically placed at gateways to networks to create a security perimeter to primarily protect an internal network from threats originating from outside, particularly, from the Internet. However, the most important issue in effectively using firewalls is developing a firewall policy. A firewall policy is a statement of how a firewall should work through a set of rules by which incoming and outgoing traffic should be authorized or rejected. Therefore, a firewall policy can be considered as the specification document for a security solution that has to protect in a convenient way the organization's resources and processes.

Firewall products have improved considerably over the past several years and are likely to continue to improve. The firewall technology is evolving towards distributed and high-speed organization in order to cope with the growing application needs and sophistication of networks.

Access, authorization, and authentication tools

Access control mechanisms are used to enforce a policy of limiting access to a resource to only authorized users. These techniques include the use of access control lists or matrices, passwords, capabilities, and labels, the possession of which may be used to indicate access rights. Authorization involves a conformance engine to check credentials with respect to a policy in order to allow the execution of specific actions. Authentication attempts to provide a proof of the identity of a user accessing the network (Obaidat, 1997; Obaidat, 1999).

Intrusion detection systems

Intrusion detection is the technique of using automated and intelligent tools to detect intrusion attempts in real-time manner. An *Intrusion Detection System* (IDS) can be considered as a *burglar alarm for computer systems and networks*. Its functional components include an analysis-engine that finds signs of intrusion from the collection of events and a response component that generates reactions based on the outcome of the analysis engine. Chapter 11 will discuss the IDS's concepts, usage, and promises.

Two basic types of intrusion detection systems exist: rule-based systems and adaptive systems. Rule-based systems rely on libraries and databases of known attacks or known signatures. When incoming traffic meets specific criteria, it is labeled as an intrusion attempt. There are two approaches for rule-based IDSs: peremptory and reactionary. In the peremptory approach, the intrusion tool actually listens to the network traffic. When suspicious activity is noted, the system attempts to take the appropriate action. In the reactionary approach, the intrusion detection tool watches the system logs instead, and reacts when suspicious activity is noticed.

Adaptive/reactive systems use more advanced techniques including multi-resolution operational research, decision support systems, and learning processes to achieve its duties.

1.7   Security planning

Security planning begins with risk analysis, which is the process that determines the exposures and their potential harms. Then it lists for all exposures possible controls and their costs. The last step of the analysis is a cost-benefit analysis that aims at answering questions like:

Does it cost less to implement a control or accept the expected cost of the loss?

Risk analysis leads to the development of a security plan, which identifies the responsibilities and the certain actions that would improve system security (Swanson, 1998).

1.7.1 Risk analysis

Risk analysis is concerned with setting up relationships between the main risk attributes: assets, vulnerabilities, threats, and countermeasures. Since the concept of risk is commonly considered as simply part of the cost of doing business, security risks must be taken as a component of normal operation within an enterprise. Risk analysis can reduce the gravity of a threat. For example, an employee can perform an independent backup of files as a defense against the possible failure of a file storage device. Companies involved in extensive distributed computing cannot easily determine the risks and the controls of the computer networks. For this reason, a systematic and organized approach is needed to analyze risks.

For companies, benefits of careful risk analysis include: (a) the identification of assets, vulnerabilities, and controls, since a systematic analysis produces a comprehensive list that may help build a good solution; (b) the development of a basis for security decisions, because some mechanisms cannot be justified only from the perception of the protection they provide; and (c) the estimation/justification for spending on security since it helps to identify instances that justify the cost of a major security mechanism.

Risk analysis is developed extensively in Chapter 14. It is a methodical process adapted from practices in management. The analysis procedure consists of five major steps: (a) assets identification; (b) vulnerabilities determination and likelihood of exploitation estimation; (c) computation of expected annual loss; (d) survey of applicable controls and their costs; and (e) annual savings of control projection. The third step in risk analysis determines how often each exposure will be exploited. Probability of occurrence relates to the severity of the existing controls and the probability that someone or something will elude the existing controls.

Despite its common usage, there are arguments against risk analysis management. First, the lack of precision of risk analysis methods is often mentioned as a deficiency. The values used in the method, the likelihood of occurrence, and the cost per occurrence are not precise. Second, providing numeric estimation may produce a false impression of precision or security. Third, like contingency plans, risk analysis has an inclination to be stored and immediately forgotten.

1.7.2 Security plans

A security plan is a document that describes how an organization will address its security needs. The plan is subject to periodic review and revision as the security needs of the organization change and threats increase. The security plan identifies and organizes the security activities for a networked information system. The plan is both a description of the current situation and a plan for change. Every security plan should be an official documentation of current security practices. It can be used later to measure the effect of the change and suggest further improvements.

A security plan states a policy on security. The policy statement addresses the organization's goals on security (e.g., security services, protection against loss, and measures to protect business from failure), where the responsibility for security lies (e.g., security group, employees, managers), and the organization's commitment to security. The security plan should define the limits of responsibility (such as which assets to protect and where security boundaries are placed). It should present a procedure for addressing a vulnerability that has not been considered. These vulnerabilities can arise from new equipment, new data, new applications, new services, and new business situations.

If the controls are expensive or complicated to implement, they may be acquired from the shelf and implemented gradually. Similarly, procedural controls may require the training of the employees. The plan should specify the order in which the controls are to be implemented so that most serious exposures are covered as soon as possible. A timetable also should be set up to give milestones by which the progress of the security program can be evaluated. An important part of the timetable is establishing a date for evaluation and review of the security solution.

As users, data, and equipment change, new exposures develop and old means of control become obsolete or ineffective. Periodically the inventory of objects and the list of controls should be updated, and the risk analysis should be reviewed. The security plan should set a time for this periodic review. After the plan is written, it must be accepted and its recommendations be carried out. A key to the success of the security plan is management commitment, which is obtained through understanding the cause and the potential effects of security lack, cost effectiveness, and presentation of the plan.

1.8   Legal issues in system security

Law and information systems are related in different manners. First, laws affect privacy and secrecy and typically apply to the rights of individuals to keep personal information private. Second, laws regulate the development, ownership, and use of data and applications. Copyrights and trade secrets are legal devices to protect the rights of developers and owners of information applications and data. Third, laws involve actions that can be taken by an organization to protect the secrecy, integrity, and availability of information and electronic services. Laws do not always provide an adequate control. They are slowly developing behind information technologies. They also do not face all irregular acts and cyber crimes committed on information and communication systems.

Progressively security services are becoming measurable like any tangible service, or a software package. Security services do have prices, but unlike tangible objects, they can be sold again and again and since the costs to reproduce them are very low, they often are transferred intangibly. However, security policy building, risk analysis, solution configuration, and personal training should be addressed appropriately with each enterprise. This shows that security services measurability is a hard task. In addition, security services can be hacked; meanwhile they should be able to be used for tracing and investigating evidences of attacks.

All these properties of information have an essential effect on the legal treatment of information. In that, there is a need for a legal basis for the protection of information.

Software privacy is the first example in which the value of information can be readily copied. Several approaches have been attempted to ensure that the software developer or publisher receives just compensation for use of his/her software. None of these approaches seem ideal. Therefore, it is likely that the availability of a complete legal frame will be needed, by any country, in addition to the technological issues.

Electronic publishing, protection of networked databases, and electronic commerce portals are important areas where legal issues must be solved as we move into an age of electronic business/economy. Ensuring for example that the electronic publisher receives fair compensation for his/her work, means that cryptographic-based technical solutions must be supported by a legal structure capable of defining the acceptable technical requirements to perform legal signature or encrypt documents.

The law managing electronic contracts and e-employment is hard to address, even though employees, objects, and contracts are practically standard entities for which legal precedents have been developed. The definitions of copyright and patent law are stressed when applied to computing and communication because old forms must be transformed to fit new objects. For these situations, however, cases and prototypes have been decided. Now they are establishing legal precedents and studies. Nevertheless, cyber crimes represent an area of the law that is less clear than other areas.

The legal system has explicit rules of what constitutes property. Generally, property is tangible, unlike sequences of bits. Getting a copy of (part of) a sensitive file or accessing unauthorized a computing system should be considered as a cyber crime. However, because "*access and copy*" operations are not physical objects, some courts in various countries are unenthusiastic to consider this as a theft. Adding to the problem of a rapidly changing technology, a computer can perform many roles in a cyber crime. A computer can be the subject or the medium of cyber crime since it can be attacked, used to attack, and used as a means to commit a cyber crime.

Cyber crimes statutes have to include a large spectrum of cases. But, even when it is acknowledged that a cyber crime has been committed, there are still several reasons why a cyber crime is hard to prosecute. This includes the need of understanding the cyber crime (computer literacy), finding evidences (e.g., tracing intrusions, and investigating disks), and evaluating assets and damages.

1.9 Summary

Security in business activity is an important issue that should be addressed appropriately by skilled people. This chapter provides guidance on the generic activities to be addressed when designing a security solution. In particular, this chapter helps organizations define, select, and document the nature and scope of the security services to be provided in order to protect business activity. Moreover, it discusses the basic functions needed to build up security services, how these functions work and interrelate, as well as the procedures necessary to implement these services.

This chapter is intended first to provide a self-contained resource to understand the role of policy security and risk management. Second, it provides managers responsible for the implementation, operation of service, and response to security incidents with a high-level description of the main procedures to protect assets. Third, this chapter attempts to demonstrate the need for security services, skilled persons, and documented policies.

References

Allen, J. H. (2001). *CERT Guide to System and Network Security Practices*, The SEI Series in Software Engineering, Addison Wesley Professional.

Australian Computer Emergency Response Team. (2004). *2004 Australian Computer Crime and Security Survey* (available at www.auscert.org.au/download.html?f=114).

Gordon, L. A., M. P. Loed, W. Lucyshin, and R. Richardson. (2004) *2004 CSI/BFI Computer crime and security survey*, Computer Security Institute publications (available at www.gosci.com/forms/fbi/pdf.jhtml).

Hare, C. Policy development. In *Information Security Management Handbook*, volume 3, Tipton, H. F. and Krause M. (eds.). Auerbach, pp. 353–89.

Holbrook, P. and J. Reynolds. (1991). *Site Security Handbook* (available at www.securif.net/misc/Site_Security_Handbook).

Internet Engineering Task Force. (1997). *Site Security Handbook*, RFC 2196. IETF Network Working Group. Available at www.ietf.org/rfc/rfc2196.txt (date of access: Aug. 24th, 2004).

Obaidat, M. S. (1993b). A methodology for improving computer access security, *Computers Security Journal*, Vol. **12**, No. 7, 657–62.

Obaidat, M. S. and D. Macchairllo. (1993a). An on-line neural network system for computer access security. *IEEE Transactions on Industrial Electronics*, Vol. **40**, No. 2, 235–42.

Obaidat, M. S. and D. Macchairllo. (1994). A multilayer neural network system for computer access security, *IEEE Transactions on Systems, Man, and Cybernetics*, Vol. **24**, No. 5, 806–13.

Obaidat, M. S. and B. Sadoun. (1997). Verification of computer users using keystroke dynamics. *IEEE Transactions on Systems, Man and Cybernetics*, Part B, Vol. **27**, No. 2, 261–9.

Obaidat, M. S. and B. Sadoun. (1999). Keystroke dynamics based identification. In *Biometrics: Personal Identification in Networked Society*, Anil Jain *et al.* (eds.), Kluwer, pp. 213–29.

Stallings, W. (2001). *Cryptography and Network Security*, 3rd edn. Prentice Hall.

Swanson, M. (1998). *Developing Security Plans for Information Technology Systems*, NIST Special Publication 800–18.

West-Brown, M. J., D. Stikvoort, and K. P. Kossakowski. (1998). *Handbook for Computer Security Incident Response Teams (CSIRTs)* (CMU/SEI-98-HB-001). *Software Engineering Institute*, Carnegie Mellon University.

2 Public key cryptosystems

Public key cryptosystems represent a basic tool for the implementation of useful security services that are able to protect the resources of an organization and provide an efficient security for the services and Web sites that an enterprise may offer on the Internet. This chapter describes the main components, functions, and usage of a public key cryptosystem. It also discusses some major attacks that have been developed to reduce cryptosystem efficiency.

2.1 Introduction

A text containing data that can be read and understood without any special measure is called *plaintext*. The method of transforming a *plaintext* in a way to hide its content to unauthorized parties is called *encryption*. Encrypting a plaintext results in unreadable text called *ciphertext*. Therefore, encryption is used to ensure that information is hidden from anyone for whom it is not intended, including those who can capture a copy of the encrypted data (while it is flowing through the network). The process of inversing the ciphertext to its original form is called *decryption*. Cryptography can be defined as the science of using mathematics to encrypt and decrypt data. Cryptography securely provides for the storage of sensitive information and its transmission across insecure networks, like the Internet, so that it cannot be read (under its original form) by any unauthorized individual (Menezes *et al.*, 1996).

A cryptographic algorithm, also called *cipher*, is a mathematical function used in the encryption and decryption processes. A cryptographic algorithm works in combination with some secret information called *key*, which is a word, number, or phrase, to encrypt the plaintext. The same plaintext can be encrypted into different ciphertext using different keys. Depending on the mechanism used, the key can be used for both encryption and decryption (in that case, the encryption is called secret key or symmetric key encryption), while other mechanisms use different keys for the encryption and the decryption process (they are called public key or asymmetric key encryption). The security of encrypted data is entirely dependent on two parameters: the strength of the cryptographic algorithm and the secrecy of the key.

A *cryptosystem* can be defined as the aggregation of a cryptographic algorithm, the set of all possible keys, and the set of protocols that make the algorithm work between the networked systems.

Often, a key is a value that is used by a cryptographic algorithm to produce a specific ciphertext. Keys are basically large numbers. Key size is measured in number of bits; the number representing a 1024-bit key is of common usage in public key cryptosystems (i.e., in RSA cryptosystem (Rivest, 1978)), where experts agree that the larger the key is, the more secure the ciphertext is. Public key size and secret key size appear to be totally unrelated. However, an 80-bit secret key has the equivalent strength of a 1024-bit public key and a conventional 128-bit key is equivalent to a 3000-bit public key. While the public and private keys are mathematically related, it should be extremely difficult to derive the private key given only the public key (and vice versa). Nevertheless, deriving the private key is always mathematically possible given enough time and computing power. This makes it very important to choose keys large enough to be secure, but small enough to be applied fairly fast. Therefore, key security appears to be highly dependent on the time duration of its use and the technique used to generate it. Larger keys, for example, will be cryptographically secure for longer periods of time.

Finally, it is important to emphasize that cryptographic algorithms should be well-designed and efficiently tested before they are used. This, however, will cost a lot of time and money since testing and more generally attacking these algorithms can take years. Opting, however, for fewer requirements will lead to a false feeling of security. The best cryptographic algorithms are the ones that have undergone thorough tests. According to the National Institute of Standards and Technology, (NIST), cryptographic algorithms should have the following features:

- A cryptographic algorithm must be completely specified and should be written in an unambiguous manner. It must be robust and should cope with a large number of users.
- A cryptographic algorithm must provide an acceptable level of protection, normally expressed in length of time or number of operations required to recover the key, in terms of perceived threats.
- A cryptographic algorithm must provide methods of protection based only on the secrecy of the related keys; it must be maintained.
- A cryptographic algorithm must not discriminate against any user or system supplier. It must be utilized in various applications and environments.

According to the NIST, the Data Encryption Standard, DES (NIST, 1999) algorithm was the first secret key algorithm and it was found acceptable for these requirements. However, DES users should be aware that it is theoretically possible to derive the key in fewer trials and, therefore, the key should be changed as often as possible. Users must change their key and provide it a high level of protection in order to minimize the potential risks of its unauthorized computation or acquisition.

2.2 Symmetric encryption

Conventional cryptography is based on the fact that both the sender and the receiver of a message (or a stream of messages) share the same secret key. Conventional algorithms are also known as secret key algorithms or symmetric key algorithms. They have constituted the only type of encryption before the development of public key encryption. Secret key encryption assumes that it is computationally unfeasible to decrypt a message given the ciphertext and knowledge of the encryption/decryption algorithm. This feature is what makes the system feasible for widespread use, since keeping the secret key algorithm public means that manufacturers can develop low-cost cryptosystems based on the given algorithm.

2.2.1 Secret key encryption features

A symmetric key encryption scheme is given when two procedures are provided to perform encryption and decryption using shared keys. Formally speaking, a symmetric-key encryption scheme is given by four components: (a) a message space M, which defines all the messages that can be encrypted; (b) a key space K, which contains all sequences of bits, integers, or strings, acceptable by the algorithm; (c) a cipher space C, whose elements are possible ciphertexts; and (d) two maps:

$$\phi : M \times K \to C \quad \text{and} \quad \gamma : C \times K \to M.$$

These functions represent the encryption and decryption processes. Both functions must satisfy:

$$\phi(\gamma(c, k), k) = c \quad \text{and} \quad \gamma(\phi(m, k), k) = m$$
$$\phi(-, k) : M \to C \text{ is a one-way function for all } k$$
$$\phi(m, -) : K \to C \text{ is a one-way function for all } m$$

The term *one-way function* means that the function is easy to compute (i.e., computable at a polynomial-time complexity); the term infeasible means that the function is not computable in polynomial-time. The above expressions state the exact relationship between the encryption and decryption processing. The adjective symmetric finds its origin in the fact that the key used for encryption is the same as the one used for decryption. Hence, the secret key must be communicated prior to its use, through a completely secure channel in order to reach full security. Even though the secret cryptosystem scheme is attractive, thanks to its security and ease of encryption and decryption, it has the major disadvantage of requiring a key that must be communicated securely. Key distribution is a major concern for such systems.

Figure 2.1 presents a model that specifies the conventional cryptosystem. It represents a source providing a message (or plaintext m), another source providing key k to both the message source and destination. The figure shows also the destination receiving the ciphertext c. Two secure channels are used to deliver the shared key. The primary symmetric encryption algorithm widely used in the computer industry for the last thirty years is the data encryption standard (DES). The latter is a block-based encryption technique that takes a 64-bit fixed block of text, and transforms the block into a 64-bit block of ciphertext. DES symmetric key is 64-bit long, of which only 56 bits are used to encrypt and decrypt the message (NIST, 1999). The remaining 8 bits are used for the parity checks of the previous 56 bits. The total number of keys is thus equal to $2^{56} (= 7.2 \times 10^{15})$.

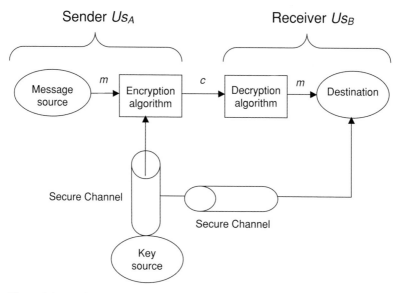

Figure 2.1   Model of secret-key cryptosystem.

The DES algorithm essentially consists of a series of permutations and substitutions. A block that is going to be encrypted is first subjected to an initial permutation (denoted by IP); then to a complex series of key-dependent cycles involving 16 sub-keys and special functions. Finally, it is submitted to another permutation, which is the inverse of the initial permutation (that is IP^{-1}). Figure 2.2 depicts the main operations executed at each cycle. Each right half is expanded from 32 bits to 48 bits. The expansion aims to make intermediate halves of the ciphertext comparable in size to the intermediate keys and to provide a longer result that can be reduced by special non-linear functions called S-boxes.

DES has several known weaknesses including complements and semi-weak keys. Notice that two messages m and $\underline{m}$ are called complements if $m \oplus \underline{m} = 11 \ldots 1$. The weakness related to complements shows that if a message m is encrypted with a particular key k, the complement of the resulting ciphertext will be the encryption of the complement $\underline{m}$ using the complement key $\underline{k}$. Semi-key weaknesses mean that there are specific pairs of keys having identical decryption, implying that two keys decrypt the messages encrypted by the other key. Other common symmetric algorithms likely to be used (and therefore replace DES) in the computer industry are triple DES, and the advanced encryption standard (AES). Triple DES is a variant of DES that decreases the risk from brute force attack by using longer keys. This gives an effective key length of roughly 168-bit. AES serves as an eventual successor to DES, as selected by NIST. The latter has conducted a selection process involving 13 secret key algorithms. AES is based on a powerful substitution-linear transformation network with 10, 12, or 14 rounds, depending on the key size and with variable block sizes of up to 256 bits (NIST, 2001).

An example of secret-key cryptosystem, which is called the one-time pad, works as follows. The system uses a random key that is as long as the plaintext (with no repetitions, if needed) and produces a random output (or ciphertext) that has no mathematical relationship

Table 2.1 *Ciphertext obtained by XOR of plaintext and key*

Ciphertext	s	e	c	u	r	i	t	y		o	f		e	−	s	e	r	v	i	c	e	s
Key	x	k	l	p	o	r	w	x		r	u		g	e	s	m	y	a	w	f	h	v
Ciphertext	k	n	o	e	2		c	a	−	2	S	−	b	e	−	n	s	w	3	e	b	e

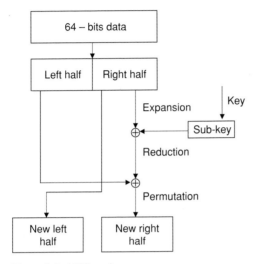

Figure 2.2   DES cycle.

to the plaintext to be encrypted. Therefore, one can deduce that there is no way to break the system. To illustrate this, we consider that binary codes between 00000 and 11111 are used to represent the set of characters:

$$\{\text{``} - \text{''}, a, b, \ldots, y, z, \text{``blank''}, 1, 2, 3,\}.$$

We assume that the ciphertext is given by: $c_i = m_i \oplus k_i$, for all $i \leq \text{length}(m)$, where:

$\oplus$ is the Exclusive OR operation;
m_i is the ith binary digit of plaintext m;
k_i is the ith binary digit of plaintext k; and
c_i is the ith binary digit of plaintext c.

Therefore, the ciphertext is obtained by performing a bitwise Exclusive OR (XOR) operation of the plaintext m with key k. The process uses the identity $m_i = c_i \oplus k_i$. An application of the system gives the example described by Table 2.1.

2.2.2   Secret key distribution

In order to communicate securely using a secret-key cryptosystem, the two parties must share the same key. Moreover, no third party can share this key with them. In addition to that, frequent key changes must be performed to limit the probability of data compromise and reduce the possibility that an attacker gets control of the key. Nevertheless, the security strength of any symmetric cryptosystem seems to rely heavily on the security of the key distribution process. Key distribution refers to the set of means and functions that are used to deliver a key to any pair of parties who wish to communicate securely. Key distribution can be realized between two parties, denoted by Us_A and Us_B, following three major methods:

1. The secret key is generated by one party and delivered manually or through a trusted party to other party.
2. The secret key is generated by one party and delivered to the other party using a shared secret previously established between the two parties.
3. The secret key is created by a third trusted party and manually delivered to Us_A and Us_B. The third party can also deliver the key on secure channels, if there is any (as presented by Figure 2.1).

Methods 1 and 2 are sufficiently secure and can be applied efficiently within a small-size organization. Meanwhile, method 3 is not appropriate for an organization's activity since recoverability of the key may be of challenging importance to the continuity of the business activity of an organization due to internal attacks. The availability of secure channels, however, appears to be a reasonable requirement, since each channel encryption is going to be exchanging data only with the trusted third party, which can be a unique entity (that is legally recognized). Nevertheless, in a distributed environment, any user may need to establish a secure communication with many other users over time, and therefore may need to share a large number of secret keys with these users. Coping with this large number would therefore represent a shortcoming of secret key cryptosystems. The problem becomes especially difficult in a wide-area distributed environment.

The scale of the distribution problem depends on the number of communicating pairs that must be supported. This may require the distribution of up to $n \cdot (n-1)/2$ keys for n users, if encryption is done at the user level. The use of a Key Distribution Center (KDC), as a third party responsible for distribution of the key, can be an appropriate solution to overcome the abovementioned shortcoming. This solution can use the concept of "hierarchy of keys" as follows. A typical scenario between users Us_A and Us_B involving the KDC would work according to a four-step process:

1. First, Us_A issues a request to the KDC for a session key to protect a specific logical connection to user Us_B. The message includes the identity of Us_A and an identifier, say ID_A, of the connection. The identifier ID_A can include a timestamp, a counter, or a random number, and a reference to the communication protocols to be used with the connection.
2. The KDC creates a special secret key Ks for the required session and responds with a message encrypted using a pre-established secret key k_A between Us_A and KDC (known only by Us_A and KDC). The message should contain: (a) the session key Ks; (b) the original request message for verification and session identification; and (c) an encrypted sub-message, encrypted by a pre-shared key k_B between KDC and Us_B, that Us_A should relay to Us_B. The sub-message contains Ks and the identity of Us_A, ID_A.

3. User Us_A can verify that its original request was not altered during its transfer and then sends a message to Us_B including its identity, the session identifier ID_A, the encrypted sub-message made by KDC to be relayed to Us_B, and an encryption of ID_A to support the authentication of Ks.

4. Upon arrival of the message, user Us_B can check the identity of Us_A and the integrity of session key Ks. At the end of this point, one can deduce that a secret key has been securely delivered to Us_A and Us_B, so that they can share it and use it during the connection.

Another scenario to deliver a session key to Us_A and Us_B using a KDC is to allow Us_A and Us_B to cooperate together to have their secret key and have the KDC realize the completion and protection of this process. This scenario can use the *Diffie–Hellman key agreement*, which is based on the discrete logarithm problem. Using the scenario, Us_A and Us_B can perform the following tasks, assuming that a prime p and an integer g are made publicly available:

1. Users Us_A and Us_B generate independently two random integers x_A and x_B, respectively.
2. User Us_A computes $y_A = g^{x_A} (\bmod\ p)$ and sends it to Us_B via KDC. Meanwhile, Us_B creates $y_B = g^{x_B} (\bmod\ p)$ and sends it back to user Us_A via KDC.
3. Us_A and Us_B compute:

$$z_A = (y_B)^{x_A} \bmod p \quad \text{and} \quad z_B = (x_A)^{x_B} \bmod p.$$

The two users assert that:

$$Ks = z_A \quad \text{and} \quad Ks = z_B.$$

The values z_A and z_B are equal. Therefore, $K_S = z_B$ is the shared secret that is only known by Us_A and Us_B. The underlying security problem behind the robustness is equivalent to solving the discrete logarithm-like problem, which can be stated as follows:

$$\text{Given: } y_A = g^{x_A} (\bmod\ p) \quad \text{and} \quad y_B = g^{x_B} (\bmod\ p),$$
$$\text{Find: } z = z_A = z_B = g^{x_A x_B} (\bmod\ p) = g^{x_B x_A} (\bmod\ p).$$

The role of the KDC in the above scenario is to guarantee the security of the exchange of y_A and y_B between the two users, and thus avoid attacks such as the *man-in-the-middle* attacks.

2.3   Public key cryptosystems

As mentioned in the previous section, in conventional encryption schemes, the key has to be shared by the communication entities and must remain secret at both ends. This does not appear to be a necessary condition to guarantee a secure communication. Instead, it is possible to develop a cryptographic algorithm that relies on one key for encryption and another key for decryption. These algorithms have the following characteristics. It is computationally infeasible to determine the decryption key given only knowledge of the cryptographic algorithm, a set of ciphertexts, and the encryption key.

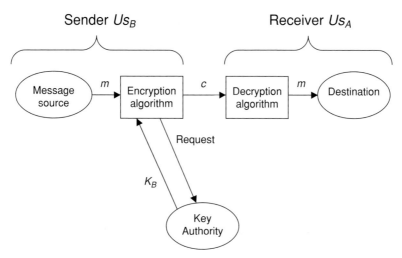

Figure 2.3   Model of public-key cryptosystem.

Public key algorithms are examples of these algorithms (Menezes, 1996). They use two different keys: an encryption key that is kept secret and a decryption key that is made public. Any user, say Us_B, that wants to send a message to user Us_A, who is in possession of a private key, can also use the related public key to encrypt the message and send it to the user. Only user Us_A, who knows his private key, can decrypt the encrypted data. A key authority guarantees the authentication of users' public keys by sending the requested public key of a recipient to a requesting sender. Figure 2.3 shows a public key cryptosystem scheme, where the key authority plays a particular role that ensures binding the public key with the related private key and the identity of its owner.

A formal description of a public key cryptosystem that differs from the one presented for a secret key cryptosystem is provided in Subsection 2.2.1 by the first statement related to the characteristics of the encryption and decryption functions ϕ and γ. This is modified as follows:

There is a function $\alpha : C \rightarrow C$ *such that* $\phi(\gamma(c, k), \alpha(k)) = c$ *and* $\gamma(\phi(m, \alpha(k)), k) = m$

where $(\alpha(k), k)$ represents the pair of private and public keys, and α characterizes the mathematical link between the private key $\alpha(k)$ and the related key owned by the user.

2.3.1   Trapdoor function model

A *one-way trapdoor function* is a one-way function, $f : X \rightarrow X$, defined from a message space X to a ciphertext space Y with the additional property given by some specific information, called trapdoor. It becomes computationally feasible to find y in the image $\text{Im} f$ of f, and an x in X such that $y = f(x)$. Most public key cryptosystems rely on the concept of one-way trapdoor function. The notion of one-way trapdoor function allows public key transmission and digital signature.

For public key transmission, if f is a trapdoor one-way function known only by one user Us_A, then one can choose:

$$Public\ key : f \quad and \quad Private\ key : f^{-1}$$

Assuming that f^{-1} is the inverse function and that f and f^{-1} are generated together. The related public key encryption algorithm should work as follows:

1. User Us_B wants to send message m to user Us_A privately. He/she encrypts m using Us_A's public key f (i.e., computes $c = f(m)$) and sends c to Us_A.
2. On receiving c, Us_A computes $m = f^{-1}(c)$, and therefore decrypts c using his/her private key, f^{-1}.

While Us_A's and Us_B's work are easy during this process, an attacker's work is hard since the attacker has first to invert the one-way function f.

The digital signature scheme is processed as follows assuming that user Us_A has f and f^{-1} as his/her *Public key* and *Private key*, respectively:

1. User Us_A wants to sign a message m and send it to user Us_B. He/she encrypts m using his/her private key (i.e., computes $s = f^{-1}(m)$) and sends the pair $\langle s, m \rangle$ to Us_B.
2. On receiving $\langle s, m \rangle$, Us_B computes $f(s)$, checks whether $m = f(s)$. If the identity is valid, then user Us_B is convinced that Us_A is the source of the message and that the message has not been modified during its transmission.

While Us_B's and Us_B's work are easy during the signature process, the attacker's work is hard to break the process since the attacker has first to invert the one-way function f.

Although no one knows whether one-way functions exist really, a candidate one-way function is exponentiation modulo prime numbers. The inverse function, the discrete logarithm, may not have a polynomial-time bounded algorithm.

2.3.2 Conventional public key encryption

The idea of a public key cryptosystem was proposed by Diffie and Hellman in their pioneering paper in 1976 (Diffie and Hellman, 1976). Their major idea was to enable exchange of secure messages between sender and receiver without having them to meet in advance and agree on a common secret key they can share to exchange messages. They proposed the concept of a trapdoor function and how it can be used to achieve a public key cryptosystem. Shortly thereafter, Rivest, Shamir and Adleman proposed the first candidate trapdoor function, the RSA. The essential steps of the public key encryption process are described below:

• Each end-system in a network generates a pair of private and public keys to be used for encryption and decryption of messages that it will receive.
• Each end-system publishes its encryption key by placing it in a public register or file, which is accessible for reading the public keys.
• A user wishing to send a message to a second user encrypts the message using the second user's public key.

- When the intended user receives the message, it decrypts it using its private key. No other recipient can decrypt the encrypted message.

The robustness of such a process is mainly based on the level of protection of the public directory and the algorithm strength. Aside from the obvious requirement that an adversary should not be able to easily derive the original message from its cipher text, the process should guarantee the following requirements:

- Computing even partial information about the original message given its related cipher-text should be very difficult.
- Detecting simple features about the traffic of messages handled by the same pair of keys (e.g., when the same ciphertext is sent twice) is a difficult task.
- The encryption algorithm should be constant in length. This means that the length of the encryption algorithm should not depend on the length of the message.

Note that these requirements are not easy to satisfy. The third requirement, for example, automatically disqualifies the use of any deterministic encryption scheme (e.g., RSA scheme). Various encryption schemes, including RSA, fall on the second requirement (Rivest, 1978).

2.4   Comparing cryptosystems

Symmetric key and public key encryption schemes have various advantages and disadvantages, some of which are common to both schemes. This section highlights a number of differences, advantages, and disadvantages.

Advantages of secret key cryptosystems

Advantages include, but are not limited to, the following main characteristics:

- Secret key algorithms can be designed to have high performance of data processing. It uses relatively short keys.
- Secret key algorithms can be employed as primitives to construct various cryptographic mechanisms including pseudo-random number generators, hash functions, and computationally efficient digital signature schemes.
- Secret key algorithms can be composed to produce stronger ciphers. Simple transformations, which are easier to analyze, can be used to construct strong product ciphers, even though they are weak when they are considered on their own.

Disadvantages of secret key cryptosystems

Such disadvantages include the following:

- In a two-party connection, the key must remain secret at both ends. This may present a vulnerability of the connection if the keys used last for longer periods.

- In a large-scale network, effective key management requires the use of an unconditionally trusted key distribution function, which unfortunately cannot be usually provided in wide area networks.
- In a two-party communication, a sound cryptographic practice requires that the key be changed frequently. Establishing a secret key per communication session would seriously increase the security strength of the cryptosystem.
- In digital signature practice, the mechanisms arising from secret-key encryption typically require either the use of large keys or the use of a trusted third party (such as a KDC).

Advantages of public key cryptography

Public key schemes have the following major advantages:

- The number of keys necessary for the use of public key cryptosystems in a network may be considerably smaller compared to the number required by secret key cryptosystems for the same environment.
- The private key is kept secret by its owner personally. No other party can have access to it. The authenticity of the public keys must, however, be guaranteed continuously by trusted entities in wide area environments.
- The administration of public keys, on use in a network, requires the presence of only a functionally trusted third party as opposed to an unconditionally trusted third party.
- A private key/public key pair may remain in use unchanged for considerably longer periods of time (and may even last several years). The mode of usage, however, should impact their durations.
- Public-key schemes provide relatively efficient digital signature mechanisms. The key used to describe the public verification function typically can be made smaller than for the symmetric-key counterpart.

Disadvantages of public key encryption

The main disadvantages of the public key scheme are:

- Performance rates for public key encryption methods are several orders of magnitude slower than the best-known symmetric-key schemes (at least the most popular methods).
- Key sizes are typically larger than those required for symmetric-key encryption, and the size of public-key signatures is larger than the tags used in secret key techniques in order to provide data authentication.
- No public key scheme has been proven to be secure. The most effective public key encryption schemes found to date have their security based on the presumed difficulty of a small set of number-theoretic problems.

In conclusion, one can say that the secret key and public key algorithms have a number of complementary advantages. Public key encryption techniques may be used to establish a key for a secret key cryptosystem being used by two communicating entities.

Combining symmetric and asymmetric cryptography

It is clear that the public key cryptographic approach has its biggest drawback in being slow. Typically, public key cryptography is not an alternative or replacement of symmetric cryptography, rather it supplements it. RSA, for example, allows two important functions that are not provided by secret key algorithms: secure key exchange without prior exchange of secrets and digital signatures for the purpose of authentication. This has encouraged the development of a large spectrum of solutions that combine the two technologies to provide a more efficient, effective and secure solution to real-life applications; especially, those applications developed in electronic administration and the electronic marketplace.

In this scenario, the two entities can take advantage of the long-term nature of the public/private to establish short-life session keys and benefit from the performance efficiencies of the secret key scheme to provide data encryption. The computational performance of a known public key encryption is inferior to that of symmetric key encryption. There is, however, no major proof that this must be the case for all public key encryptions. The important points to infer in practice are:

- Public key cryptosystems facilitate the efficient signature-based services, particularly the services which provide non-repudiation and key management.
- Secret key cryptosystems are efficient for encryption-based services such as data confidentiality.

For encryption of messages, public key cryptography and secret key cryptography are often combined in the following way (as depicted by Figure 2.4). A user, say Us_A, that wishes to send directly and securely a message to user Us_B, first generates a random secret key k_s (for algorithm AES for example) and encrypts the message with secret key k_s. Second, Us_A concatenates the resulting encrypted message m with key k_s to form a new message $m \| k_s$. Third, Us_A encrypts $m \| k_s$ using Us_B's public key and sends it to Us_B over the insecure communication channel. This protocol is sometimes known as an *RSA digital envelope*.

2.5 Public key main algorithms

Various public key algorithms have been developed. We describe in this section the two most important and widely used schemes. They are the RSA algorithm and the ElGamel algorithm.

2.5.1 *RSA algorithm*

The Rivest–Shamir–Adleman scheme is one of the most used public key algorithms. This scheme was introduced in 1978 and, nowadays, it remains secure and widely accepted. It incorporates results from number theory combined with the difficulty to determine the prime factor of large integers. It is a bloc-based encryption in which the plaintext and ciphertext are integers between 0 and $n - 1$ for some integer n. The encryption scheme makes use of an expression with exponents. Two interchangeable keys d (called private key) and (n, e)

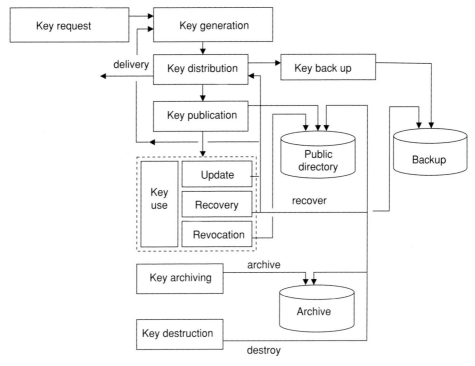

Figure 2.4 Key management life-cycle.

(called public key) are used for encryption and decryption. The plaintext block m (having a binary value smaller than n) is encrypted as:

$$c = m^d \bmod n.$$

Because of the exponentiation nature retrieving m from $m^d \bmod n$ is hard. The decrypting key e is carefully selected so that $(m^e)^d \bmod n = m$. Therefore, knowing d, the legitimate destination simply computes $c^e \bmod n$ to recover m. Both sender and receiver must know the value n. The sender knows the value d, and the receiver only knows the value e. Thus, this is a public key encryption algorithm using the following keys:

$$Public\ key : (n, e) \quad and \quad Private\ key : d$$

For this algorithm to be satisfactory for public-key encryption, the following requirements need to be fulfilled:

- It is possible to find integers n, e, and d such that:

$$m^{ed} = m \bmod n \quad \text{for all} \quad 0 \le m < n.$$

- It is relatively easy to compute $m^d \bmod n$ and $c^e \bmod n$ for all values of m smaller than n.
- It is computationally infeasible to compute d given e and n.

Traditional number theory shows that if integer n is a product of two primes p and q ($n = p.q$), d is chosen relatively prime modulo $(p - 1)(q - 1)$ (i.e., integer d has no factors in common with $(p - 1)(q - 1)$, and $m < n$, then the requirements are met for a unique e satisfying the following:

Let $\phi(n)$ be a function, called *Euler function*, given by $\phi(n) = (p - 1)(q - 1)$. The Euler function $\phi(n)$ is the number of positive integers smaller than n and relatively prime to n. Then there is an integer e such that:

$$
\begin{aligned}
e.d &= k\phi(n) + 1 & \text{for some } k, \\
m^{ed} &= m \bmod n & \text{for all } 0 \le m < n.
\end{aligned}
$$

Let us notice now that n, p, and q should be chosen to be very large. Typically, nowadays p and q are chosen to exceed 1024 bits each, so that n can exceed 2048 bits long. A large value of n effectively prohibits factoring n and therefore providing primes p and q. Moreover, a relatively large value of d can make the method robust.

Illustrative example

An example illustrating the RSA algorithm using relatively small n, p, and q gives the following:

- $p = 17$, $q = 13$, $n = pq = 221$
- $\phi(n) = 192$, $d = 5$, $e = 77$ (since $5 \times 77 = 1 + 2 \times 192$).

Choosing a message m such that $m = 6$, implies the ciphertext c is given by:

$$
c = m^d = 6^5 = 146 \bmod 221.
$$

Then the ciphertext associated with message m is given by 146.

2.5.2   ElGamel algorithm

The ElGamel public-key encryption algorithm was proposed in 1984 (ElGamel, 1985). Even though this algorithm is not widely used, it represents a US digital signature standard. The algorithm relies on the difficulty to solve the discrete logarithms problem. The ElGamel algorithm consists of two steps: (a) generation of the private and public keys and (b) encryption and decryption process.

During generation, user Us_A generates a large prime number p and two integers g and x, such that:

$$
\begin{aligned}
0 &< g < p \quad \text{and} \quad 1 < a < p - 1, \\
(p &- 1) \text{ has a large prime factor.}
\end{aligned}
$$

The user then computes $y = g^a \bmod p$. Then he/she states that a is his/her private key and (p, g, y) is the related public key. The encryption/decryption process works as follows:

To encrypt a message to Us_A, user Us_B obtains Us_A's public key $\langle p, g, g^a \bmod p \rangle$ then takes a message (or the blocks of a message) as an integer m that is in the range $\{0, \ldots, p^{-1}\}$.

He/she selects a random integer k, $1 \leq k \leq p - 2$, and computes:

$$\gamma = g^k \bmod p \quad \text{and} \quad \delta = m \times (g^a)^k \bmod p.$$

Finally, user Us_B sends the ciphertext $c = (\gamma, \delta)$ to user Us_A.

To recover the plaintext m from ciphertext c, user Us_A should perform the following actions: (a) Us_A uses the private key a to compute $\gamma^{p-1-a} \bmod p$ (this is equal to $\gamma^{-a} \bmod p = g^{-ak} \bmod p$) and (b) recover m by computing:

$$(\gamma^{-a}) \times \delta \bmod.$$

The decryption process guarantees the recovery of m, since we have:

$$(\gamma^{-a}) \times \delta = g^{-ak} \times m \times g^{ak} = m \bmod p.$$

Example

To illustrate the ElGamel cryptosystem, let us consider this hypothetical example. User Us_A chooses $p = 89$, $g = 2$, and $a = 31$; then he/she generates the public key:

$$(89, 2, 2^{31} \bmod 89) = (2, 89, 67).$$

Decrypting message $m = 91$, user Us_B chooses $k = 43$ and performs the following computations:

$$\gamma = g^k \bmod p = 2^{43} \bmod 89 = 33$$
$$\delta = m \times (g^a)^k \bmod p = 91\,(2^{31})^{43} \bmod 89 = 26.$$

The decryption scheme computes:

$$(\gamma^{-a}) \times \delta = (2^{43})^{-31} \times 91 \times 2^{31})^{43} = (33)^{58} \times 26.$$

Finally, this gives m.

2.6   Public key management

A large degree of the efficiency of public-key cryptosystems can be achieved by providing a secure management of the public keys. This function includes the key distribution, key recovery and key generation. This section will develop the main features of a public key management system. It also discusses the public distribution scheme provided in the literature.

2.6.1   *Key management life cycle*

It is clear that the key management is easier when all cryptographic keys are made fixed over time. However, the set of states through which a key evolves during its existence, called the *key life cycle*, requires a lot of attention and controls to protect the active keys during their creation, usage, backup, and deletion. The life cycle processes (as depicted by Figure 2.4) operate as follows:

1. *Key generation*   The generation process starts with the acquisition of initial keying material (e.g., information provided by the user through a request). The user requesting a pair of keys, can generate the pair or acquire it from trusted system components. Key generation applies methods that ensure the appropriate properties and randomness (unpredictability of the key) for the intended application.

2. *Key distribution*   The distribution process guarantees the delivery of the keying material to the user that has requested the generation of a pair of keys. The process includes the installation for operational use within the user's system. The installation may utilize different means and techniques to achieve distribution (e.g., read-only-memory component, hardware tokens, and tools for the transfer of disk content).

3. *Key publication*   In association with key distribution, the keying material is officially stored (by a registration trusted authority) for appropriate and well-specified needs, including making the public key publicly available for access through a public directory or other means. An official storage of the private key can be securely performed under the authorization of the user for recovery needs, if any.

4. *Key operation*   The operation includes five functions: key use, key update, key storage, key recovery, and key revocation. Key use facilitates the operational availability of the private/public key for cryptographic purposes. Key update ensures that prior to the expiring life-time the keying material is replaced by a new keying material. Key storage provides the short-term backup of keying material in secure media and may provide a data source for key recovery. Key recovery restores the keying material in case of key loss or key compromise. Finally, key revocation removes the public key from the operational use prior to its expiring life-time.

5. *Key archiving*   Pairs of keys that are no longer in operational use should be archived to provide a long-term source for key retrieval and security proofs under specific conditions. This includes archiving of official material for legal periods.

6. *Key destruction*   This process allows the removal of a pair of keys from all official storages, provided that there are no further needs to maintain the keying material and its association with an entity, which has used it during its life-time.

Figure 2.4 shows that the key life-cycle process integrates the use of three containers (files, databases or repositories). They are: the public directory, backup container, and key archive.

2.6.2   Key distribution

Different protocols have been developed for key distribution within different frameworks and techniques. The available protocols often depend on how they guarantee the level of trust a user is expecting from them, the amount of information the sender and receiver already share, the degree of on-line interactions that the user should have with the key repository, and the type of standards they allow. In what follows, we consider typically the distribution using a publicly available directory of public keys, where maintenance and distribution of the public directory would have to be the responsibility of some trusted third party (TTP).

The TTP has to perform the following generic tasks, no matter how the distribution scheme realizes its tasks:

1. The TTP manages, maintains, and updates a directory where the entries should (at least) have the form $\langle ID_{Use}, Id_{Ent}, Public\text{-}key, Date \rangle$, where ID_{Use} is the user identification, Id_{Ent} is the entry identification within the directory, and *Date* stands for the date of insertion of the entry in the directory and its validity period (eventually).

2. Users register their public keys with the TTP through well-established procedures that may have to indicate whether the registration should have to be in person or by means of secure authenticated communication. It also indicates the type of verification the TTP should perform before the publication of the public key.

3. The TTP allows users to access the directory electronically. It is responsible for keeping the directory publicly available. To this end, TTP should provide the appropriate level of security to: (a) the communication between the directory and the accessing user; (b) public key repository; and (c) different tools used at the TTP site to access and maintain the repository.

The above scheme is definitely more secure than individual public announcements; however, it presents different shortcomings. First, the public key archive may constitute an important vulnerability, if the level of protection provided by the TTP can be hacked. Second, the level of verification may be unsatisfactory to provide trustworthiness. Finally, on-line interaction with registrees, if any, may be hacked and therefore be subject to inappropriate data insertion. Stronger security for public-key distribution can be achieved by providing stronger control over the distribution of public keys from the directory. A typical scenario assumes the existence of a legally established key distribution center (KDC) that maintains the directory of public keys of all based on a legally approved policy.

The access protocol provided by the KDC operates as described by the following steps:

- A user, named Us_A, sends a message to the KDC requesting the public key of user Us_B that he/she wants to communicate with. The message can be in a clear form or encrypted with the public key of the KDC. It also can be time-stamped.

- The distribution center responds with Us_B's public key and his/her identifier $ID_{Use(B)}$, encrypted using the distribution center's private key. There is no confidentiality issue with this information, but only integrity. User Us_A is able to decrypt the message using the KDC's public key. Therefore, Us_A is guaranteed that the message originated from the KDC has a valid content, including Us_B's public keys and identity, the information that the original request may require, and the timestamp, if any.

- User Us_A can store Us_B's public key and also can start communicating directly with Us_B using the key to encrypt any message he/she wants to send confidentially to Us_B. User Us_A is sure that only Us_B can read it.

- User Us_B can use the identity of Us_A to request Us_A's public key from the KDC. On receiving the required public key, Us_B can use it to respond securely to Us_A.

The registration process with KDC can be provided by different means. First, the KDC can publish widely its own public key, and any user who wants to register with that center simply sends his/her public key and whoever required information to the KDC encrypted under the center's public key. Second, the KDC can provide a secure shell to any user who is willing to register to allow the user delivering his/her public key and identity. The KDC should have adopted a procedure for verifying the delivered information. The verification uses a set of methods that range from a simple check of the link relating the user's public key and private key to the presentation of official proofs individually.

This security solution may be attractive. However, it still represents some shortcomings. The KDC could reduce the performance of the KDC server, since a user must appeal to the KDC for the public key for every other user that he/she wishes to communicate with. The key distribution center is an essential component in the distribution scheme: it must be available anytime any user needs a public key. Finally, key authentication may be questionable; that is, that they belong to the users whose identities are associated.

An alternative approach is to create what are called digital certificates, which can be used by users to exchange keys without contacting a KDC. A public key and the user's identity are bound together in a digital certificate, which is then signed by a third party certifying the accuracy of the binding. The third party is called the certification authority. A user wishing to communicate directly with another user conveys directly his/her public key information to the other by transmitting his/her digital certificate. The receiving party can verify that the certificate is signed by the authority. The following requirements are essential to the success of this scheme:

*   Any user should be able to read the content of a certificate to determine the name and public key of the certificate's owner.
*   Any user should be able to verify that the certificate originated from the certificate authority and is not counterfeit.
*   Any user should be able to access a public directory, which is protected by the KDC in order to check the non-revocation of a received certificate.

Certificate authority function, certificate content, and certificate management will be discussed in detail in Chapter 4. Different variations on the certificate concept will also be discussed in what follows.

2.6.3   Key recovery

Ciphertexts and private keys do not enjoy physical security since they are intended for communication over non-secure channels. Thus, there are no *crypto locksmiths* that can retrieve lost or forgotten keys! This may generate unacceptable losses to organizations if the keys have been used to encrypt files, which need to be accessed for business purposes. Key recovery allows enterprises and government agencies (subject to legal controls) to access copies of the cryptographic keys being used to protect information exchange and storage. Key escrow for instance provides a means to allow other legitimate users access, on an emergency basis, to adequately secured data (Denning, 1996; NIST, 1994). The Clipper Key Escrow (CKE) is a special example of key escrow methods. The main idea of CKE is to divide a key into n pieces and distribute them among n individuals in such a way that any k individuals, among them, are able to recover the key, but no $(k-1)$ individuals can do so (Shamir, 1995). CKE uses an encryption protocol and key exchange protocol.

However, the concept of key escrow is seen by many experts as a step to imposing controls on cryptography and limitation of privacy.

Meanwhile, it was argued that end users needed an emergency means for key recovery to protect against the possibility that their primary keys were lost, stolen, or damaged. Key recovery represents a capability to recover the cryptographic keys being used to protect information when the primary means for obtaining these keys is unavailable. For businesses protecting their data using encryption utilizing employee-based keys, it was also demonstrated that it would be important for the enterprises to maintain access to the encrypted data in the event of problems with the keys held by employees.

Key recovery for stored data

If cryptography is used for stored data, there is an evident risk that key loss or damage might induce a loss of the data itself. Such limitations show that the benefits of backing up encryption keys will almost always outweigh the additional risks that this will induce. Therefore, it appears normal in a business environment to provide for emergency key recovery when encryption is used for stored data.

Key recovery for communications

If cryptography is used for protecting communications channels, there is no end user interest in key recovery since the unencrypted data streams will be available to all the parties involved in the communication. If an encryption key is lost or damaged it does not need to be recovered since the data can be sent again using a new encryption key. It is sometimes suggested that enterprises have a need for communications key recovery in order to check what their employees are doing by decrypting at the communications level. While this appears understandable at a superficial level, a careful analysis suggests that such a requirement is unlikely to exist.

When an enterprise wants to secretly intercept its own encrypted communications data, it is hard to accept why it would do it on the encrypted side of a communication link rather than on the unencrypted side. If the argument is that this has to be hidden and not visible to any employees then it will have to intercept this traffic outside the company domain and this will be filled with many technical and legal difficulties. For such reasons, any interception will almost certainly have to be done within the enterprise boundary. Another argument often used for enterprise communications key recovery is the need for the enterprise to audit all its transactions. In that case, it is hard to understand why this would be done after the communication-level encryption is employed rather than prior to it.

Typically, key recovery systems work following different paradigms. Early key recovery techniques were based on the storage of private keys by a trusted authority. More recent systems allow key recovery agents to maintain the ability to recover the keys related to an encrypted communication session, provided that such session keys are encrypted by a master key known by the agent. However, all these systems share two essential and negative features: (a) the use of an external mechanism to the primary means of encryption/decryption and (b) the existence of a highly secured key that should be protected for a longer period.

On the other hand, key recovery systems are by nature less secure and have a high cost. Key recovery degrades the protections that are made available using primary encryption.

Finally, we notice that the failure of key recovery mechanisms can damage the proper operation and encryption-based services. Threats include disclosures of keys, theft of valuable key information, or security service failure. Moreover, new vulnerabilities are introduced by key recovery: inherent guarantees of security are removed and new concentrations of decryption information are created.

2.7 Attacks against public key cryptosystems

Cryptanalysis is the science that aims to recover plaintexts from the related ciphertexts, discover private keys, and break cryptographic algorithms. Cryptography and cryptanalysis are considered complementary branches of science. Advances in one are usually followed by further advances in the other. In this section, we give a brief description of the main methods used to attack the RSA cryptosystem. These include the factoring methods, attacks on the underlying mathematical function, and attacks that exploit implementation details and flaws. The choice of RSA is made for two reasons. First RSA is among the most commonly used cryptosystems. Second, twenty years of research has led to a number of fascinating attacks over RSA, but none of them is considered devastating (Boneh, 1999; Nicopolitidis, 2003).

Factoring

The security of the RSA encryption is intimately linked to the integer factorization problem, which aims at finding prime factors of composite numbers. The integer factorization problem is widely believed to be a hard problem (i.e., there is no polynomial-time algorithm that solves the problem). If an adversary can factor the integer n occurring in a public key (n, e), he/she can recover any plaintext since it is possible to deduce the private key easily. Different factoring methods have been developed. We give below an outline of some of them.

Trial division attempts to divide n by all successive primes until one factor is found. Although this method is effective when factoring relatively small composite numbers and numbers having relatively small factors, it is useless against the types of numbers dealt with in RSA.

Pollard's $(p - 1)$ method relies on a special property of a divisor of n. Let p be an unknown factor, for which $(p - 1)$ is a product of a number of small primes (smaller than a given real number α). Let us now define β as the product of large enough powers of all primes smaller than α. Then one can state that β is a multiple of $p - 1$; and therefore $2^\beta = 1 \bmod p$ (this is due to Fermat's Theorem). Thus p divides the greatest common divisor of $2^\beta - 1$ and n. Since β can be large, the number $2^\beta - 1$ can be computed modulo n. Therefore, using Pollard's method, one can be assured to find a prime factor of n in time proportional to the largest prime factor of $p - 1$. But since p is unknown, we have to choose various increasing values for the bound β.

A third method that we consider here is called the Quadratic Sieve Method (or QSM). It attempts to find integers x and y such that:

$$x^2 = y^2 \bmod n \quad \text{and} \quad x \text{ different from } y \bmod n \text{ and } -y \bmod n.$$

On the existence of x and y, we conclude that:

$$n \text{ is a divisor of } x^2 - y^2 = (x - y)(x + y) \text{ and}$$
$$n \text{ does not divide } (x - y) \text{ nor } (x + y).$$

Therefore, it follows that the greatest divisor of $x - y$ and n is a non-trivial factor of n. If $n = pq$, a random solution (x, y) of the equality $x^2 = y^2 \bmod n$ would provide a factor of n with probability higher than 0.5.

Attacks on the RSA function

The following attacks exploit special properties of the RSA function. They take advantage of a bad choice of the private exponent d or public exponent e, partial key exposure, or a relation between encrypted messages (Boneh, 1999).

The low private exponent attack shows that the choice of small values for the private key d leads to a break of the system. More specifically, the attack shows that if the condition $d < 1/3(n)^{1/4}$ is satisfied then the public key (n, e) can efficiently recover the secret key d. Therefore, a typical constraint for d, when n is 1024-bit modulus, is to have at least 300-bit length.

The partial key exposure attack applies when the modulus n is k bits long and $(k/4)$ bits of it at least are exposed. In that case, a malicious adversary can recover the value of d in time proportional to $elog(e)$. This shows that if exponent e is small, the exposure of a quarter of the bits of d can allow the recovery of d. Therefore, the issue of protecting RSA private keys is an essential task. Unfortunately, protecting private keys is overlooked in an organization's environments.

Implementation attacks

These attacks, also named side-channel attacks, target implementation details of the RSA cryptosystem or exploit faults observed in the implementation.

Timing attacks against RSA cryptosystems take advantage of the relation between the private key and the runtime of the cryptographic operation. By measuring the runtime of the private operation on a large number of messages, a timing attack allows an adversary to recover the bits of d one at a time, beginning with the least significant bit.

Various other interesting attacks against the RSA function have been developed. All these attacks have taken advantage of RSA features to reduce the complexity of breaking down the cryptographic system in particular cases; but did not create real breaches in the RSA cryptosystem. RSA attacks available on the Internet include the short pad attack, broadcast attack, power analysis attack, and failure analysis attack. Therefore, there is a set of issues that deserve particular attention in order to provide a high level of security to public key cryptosystems. They are: key size, implementation details, and features of cryptographic parameters. Rules to observe for the keys impose that keys should be long enough to make the attacks unachievable and that deciding the key size should take into consideration the value of the data to encrypt, its life time, and the attacks that can be involved.

Issues to consider for the security of the RSA cryptography include the following requirements:

- Public key: The values 3 and $2^{10} + 1$ are commonly used for the public exponent e. Since some of the attacks developed in the literature have achieved success with value 3, it is important to use the other value.
- Prime factors: Prime factor of the modulus n should be large enough so that $p - 1$ has large prime factors.

Moreover, during the implementation phase, particular attention should be paid by the designers and developers of cryptographic applications to the implementation details.

2.8   Conclusion

Secret key cryptosystems and public-key cryptosystems appear to be very efficient as well as complementary tools to provide efficient security services. While the first class provides efficient means for encrypting messages, the second class provides interesting mechanisms for digital signatures, key establishment, and session keys delivery.

This chapter describes the main features of the two classes, addresses the advantages and limits of each class, and particularly highlights the role of one-way functions and key distribution centers. Problems related to key recovery are also discussed from different points of view.

References

Boneh, D. (1999). Twenty years of attacks on the RSA. *Notices of the ACM*, Vol. **44(2)**, 203–13.

Denning, D. (1996). A taxonomy for key escrow encryption systems, *Communications of the ACM*, **39**(3), 34–40.

Diffie, W. and M. Q. Hellman (1976). New Directions In Cryptography. *IEEE Transactions on Information Theory*, Vol. **22(6)**, 644–54.

ElGamel, T. (1985). A public key cryptosystem and a signature scheme based on discrete logarithms. *IEEE Transactions on Information Theory*, IT **31**, 4, 469–72.

Menezes, A., P. van Oorschot, and S. Vanstone (1996). *The Handbook of Applied Cryptography*. CRC Press.

National Bureau of Standards (1993). *Data Encryption Standard (DES)*, FIPS Publications, 46-2, (available at http://www.itl.nist.gov/fipspubs/fip46-2.htm).

NIST (1994). *Escrow encrypted standard*, FIPS Publications. 185.

NIST (1999). *Data Encryption Standard (DES)*. Federal Information Processing Standards Publications, FIPS 46-3.

NIST (2001). *Advanced Encryption Standard (AES)*. FIPS Publications, 197. http://csrc.nist.gov/publications/fips/fips197/fips-197.pdf, November 26, 2001.

Nicopolitidis, P., M. S. Obaidat, G. I. Papadimitriou, and A. S. Pomportsis (2003). *Wireless Networks*. Wiley.

Rivest, R., A. Shamir, and L. Adleman. (1978). A method for obtaining digital signatures and public-key cryptosystems. *Communications of the ACM*, **21**(2), 120–6.

Shamir, A. (1995). *Partial Key Escrow*: A New Approach to Software Key Escrow, presentation at NIST Key Escrow Standards meeting, Sept. 15, 1995.

3 Authentication and digital signature

This chapter considers the techniques developed to provide assurance that the identity of a user is as declared and that a transmitted message has not been changed after its signature. This prevents impersonation and maintains message integrity. Weak authentication and strong authentication schemes are addressed and the most common authentication services are also elaborated on in this chapter.

3.1 Introduction

As stated in the previous chapters, entity authentication can be defined as the process through which the identity of an entity (such as an individual, a computer, an application, or a network) is demonstrated. Authentication involves two parties, a *prover* (called also *claimant*) and a *verifier* (called also *recipient*). The prover presents its identity and a proof of that identity. The verifier ensures that the prover is, in fact, who he/she claims to be by checking the proof. Authentication is distinct from identification, which aims at determining whether an individual is known to the system. It is also different from authorization, which can be defined as the process of granting the user access to specific system resources based on his/her profile and the local/global policy controlling the resource access. In the following sections, however, we will use the terms identification and authorization to designate the same concept.

Message authentication, on the other hand, provides the assurance that a message has not been modified during its transmission. Two main differences can be observed between entity authentication and message authentication, as provided by the techniques described in this chapter: (a) message authentication does not provide time-related guarantees with respect to when the message has been created, signed, sent, or delivered to destination and (b) entity authentication involves no meaningful message other than the claim of being a given entity.

An example of authenticators (or authentication tools) is the use of biometrics technology, where an individual's identity can be authenticated using, for instance, fingerprints, retinal patterns, hand geometry, voice recognition, or facial recognition. Most of these forms of authentication are being used widely. Digital signatures are beginning to replace passwords as a means of access controls to networks and information systems. Digital signatures allow users receiving data over communication networks to correctly determine the origin of the information as well as find out whether it has been changed during transmission (Obaidat, 1994; Obaidat, 1997; and Obaidat, 1999).

Because authentication has started to be such an essential component of business and consumer activity, all the various technologies being developed to provide authentication services will have to prove their efficiency and level of security. However, the functions used to produce an authenticator may be grouped into three categories: (a) message encryption-based authenticators, which use encryption so that the produced ciphertext of the entire message serves the authentication of the message (or/and encryptor); (b) cryptographic checksum-based authenticators, which use public checksum-like functions of the message and a secret key that produces a fixed-length value to serve as the authenticator; and (c) hash function-based authenticators, which use public one-way functions to map a message of any length into a fixed-length value that serves as the authenticator of the message.

An entity authentication protocol is a real-time process that provides the assurance that the entity being authenticated is operational at the time when that entity has carried out some action since the start of the protocol execution. From the point of view of the verifier, the result of an entity authentication protocol is the acceptance of the prover's identity as authentic or its non-acceptance. The authentication protocol should fulfill the following objectives:

- The probability that any third party different from the prover, using the authentication protocol and impersonating the prover, can cause the verifier to authenticate the third party as the prover, is negligible.
- The verifier should not be able to reutilize the information provided by the prover to impersonate him/her to a third party.
- A signed message cannot be recovered from the signature code during signature verification.

A digital signature can be defined as a cryptographic message enhancement that identifies the signer, authenticates the message on a bit basis (i.e., every bit is authenticated) and allows anyone to verify the signature, with the restriction that only the signer can apply. Digital signature is different from authentication. A digital signature identifies the signer with the signed document. By signing, the signer marks the text in his/her own unique way and makes it attributed to him/her. The concepts of signer authentication and document authentication encompass what is called non-repudiation service. The non-repudiation service provides proof of the origin or delivery of data in order to protect the sender against false denial by the recipient that the data has been received, or to protect the recipient against false denial by the sender that the data has been sent. Typically, a digital signature is attached to its message and stored or transmitted with this message. However, it may also be sent or stored as a separate data element since it maintains a reliable association with its message.

In addition to the above features authentication and digital signature can be classified into weak and strong techniques, while entity authentication methods can be divided into three classes, depending on which paradigm they are based on. They are:

1. *Known-object-based authentication* Methods from this class use as an input what the user presents (through challenge-response protocols) for authentication. Examples

of known object-based methods include standard password, personal identification numbers (PIN), secret keys, and private keys.

2. *Possessed-object-based authentication* Techniques in this class use physical devices to authenticate an entity. Examples of such techniques include credit card, smart card, and hand-held customized devices that provide time varying PINs.

3. *Biometric-object-based authentication* Methods in this class use human characteristics to achieve user authentication. Examples of characteristics include fingerprints, retinal patterns, voice, hand geometry, and keystroke dynamics.

Techniques used in possessed-object-based authentication are typically non-cryptographic and will not be extensively considered in this book. Techniques belonging to the third category will be discussed in Chapter 5. The current chapter discusses the techniques used by known-object-based authentication.

3.2   Weak authentication schemes

Weak authentication typically fails to provide a complete and efficient authentication. Two classes of weak authentication schemes are of particular interest since they are commonly used. They are the password-based authentication and the PIN-based authentication. We discuss in this section the main features and drawbacks of weak authentication techniques.

3.2.1   Password-based authentication

Password authentication is perhaps the most common way of authenticating a user to an electronic system. To be authenticated, the system compares the entered password against the expected response after the login of the user. Typically, password-based authentication falls into the category of secret-key methods as it uses conventional password schemes that are time-invariant passwords. Passwords as associated with an entity are strings of characters (usually with a size larger than eight), which serves as a shared secret key between the entity and the system to be accessed. To be authenticated, the entity provides a (*login, password*) pair, where *login* is a claim of identity and *password* is the shared password, which is used to support the claim. The system then checks whether the pair matches the entity's identity and the secret it shares with the entity.

Generally, password schemes differ by the technique used to perform the verification and store the information providing password verification. For non-time-varying passwords, the system can store the entity's passwords in a system file, which is protected against read and write operations. In that case, the system does not use any kind of cryptographic object. Therefore, this method provides no protection against privileged system users and subsequent accesses to the file after backup operations. To overcome such drawbacks, the verifying system can apply a one-way function to the passwords and store the resulting values. In such a situation, to verify the password provided by an entity, the system computes the one-way function on the entered data and checks whether the result matches with the entry it has stored.

Non-time-varying password schemes present various security weaknesses. An adversary, for example, can perform various attacks to get control of a user's password. He can observe it as it is introduced by the user, or during its transmission; and then he can use the captured password for subsequent impersonation. He also can perform password-guessing and dictionary attacks. Finally, in the case of the use of one-function as a means to protect passwords, the adversary may attempt to break the list of passwords by providing arbitrary passwords, one by one, and comparing their values under the one-way function to passwords in the file.

While there are substantial problems with non-time-varying password-based authentication, it should be noticed that passwords are very familiar and offer a high level of user acceptability and convenience. Added to this, administrative rules can be used within an enterprise to ensure the user-chosen passwords satisfy certain criteria for acceptability (e.g., size, use of digits) and provide a periodic modification of the passwords under use.

A natural enhancement of fixed-password schemes is given by time-varying password schemes such as what are called "one-time password" schemes, which ensures that a system user utilizes a new password at each new access. With one-time password schemes, a list of passwords is managed for each user. Each password in the list is used only once. The passwords can be in three forms:

1. Pre-written in the list;
2. sequentially updated; or
3. sequentially computed.

In the first case, the list is not used sequentially, but the entity and the system agree to use a challenge-response table containing n pairs of the form:

$$\langle i, password_i \rangle, i \leq n.$$

On access request, the system challenges the users with a value for i, and waits for the right password. The second case considers an initial unique shared password.

The user is assumed to create and transmit to the system the new password, say $password_{i+1}$, during the authentication period covered by $password_i$. The last case assumes that the user has an initial password, say pwd_0. The user utilizes a one-way function h to define the password sequence so that, for $i < n$, the ith password pwd_i is give by:

$$pwd_i = h^i(pwd) = h(h(\ldots(pwd)\ldots)).$$

On the occurrence of problems (or on reaching the value n), the system is assumed to restart with the shared initial password.

Even though they provide better security than the non-time varying schemes, the one-time schemes, however, present several drawbacks. An active attacker, for example, can attempt to intercept an unused one-time password and impersonate the system.

3.2.2   PIN-based authentication

Personal identification numbers (PIN) schemes can be classified as special time-invariant passwords since a physical device, such as a banking card or a token, is used to store the PIN.

Typically, PINs are short strings of digits (from four to ten digits). This scheme represents a vulnerability that should be covered by additional constraints. Examples of protections include invalidating the physical device when more than a pre-specified number of incorrect PINs are attempted by an adversary (or the user himself).

In an authentication system that uses PINs, a claimed identity accompanied by a user-provided PIN may be verified (on line) by comparing it with the PIN stored for that user in a system database. An alternative approach, called validation off-line, does not use a centralized database and considers that the verification is performed based on information stored on the device itself. In such a situation, the PIN may be defined to be a function of a secret key and the identity of the user associated with the PIN. Moreover, the device should contain additional information allowing the token (and therefore the associated user) to be authenticated. This however requires the user to possess the device and remember the PIN.

3.3   Strong authentication schemes

Typically, a strong authentication scheme is based on the concept of cryptographic challenge-response protocol, which works as follows. A user wishing to access a service (or use a resource) should prove his/her identity to the verifier by demonstrating knowledge of a secret information known to be sufficient to authenticate the claimant. The demonstration is usually made without revealing the secret information to the verifier. It is typically achieved by providing the right response to a time-varying question (or challenge) related to the secret information.

Typically, a cryptographic challenge-response protocol is built on secret-key cryptosystems, public-key cryptosystems, and zero-knowledge techniques. It often uses time-variant parameters to uniquely identify a message or a sequence involved in the process, and thus protect against replay and interleaving attacks. Examples of time-variant parameters include the timestamps, random numbers, and sequence numbers. Combinations of these parameters may be used to ensure that random numbers are not repeated, for example. Often, random numbers are used as follows. An entity can include a random number in a transmitted message. The next received message, whose construction has used the random number, is bound to that number, which links tightly the two messages. Various drawbacks can be observed with protocols using this technique, including the use of pseudorandom number generators and the generation of additional messages.

A sequence number, such as serial number, transaction number, and counter value serves as an identifier of a message within a series of messages. Sequence numbers must be associated with both the source and destination of a message. The association can be explicit or implicit. Parties using a sequence number scheme agree on the fact that a message is accepted only if the sequence number contained in the message conforms to a pre-defined policy. The bare minimum policy should define the starting values of the sequence, window time, and monotonicity type of the sequence. Problems can limit the use of sequence numbers including delays experienced at the verifier's side.

Timestamps can be used to provide timed guarantees and prevent message replays and can serve to implement access privileges. Timestamps work as follows. An entity originating a message inserts in it a timestamp that is cryptographically bound to it. On the receipt of the timestamp, the destination computes the difference τ between its local time and the received timestamp. The message is received if τ is within the acceptance period of time and no other message with the same timestamp has been previously received from the originating entity. However, the security of the timestamp scheme relies on the use of a synchronized clock.

3.3.1 Challenge-response by cryptosystems

Challenge-response mechanisms based on cryptosystems expect the entity requiring an access to share a secret key with the verifier (in the case of secret-key cryptosystems) and to secure his public key (in the case of public-key cryptosystems).

Secret key cryptography

The general model of a challenge-response mechanism using secret key cryptography (and a random number) can be described as follows. Let r denote a random number; $\phi(k, -)$ denotes a secret key cryptographic function parameterized by a shared key between Us_A and Us_B. Assuming that each user is aware of the identity of the other entity, the authentication of Us_A starts by having the verifier Us_B sending a random number r. On receiving r, Us_A computes $\phi(k, \langle r, m \rangle)$ and sends it back to Us_B. Message m is optional used to prevent replay attacks. Syntactically, this can be described by:

$$Us_B \rightarrow Us_A : r$$
$$Us_A \rightarrow Us_B : \phi(k, \langle r, m \rangle).$$

Then the verifier decrypts the received random number and checks whether it is the random number that was provided before.

The above mechanism may include a one-way function to provide a more efficient challenge-response mechanism. In the case of mutual authentication, three messages can be used:

$$Us_B \rightarrow Us_A : r$$
$$Us_A \rightarrow Us_B : \langle r', \phi(k, h(\langle r, r', m \rangle)) \rangle$$
$$Us_A \rightarrow Us_B : \phi(k, h(\langle r, r', m' \rangle))$$

where r is a random number generated by Us_B, r' is a random number generated by Us_A, m and m' are optional messages, and h is the one-way function.

Public key cryptography

Public key techniques can be used to provide challenge-response authentication, with the originator of a request demonstrating knowledge of his private key. The originator can use two approaches to achieve this: either he decrypts a challenge that the verifier has encrypted using the originator's public key or digitally signs a challenge.

Challenge-response based on public keys operates typically as follows. User Us_B generates a random number r. Then, he computes its value $h(r)$ using a one-way function, encrypts r along with a general message m, using Us_A's public key, and sends to Us_A the value $h(r)$ appended to m and the encryption result. User Us_A decrypts the received message to recover r' and $h(r')$. If $h(r) = h(r')$, Us_A sends r to Us_B; otherwise he exits. Formally, this is done by:

$$Us_B \rightarrow Us_A : \phi(k_A, \langle h(r), m \rangle)$$
$$Us_A \rightarrow Us_B : r \text{ if reception is valid else exit}$$

where ϕ and k_A stand for the public-key encryption function and the public key of Us_A, respectively.

Challenge-response based on digital signature typically assumes that a request originator, upon receiving a random number r from the verifier, generates a random number r' and an optional message m (which may be equal to the identity of the verifier). Then he signs $\langle r, r', m \rangle$ and sends the result to the verifier of the digital signature appended to r' and m. If Us_B is the request generator then the verifier result received by Us_A has the following form:

$$\langle \langle r, r', m \rangle, \phi(K_B, \langle h(r, r', m) \rangle)$$

where K_B stands for the private key of Us_B.

3.3.2 Challenge-response by zero-knowledge techniques

The abovementioned challenge-response mechanisms might reveal part of the secret information covered by the user wishing to be authenticated by a verifier. A malicious verifier, for example, may be able to submit specific challenges to obtain responses capable of recovering such part of information. Zero-knowledge was proposed in the literature to overcome such drawbacks by allowing a user to prove knowledge of secret information while revealing no information of some interest to the verifier whatsoever (Anshel, 1997).

Roughly, the zero-knowledge technique is a form of interactive proof, during which the originator (or prover) and the recipient (or verifier) exchange various messages and random numbers to achieve authentication. To perform such interactive proofs, the concept of proof is extended to integrate some forms of probabilistic features. The general form of a zero-knowledge scheme is given by a basic version of the Fiat–Shamir algorithm (Fiat, 1987). We present this algorithm in the following. However, we notice that more efficient versions of this algorithm are now under use within various solutions. In these versions, multiple challenges may be used.

In the basic version of Fiat–Shamir authentication, user Us_A proves to Us_B knowledge of a secret s in n executions of a 3-phase process:

Phase 1

Secret generation A trusted third party (TTP) selects two large prime numbers p and q. Then he publishes $n = p \cdot q$ while keeping secret p and q. A user, say Us_A, wishing to be authenticated by Us_B, selects a secret s relatively prime to n, $1 < s < n$, computes a public key c by:

$$c = s^2 \bmod n$$

and registers the public key with the trusted third party.

Phase 2

Exchanging messages User Us_A generates a random number r, $0 < r < n$, and participates in the following three actions:

- $Us_A \rightarrow Us_B$: $x = r^2 \bmod n$
- $Us_B \rightarrow Us_A$: random Boolean number b
- $Us_A \rightarrow Us_B$: $y = r \cdot s^b \bmod n$.

The above actions are repeated t times ($t < n$). At the end of the t rounds, one can say that: (a) a number of $3t$ messages have been exchanged; (b) user Us_A has selected t random values $r_1, \ldots, r_t$; and (c) Us_A has computed t values $x_1, \ldots, x_t$, and determined t numbers $y_1, \ldots, y_t$, while Us_B has selected t random Boolean values $b_1, \ldots, b_t$ and received all x_i, and y_i, for $i = 1, \ldots, t$, such that:

$$x_i = r_i^2 \bmod n, \; y_i = r_i \cdot s_i^{b_i} \bmod n.$$

Phase 3

Verification Us_B accepts the proof if, for all $i \le t$, the equation:

$$y_i^2 = x_i \cdot c^b \bmod n$$

is satisfied and that y is not 0. Both terms of the equality take the form $r_i^2 \cdot s_i^{2bi}$ in the case of success.

By sending challenge b, the verifier aims to first check whether Us_A is able to demonstrate (t times) that he has knowledge of the secret s, and to deny actions performed by an adversary impersonating Us_A such as selecting any random number r and sending $x = r^2/c$ to verifier Us_B. On receiving $b = 1$, for example, the adversary will only answer by sending r, which is enough to satisfy the above equality. However, this will not work for $b = 1$. The response $y = r$ (response for challenge $b = 0$) is independent of the secret s, while the response $y = r \cdot s \bmod n$ (made when $b = 1$) provides no information about s because r is a random number.

3.3.3 Device-based authentication

Normally, many user authentication protocols are susceptible to masquerading, spoofing, interception, and replays of authentication messages. Some current approaches address these exposures by using an authentication device with limited processing capability that contains a cryptographic strong key which aims to help user authentication in a hostile environment. The device's key is randomly selected out of the key space of the embedded cryptographic algorithm. Since the key is strong, the probability of success of a brute force attack is almost null. The activation of the device operation takes place directly between the user and the device and is performed by the user using a weak initial secret (e.g., password and PIN).

Current device-based authentication methods differ in many aspects. The following features characterize some dimensions of that difference including the security of the method and its deployment costs:

- **Device–workstation interface** The device and workstation need to communicate. They can do it through an electronic interface such as a card reader or via the user himself, who may enter manually the information provided by the device.
- **Clock availability** An internal clock may be needed by a device for the generation of the necessary parameters (e.g., generated random numbers, timestamps) or the computation of other useful parameters (e.g., one-way function values).
- **Storage usage** Non-volatile read-only storage may be needed to store sensitive information and computational procedures within the device (e.g., cryptographic key, random numbers generator).
- **Encryption capabilities** Specific devices implement public key cryptography, however, for the sake of performance, the devices may not have to implement the decryption algorithm or may only need to perform a one-way function value computation.
- **Exchanged information** The size and structure of the exchanged information between a device and the workstation may differ from one method to another. The way the information is passed is also an important factor to consider.

With current methods, the user-device relationships developed for the needs of authentication are based on the delegation concept, where the device performs an authentication procedure on behalf of the user. Unfortunately, a major drawback still persists with these methods including the potential of masquerading that can be performed using stolen devices.

3.4   Attacks on authentication

Nowadays, malicious adversaries can attempt to defeat authentication schemes and perform a set of damaging attacks including the following non-exhaustive list:

Impersonation attacks An adversary can attempt to capture information about a legal user and impersonates that user. Impersonation is easy to perform if the adversary succeeds to compromise the keying material of a user (password, PIN or key).

Replay attacks An adversary can attempt to capture authentication related information and replays it to impersonate the user originating the information with the same or a different verifier (Syverson, 1994). Assume, for example, that a user U authorizes the transfer of funds from a banking account to another by encrypting the request by a signature key known only to him. To get this transaction done, U sends it to the bank system, which checks the signature and executes the transactions. An adversary H, wishing to have the same request repeated without having U's authorization, would need only to produce the signed transfer, if anti-replay measures have not been provided.

Forced delay attacks The adversary executes a forced delay attack when he can intercept a message, drop it from the net, and relay it to its destination after a certain period of time. This attack is different from replay attacks. Delaying signed transactions may induce serious damages in e-business.

Interleaving attacks An adversary can involve selective combination of information from one or more authentication processes, which may be ongoing processes (Tzeng, 1999). Examples of interleaving attacks include the oracle session attack and the parallel session attack.

To explain these attacks, we consider the case of the two-way authentication protocol, which can be described formally as follows:

- $Us_A \rightarrow Us_B : \phi(k, r)$
- $Us_B \rightarrow Us_A : \langle r, \phi(k, r') \rangle$
- $Us_A \rightarrow Us_B : r'$

where k is a shared secret key, and r and r' are random numbers generated by Us_A and Us_B, respectively.

Oracle session attack This attack is performed as follows. An intruder Us_H starts a session with Us_B, in which he poses as Us_A. Intruder Us_H generates a random number, which is considered to be $\phi(k, r_1)$; Us_H has no knowledge of r_1. He sends it to Us_B, who assumes that its source is Us_A. User Us_B decrypts the "encrypted" random number with the secret key k that he shares with Us_A and he generates a random number r_2 which he encrypts with the same secret key and sends the random number r_1 and the encrypted random number $\phi(k, r_2)$ to Us_A. Then Us_H intercepts the message and starts a new session with Us_A, in which he impersonates Us_B. He sends the encrypted random number $\phi(k, r_2)$ to Us_A. User Us_A decrypts the encrypted random number, generates a new random number r_3 and encrypts it. He sends the random number r_2 and the encrypted random number $\phi(k, r_3)$ to Us_B. Intruder Us_H intercepts the message and sends the decrypted random number r_2 to Us_B. He then quits the session with Us_A. Us_B receives the random number r_2, which is the same as the one he generated, and so he thinks that he is communicating with Us_A. Finally, Us_H has falsely authenticated himself as Us_A to Us_B. He has used Us_B as an oracle to decrypt the encrypted random number that Us_B has generated.

A security measure to prevent an adversary from performing an oracle session attack on the two-way authentication protocol is to transform the second step as follows:

$Us_B \rightarrow Us_A : \langle \phi(k, r), \phi(k, r') \rangle.$

Parallel session attack This attack works as follows, when performed on the enhanced two-way authentication protocol. User Us_A wishes to start a session with Us_B. He generates a random number r_1 and sends it to Us_B. An adversary Us_H intercepts the message and starts a second session with Us_A, in which he impersonates Us_B and sends r_1 to Us_A (as if he generated it). User Us_A assumes that the source of the random number is Us_B and encrypts it with the secret key k he shares with Us_B. Then Us_A generates a random number

r_2, encrypts with k, and sends the encrypted numbers to Us_B as part of the parallel session. Intruder Us_H intercepts the encrypted random numbers and sends them back to Us_A as part of the first session. Assuming that he is receiving the encrypted random numbers from Us_B, Us_A decrypts them. The received number r_1 is the same as the one he has generated for the first session, so he believes that he is communicating with Us_B. Therefore, Us_H has now authenticated himself falsely to Us_A as Us_B.

After that authentication Us_A sends r_2 back to Us_B to complete the first session. Intruder Us_H intercepts the message and sends r_2 back to Us_A as part of the parallel session. User Us_A receives the random number r_2 that is the same as the one he has generated for the parallel session and believes that he is communicating with Us_B. Hence, Us_H has finally authenticated himself two times falsely to Us_A as Us_B. Intruder Us_H has used Us_A as an oracle to encrypt the random numbers that he generated. Hence, two authentication sessions have taken place.

The above mentioned authentication attacks can be avoided by applying several actions. Replay attacks can be stopped by challenge-response techniques using a sequence number. Interleaving attacks can be avoided by linking together all messages using a sequence number. A protection against forced delay attacks can combine the use of random numbers with reduced time-windows. Nevertheless, an efficient security solution to provide protection to authentication protocols should consider how the protocols operate and suggest actions appropriately. Moreover, the following general security guidelines must be observed to protect authentication schemes:

- When a weak authentication based on passwords is used, a password policy should be stated and should include rules related to passwords length, complexity, aging, reuse possibility, and timetable access. Such parameters depend heavily on the context and frequency of the server use, account types, and risk associated with passwords compromise.
- A security policy should be made available to describe under what conditions users' accounts are created, modified, and deleted. A set of administrative practices should be referred to specify requirements regarding users' usage of authentication materials (such as passwords).
- The implementation of a risk mitigation strategy is highly recommended in order to study the possible attack scenarios, provide efficient authentication, and reduce the cost relative to the security measures that can be applied against threats.
- If authentication needs to be valid during a connection lifecycle, authentication should be redone periodically in a way that an adversary cannot profit from the duration of the process.
- If the authentication process is linked to any active integrity service, the authentication process and the integrity service should use different keying materials.
- If timestamps are used in an authentication scheme, the working stations that are involved in the procedures of timestamps' computation and verification should be protected and tightly synchronized.
- Anonymous remote attempts of authentication should be unauthorized. The number of limited attempts of authentication that are made remotely by any user should be limited to a small threshold.

- Selecting, generating, and configuring the security parameters used by authentication schemes should take into consideration the reduction of the probability of successful attacks.
- In the case where a trust relationship is defined between servers, authentication-related configurations should be reviewed carefully. A hacker can use the trust relationship to access from a compromised host to another.
- When assessing/auditing the security of an authentication scheme, the underlying cryptographic algorithms and digital signature schemes should be checked as highly secure.
- In assessing an authentication scheme, the potential impact of compromise of keying material should be studied. In particular, the compromise of long-term keys and past connections should be considered.

3.5 Digital signature frameworks

A digital signature is a data string, in a digital form, which associates a message with its originator. A digital signature scheme uses a generation process, which produces the digital signature, and a digital signature verification process, which checks whether the digital signature is authentic and requires the original message as an input. The signature generation and verification scheme works as follows. During signature generation, an entity wishing to sign a message m selects a public-key cryptosystem and uses his/her private key to associate a value to m and encrypts the value. He then sends the message and the encrypted value to the recipient. During verification, the recipient re-computes the associate value of m, and checks if it matches the result obtained after decrypting the received encrypted value. The digital signature function includes an authentication method and attempts to satisfy the following requirements:

- The signature must depend on the message being signed and should prevent forgery.
- The signature should be relatively easy to compute and relatively easy to verify.
- The signature must be easy to store in terms of structure, size, and link to the related message.

Therefore, digital signatures involve only two communicating parties (signer and verifier) and assume that the verifier has access to (or knowledge of) the signer public key. A digital signature guarantees the integrity of the signed message. It fails, however, to provide confidentiality since the message is sent in clear text. Confidentiality can be provided by encrypting the entire message (including the encrypted value) with either a secret key shared with the destination or with the destination's public key.

The efficiency of a digital signature scheme relies on the security of the sender's private key and the guarantee of public key ownership. If a sender later wants to deny sending a message, the sender can claim that his private key was lost or that the public key used to verify the signature does not belong to him. Administrative controls related to the security of private keys and ownership of public keys can be employed to thwart this situation. One example of control is to require that: (a) every signed message should include a timestamp; (b) every loss of keys should be immediately reported to a central authority; and (c) every

3.6   Hash functions

A hash function is a one-way function, denoted by $h(-)$, that satisfies the following properties:

- Compression: h takes any input m (having an arbitrary length) and outputs the value, $h(m)$, of fixed length called *hash digest*.
- Collision resistance: Given $h(x)$, it is computationally unfeasible to find a message m such that $h(x) = h(m)$.

Therefore, it takes a message of any size and generates a small, fixed-size block of data from it (called a message digest, or hash value). Re-executing the hash function on the same source message will always yield the same message digest. Moreover, a hash is not predictable in operation. That means, a small change in the source message will have an unpredictably large effect on the final digest. Thus, the message digest calculated by a hash function can be considered a good fingerprint of the original message.

Additional properties of hash functions include: (a) non-correlation, which means input bits (occurring in m) and output bits (appearing in $h(m)$) are not correlated in any manner; (b) near-collision resistance, which states that it is computationally unfeasible to find two messages m and m' such that message digests $h(m)$ and $h(m')$ differ in a small number of bits; and (c) local one-wayness, which states that it is as difficult to determine any substring as to determine the whole message m knowing $h(m)$ (Meneze, 1996; Nicopolitidis, 2003).

Typically, a hash function is designed to take an arbitrary length input by processing iteratively its fixed-size blocks. As depicted in Figure 3.1, the general model of a hash function segments the input m into fixed-length blocks m_i of fixed size l (typically $l = 512$) and makes use of a function, called *compression function*, and an initial vector, denoted by IV. The message may involve appending (or padding) extra bits to cope with the fixed size of the blocks composing the message and the message length. If l is the message size (including the padding field) and r is the block size then the message should contain t blocks such that $t = l/r$, and the iteration lasts t loops.

At the beginning of iteration, the compression function f takes m_1 and computes. $N_1 = f(m_1, IV)$. Letting N_i denote the ith output of the iteration, the general model gives:

$$N_0 = IV$$
$$N_1 = f(m_1, IV)$$
$$N_i = f(m_i, N_{i-1}), \quad \text{for } t + 1 > i > 0,$$

The global model may require the application of an additional function, called transformation function, g, to compute the hash value of m using equality $h(m) = g(N_t)$. Therefore, hash functions can be distinguished typically by the nature of the preparation function (padding and segmenting), compression function, and transformation function.

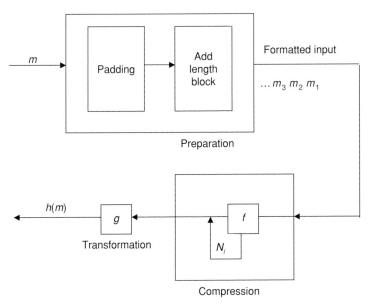

Figure 3.1 Global model for hash functions.

Preparation function

The initialization process includes padding, deciding block lengths and selecting the initial vector *IV*. The padding appends a minimal number of extra bits to reach a message length that is a multiple of the block size *r*. The padding part starts with a 1-bit and then includes as many 0-bits as needed. Thus, if the length of the original message *m* is a multiple of *r*, then the padding operation results in the generation of an extra block. By pre-agreed convention, a length field can be added to specify the bit-length of the original message.

3.6.1 Examples of hash functions

The initial vector *IV* can be of three types: fixed, random, or message dependent. In all cases, *IV* must be used at the sender and the receiver side as well. If not known prior to verification, *IV* should be transmitted along with the message.

Compression function

Various schemes can be applied to describe the compression functions used with hash functions. Generally, the most used compression functions format their input as a build on 1, 2 or 4 blocks and make use of the following constructs:

• A cryptographic function $\phi(k,-)$ parameterized by a secret key k. This gives:

$$h_i = \phi(m_i, h_{i-1}) = \phi(k, m_i) + h_{i-1}.$$

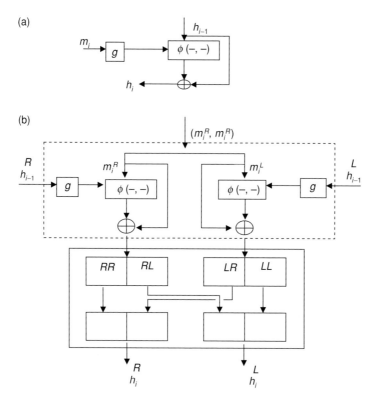

Figure 3.2   General model for compression.

- A cryptographic function $\phi(g(m_i),-)$ parameterized by a secret key $g(m_i)$. This gives:

$$h_i = \phi(m_i, h_{i-1}) = \phi(g(h_{i-1}), m_i) + m_i + h_{i-1}.$$

Figure 3.2 depicts these constructs, where function g can have different forms including identity. Figure 3.2(b) considers that all sub-messages m_i and all the intermediate hash values are formatted into two blocks:

$$m_i = \langle m_i^R, m_i^L \rangle \quad \text{and} \quad h_i = \langle h_i^R, h_i^L \rangle.$$

The permutation only reorders the four blocks occurring in h_i. Therefore, the expression of h_i of Figure 3.2 can be formally given by:

$$h_i = \langle h_i^R, h_i^L \rangle = p \left(\langle \phi(g_R(h_{i-1}^R), m_i^R) + h_{i-1}^R, \phi(g_L(h_{i-1}^L), m_i^L) + h_{i-1}^L \rangle \right).$$

Several hash functions have been specifically designed with a real consideration to the optimization of function performance. Hash functions include a long series of message digest algorithms called MD2, MD4, MD5 and the secure hash algorithm (SHA-1) as compared in Robshaw (1995). MD4 is a 128-bit hash function designed to replace MD2 in a way that brute force attack to find distinct messages with the same hash digest should last about 2^{64}

operations and finding a message knowing its hash digest should take approximately 2^{128} operations. Unfortunately, MD4 is no longer used as a hash function, since collisions have been found for it in 2^{64} operations.

MD5 was designed to strengthen MD4. However, MD5 has been also found to have weaknesses due to collisions related to the MD5 compression function. SHA-1 also extends the MD4 algorithm by fixing the bit size length to 160 (and other large values) and allows the compression functions to have four rounds instead of three rounds. SHA-1 provides increased protection against brute-force attacks.

3.6.2   Security of hash functions

Many attack methods have been developed during recent years and are available in the literature. Some attacks can be classified as general methods, while others depend on particular properties of the hash functions. Attacks include birthday attacks, attacks on the bit size length, attacks on compression functions, chaining attacks, and attacks using precomputation. Given a fixed message, a traditional way to find another message m' colliding with m is to randomly select m' and check whether $h(m') = h(m)$. Assuming that the bit size length is equal to l, the probability that given m' the above equality is satisfied is defined by 2^{-l}.

Birthday attacks are algorithm-independent attacks so that they can be applied to every hash function, using only significant parameters such as the hash-result bit size. Typically, a birthday attack considers as input a legitimate message m_1, a fraudulent message m_2, and an l-bit hash function h and delivers two new messages m'_1 and m'_2 using various modifications of m_1 and m_2 such that $h(m'_1) = h(m'_2)$. Therefore m'_1 and m'_2 would have the same hash digest and signature. To retrieve m'_1 and m'_2, the algorithm performed by the birthday attack starts by computing $2^{1/2}$ modification of m_1. Then, it computes and stores their hash digests (along with the related message). Finally it performs the same thing with message m_2 and checks, for every modified message m'_2 whether there is m'_1 such that $h(m'_1) = h(m'_2)$. Traditional mathematical results show that the probability of success of the attack is higher than $1/2$.

Attacks on the compression function attempt to find collision for the pre-image and initial vector. For a message m decomposed as $m = \langle m_1, \ldots, m_t \rangle$, a pre-image attack attempts to find a pre-image m'_1 such that the compression function does not make a distinction between $m_1, \ldots, m_t$ and $m'_1, \ldots, m_t$. The collision IV attack attempts to find a collision for the initial vector in the sense that the compression function outputs the same value for message m using two initial vectors IV and IV'.

Chaining attacks are built based on the chaining variables that constitute intermediate inputs of the iterations occurring in the compression function. Mainly, if a message is decomposed as $m = < m_1, \ldots, m_t >$ and f is the compression function taking as input m, then:

$$N_0 = IV,$$
$$N_i = f(m_i, N_{i-1}) \quad \text{for } i \leq t.$$

Suppose that an adversary picks any one among the $m_i's$, $i = 1, \ldots, t$, and replaces it with another block without modifying the hash digest of message m. The adversary could decide to replace m by a new message m' such that $h(m) = h(m')$.

3.6.3 Message authentication

Hash functions whose purpose is message authentication are called message authentication code (MAC) algorithm. Hence, a MAC aims to provide assurances regarding the source of a message and its integrity. It typically uses two independent input variables: a message input and a secret key and makes use of a cipher-block chaining. Typically, a block chaining inputs a message m, a description of the block, and a shared key (for a given cryptosystem). It outputs an n-bit MAC of m. Like hash functions, a general model for a block-chaining message authentication performs five operations: (a) a padding operation, which adds extra bits to message m; (b) a blocking operation, which divides the resulting string into blocks (denoted $m_1, \ldots, m_t$); (c) a block-chaining processing, which generates a finite sequence of encryption values given by:

$$h_1 = f(k, , m_1)$$
$$h_i = f(k, , h_{i-1} + m_{i-1}), \quad i = 2, \ldots, t$$

where $f(k, , -)$ is the encrypting function; (d) MAC value determination, which defines h_t as MAC(m); and (e) a fifth process can be used to strengthen the block-chaining message authentication and reduces the threat of exhaustive key search without affecting the efficiency of the intermediate computation. The process develops a new sequence $\{g_i\}_{i \leq t}$ using a second secret key, say k'. The sequence is given by:

$$g_i = f(k', h_i), \quad i = 1, \ldots, t.$$

Another variant of the block-chaining message authentication, called RIPE-MAC, has been developed. It mainly differs in the way the internal encrypting function is performed (using DES algorithm), the number of secret keys it uses, and the expression defining the intermediate and final MAC values h_i, which is given by the following scheme (if only one key is used):

$$hi = f(k, , h_{i-1} + m_{i-1}) + m_{i-1}, \quad i = 2, \ldots, t.$$

It also uses an additional key k' derived from k by complementing it in a predefined way.

3.7 Authentication applications

Various authentication functions have been developed to support application-level authentication. Among those services one can mention the X.509 directory-based authentication service and the Kerberos application.

3.7.1   X.509 authentication service

The ITU-T recommendation X.509 defines a directory service that maintains a database of information about users for the provision of authentication services. Information includes: user identification, email address, public key, the algorithm that the user invokes when signing digitally, and the identity of the authority that guarantees all the information. Entries in the database are called certificates. The certificate structure and the authentication protocols implemented based on this paradigm will be addressed in the following chapter. The efficiency of the database is mainly based on the fact that the certificates are unforgeable and the container is well protected.

X.509 also contains three authentication procedures: the one-way authentication, the two-way authentication, and the three-way authentication. The first procedure performs a unique transfer of information from the claimant to the verifier in order to establish that the message was generated by the claimant, that the message was intended to the verifier, and that it has not been modified or replayed. During the one-way procedure, only the identity of the claimant is verified. The message includes at minimum: (a) a random number *rand*, whose role is to protect from replay attacks; (b) a timestamp *tmp*, which includes the message generation time along with a validation procedure; (c) the verifier's identity, which accesses the X.509 directory for the purpose; (d) a signed data, which transports the proof of claimant's identity (including *rand* and *tmp*) along with a certificate of the claimant; and (e) an optionally encrypted data (using the public key of the verifier) in which the claimant can insert a session key to be used after authentication. The random number should be unique during the validation period.

In addition to the guarantee provided by the one-way authentication, the two-way authentication includes the transmission of a reply message from the verifier to the claimant that establishes the identity of the verifier (or verifier's certificate). Therefore, the two-way procedure allows both parties to authenticate the identity of the other. The reply should include several attributes including the random number *rand* (as generated by the claimant) in order to validate the reply; a new random number *rand'* that aims to protect the reply from replay attacks, and a signed data that is able to authenticate the verifier involving *rand, rand'*, and the identity of the claimant.

The three-way authentication procedure adds to the messages exchanged during the two-way authentication a third message sent by the claimant to the verifier sending the random number *rand'* and can include signed data that can help synchronizing clocks.

Finally, the verification applied during the authentication of both parties is performed online by accessing the appropriate X.509 directory to check whether the certificate provided by the other party is the one stored on the database for that entity.

3.7.2   Kerberos service

Kerberos is an authentication service developed at MIT. It allows a distributed system to be able to authenticate requests for service generated from workstations (Kohl, 1994). Kerberos authentication protocol (version 5) operates following a 3-step process: (a) when a user requests to login to a workstation, the workstation informs a key distribution centre (KDC) about the request; (b) the KDC generates a *ticket granting ticket* (TGT), encrypts it with a user key, say k_u, and sends it back to the workstation; (c) on receiving the encrypted ticket, the workstation asks the user to provide his password that it hashes to determine k_u and uses the key to decrypt the ticket. If the decryption works properly, the user is authenticated and his request is accepted.

Several vulnerabilities can be observed with Kerberos since it is a password-based authentication protocol. Vulnerabilities include (Bellovin and Merrit, 1991): unworthiness of workstations for decrypting the TGT and storing users' keys, as well as vulnerability to dictionary attacks. Some of these problems have been solved by employing special-purpose hardware including the use of smartcards, which typically changes these five operations as follows (Itoi and Honeyman, 1999):

1. On receiving a request from a user to login to the workstation, the workstation informs the KDC.
2. The KDC generates a *ticket granting ticket* (TKT), encrypts it with the user key, say k_u, and sends it back to the workstation.
3. On receiving the encrypted ticket, the workstation sends it to the smartcard provided by the user.
4. The smartcard decrypts the TGT and returns back the TGT in clear text to the workstation.
5. The workstation checks whether the TGT is valid. If that is the case, it authenticates the user.

While the use of passwords in Kerberos provides a proof of "*what the user knows,*" using a smartcard a user can prove "*what the user has.*" The extension of Kerberos provided by the preceding five operations overcomes different vulnerabilities since k_u never leaves the smartcard. This process eliminates the threat of dictionary attack and states that neither a malicious workstation nor a malicious system administrator can steal a user key since it is stored on an independent system. In addition, by providing a PIN, a user can strengthen the process since in order to impersonate the user, an adversary would have to steal the smartcard and the related PIN. However, in a compromised workstation, a user's TGT is unfortunately exposed and can be used to obtain service tickets TGT. Therefore, a dictionary attack cannot be operated.

3.8   Authentication network services

Various authentication services can be offered to support e-commerce, e-government, and e-banking. Moreover, network-based authentication can be provided to support the authentication of network subscribers. While the first three authentication services will be considered in the following chapters, we give in what follows two examples of network-based authentication services: IP-based authentication and GSM authentication.

3.8.1   IP authentication header protocol

Authentication Header Protocol (AH) ensures connectionless data integrity, data origin authentication, and protection against replay for the entire IP packet. In fact, what is really authenticated is being in possession of a shared secret, and not the IP address or the user having generated the packet. With AH, the data is readable, but is protected from modification.

AH integrity and authentication are provided jointly, using an additional block added to protect the message. This block is called an *Integrity Check Value* (ICV). Thus, it is a generic term used to indicate either a Message Authentication Code or a Digital Signature. Currently, all the proposed algorithms are obtained using a one-way hash function and a secret key. A generation of a digital signature is also possible by the use of public key cryptography. Protection against replay is provided by a sequence number and its use can be selected by the recipient during the negotiation process option.

The authentication is computed over the whole IP packet in both tunnel and transport modes. This can lead to some problems since there are fields in the IP header (like the TTL field) that may be modified while flowing through the network. To solve this problem, IPSec proposes to reset the fields included in the authentication computation to zero. IPSec also sets up *security associations* (SAs), which are simplex (or unidirectional) connections that provide security services to the traffic it carries, including the authentication header protocol.

In order to establish an SA, IPSec relies on the Internet security association and key management protocol (ISAKMP, (Maughan *et al.*, 1998)), which defines protocol formats and procedures for security negotiation. An SA defines the used cryptographic algorithms and keys. It must be established between two peers before using IPSec and be defined for AH protocol. In addition to the SA concept, another notion characterizes the IPSec response. It is the security policy, which defines the rules used by the protocol to establish the behavior of the IPSec policy over the traffic that flows to/from/via a node where IPSec is implemented. An IPSec rule consists of a series of data that determine a certain treatment to a specified traffic (e.g., allow, block, or negotiate security).

3.8.2   Authentication in wireless networks

GSM subscribers are traced during intra- and inter-domain mobility. Every mobile station MS should inform the network of its location. This information is used to update the visitors location register (VLR) and the home location register (HLR) of the mobile. The establishment of a connection is controlled by the Authentication Center (AC). For each communication, a real-time authentication of the caller and the called is realized in order to protect against attacks.

Each MS has a smartcard (SIM) containing a secret key k_{SM}, shared between MS and the HLR only. When a MS enters the coverage of a VLR, it provides its identity (International MS identifier, IMSI) to the VLR. The IMSI is relayed to the HLR along with the VLR identity, and the position of the MS. The HLR asks the AC for a pair (*rand, sres*) containing a challenge (or a random number) and a signed response. The pair is then sent to the VLR and is used only once. Parameter *sres* is computed using a classified algorithm, denoted by A_3 as function on *rand* and k_{SM} (i.e., $sres = A_3 (rand, k_{SM})$).

Subsequently, that challenge *rand* is delivered to MS. Knowing A_3, k_{SM}, MS can recompute *sres* on receiving *rand*. Therefore, the interaction between MS and VLR consists of a challenge-based authentication of MS by VLR. Furthermore, if confidentiality is required the VLR can receive a key session *ks* computed by the AC, using algorithm A_8, and delivered with the pair (*rand, sres*). Formally, we have $k_s = A_8(rand, k_{SM})$. The major concern with the GSM authentication scheme is the assumption that the internetwork involving HLR, VLR and AC is completely secure during the transmission of the pair (and the key session, in the case of confidentiality).

3.9   Conclusion

Authentication and digital signatures are vital means by which e-based systems and computer networks can be protected from being accessed by unauthorized individuals, and intruders.

This chapter is meant to shed some light on the various schemes and techniques of authentication and digital signatures as applied to such systems. We reviewed the main aspects of any digital signatures framework, hash functions, and types of digital signatures. We also reviewed the possible attacks on authentication schemes, authentication applications, and authentication services.

References

Anshel, M. and D. Goldfeld (1997). Zeta functions, one-way functions, and pseudorandom generators. *Duke Math Journal*, **88**, 371–90.

Bellovin, S. M. and M. Merrit (1991). Limitations of Kerberos authentication system. In *Proceedings of the Winter 1991 Usenix Conference* (available at http://research.att.com/dist/internet_security/kerblimit.usenix. ps).

ElGamel, T. (1985). A public key cryptosystem and a signature scheme based on discrete logarithms. *IEEE Transactions on Information Theory*, IT **31**, 4, 469–72.

Fiat, A. and A. Shamir (1987). How to prove yourself: practical solutions to identification and signature problems. In *Advances in Cryptology (CRYPTO86)*, *Lecture Notes in Computer Science* 263, Springer-Verlag, 186–94.

Itoi, N. and P. Honeyman (1999). Practical Security Systems with Smartcards. In *Workshop on Hot Topics in Operating Systems 1999*. IEEE Computer Society Press, 185–190.

Kohl, J., B. Neuman, and T. Ts'o (1994). The evolution of Kerberos authentication service. In *Distributed Open Systems*. IEEE Computer Society Press, pp. 78–94.

Maughan, D., M. Schertler, M. Scheider, and J. Turner (1998). Internet Security Association and Key Management Proto (ISAKMP) RFC 2408. Retrieved October 25, 2006 from http.//dc.qut.edu.au/rfc/rfc2408.txt.

Menezes, A., P. van Oorschot, and S. Vanstone (1996). *The Handbook of Applied Cryptography*. CRC Press.

Nicopolitidis, P., M. S. Obaidat, G. I. Papadimitriou, and A. S. Pomportsis (2003). *Wireless Networks*. Wiley.

Obaidat, M. S. and D. Macchairllo (1994). A multilayer neural network system for computer access security. *IEEE Transactions on Systems, Man, and Cybernetics*, Vol. **24**, No. 5, 806–13.

Obaidat, M. S. and B. Sadoun (1997). Verification of computer users using keystroke dynamics. *IEEE Transactions on Systems, Man, and Cybernetics*, Part B, Vol. **27**, No. 2, 261–9.

Obaidat, M. S. and B. Sadoun (1999). Keystroke dynamics based identification. In *Biometrics: Personal Identification in Networked Society*, Anil Jain *et al.* (eds.), Kluwer, pp. 213–29.

Robshaw, M. (1995). *MD2, MD4, MD5, SHA and other hash functions*. RSA laboratories Technical Report TR-101 (available at www.rsasecurity.com/rsalabs/index.html).

Syverson, P. (1994). A taxonomy of Replay attacks. In *Computer Security Foundations Workshop VII*. IEEE Computer Society Press, pp. 131–6.

Tzeng, Z. G. and C. M. Hu (1999). Inter-protocol interleaving attacks on some authentication and key distribution protocols. *Information Processing L.*, **69**(6), 297–302.

Part II

E-system and network security tools

Introduction to Part II

Data to be accessed via communication networks or transmitted over public networks must be protected against unauthorized access, misuse, and modification. Security protection requires three mechanisms: enablement, access control, and trust management. Enablement implies that a cohesive security policy has been implemented and that an infrastructure to support the verification of conformance with the policy is deployed. Perimeter control determines the points of control, the objects of control and the nature of control to provide access control and perform verification and authorization. Trust management allows the specification of security policies relevant to trust and credentials. It ascertains whether a given set of credentials conforms to the relevant policy, delegates trust to third parties under relevant conditions, and manages dynamically, if needed, the level of trust assigned to individuals and resources in order to provide authorization.

Public key infrastructures (PKIs) represent an important tool to be used in enablement, while biometric-based infrastructures are gaining an important role in providing robust access control. Biometrics are automated methods of recognizing a person based on a physiological or behavioral characteristic. Biometric-based solutions are able to provide for confidential financial transactions and personal data privacy. The need for biometrics can be found in electronic governments, in the military, and in commercial applications. In addition, trust management systems start to be used in a large set of environments such as electronic payment and healthcare management, where transactions and accesses are highly sensitive.

The chapters included in the second part of this book aim at presenting the techniques and major principles driving the operation and proof of the infrastructures related to access control, authorization and authentication. Three systems are considered: public key infrastructure systems, biometric-based security systems, and trust management systems.

Chapter 4 describes PKI functions and develops the use and management of digital certificates as a major tool for identification, signature, and conformance. A particular emphasis is put on the definition of certificate generation, certificate verification, and certificate revocation. Several other related issues are discussed including cross-certification, PKI operation, PKI assessment and PKI protection.

Chapter 5 discusses the major biometric techniques that are used to secure and authenticate access to e-based systems. These techniques have received a lot of attention in recent years.

Chapter 6 presents a comprehensive approach to trust management. It considers the major techniques and functionalities of a trust management system. Moreover, it describes two among the well-known trust management systems: PolicyMaker and Referee. The chapter also discusses three applications in medical information systems, electronic payment and distributed firewalling.

4 Public key infrastructure (PKI) systems

Data that can be accessed on a network or that are transmitted on the network, from one edge node to another, must be protected from fraudulent modification and misdirection. Typically, information security systems require three main mechanisms to provide adequate levels of electronic mitigation: enablement, perimeter control, and intrusion detection and response. Enablement implies that a cohesive security plan has to be put in place with an infrastructure to support the execution of such a plan. The public key infrastructure (PKI) being discussed in this chapter falls under the first approach.

4.1 Introduction

One of the most decisive problems in business transaction is the identification of the principal (individual, software entity, or network entity) with which the transaction is being performed. As the traditional paperwork in business is moving to electronic transactions and digital documents, so must the reliance on traditional trust objects be converted to electronic trust, where security measures to authenticate electronic business actors, partners, and end-users before their involvement in the exchange of information, goods, and services are provided. Moreover, the obligation to provide confidentiality and confidence in the privacy of exchanged information is essential. Extending this list of security services should include the necessity to establish the non-repudiation of transactions, digitally attest the validity of transactions by trusted third parties, or securely time-stamping transactions.

As world activities (e.g., economy, industry, and society) become increasingly dependent on the electronic accessibility, storage, promotion, and delivery of sensitive information, the problem of maintaining a level of trust in all business processes becomes essential. A large spectrum of security services must be provided to maximize the advantages of the new economy, make electronic exchange easier and more secure, reduce protection costs, and optimize business profit.

Public key infrastructures (PKIs) provide well-conceived infrastructures to deliver security services in an efficient and unified style. PKI is a long-term solution that can be used to provide a large spectrum of security protections. PKI's outcome can be estimated through the ongoing increase of business applications, which enables enterprises to conduct business electronically and safely. From the cryptographic building blocks (that we have developed in the previous chapters) through the system architecture, certificate life-cycle management, and deployment issues, this chapter aims to give a vendor-neutral introduction to the PKI technology. It also aims to explain how security is demonstrated through the use of chains of certificates.

Public key cryptography, as explained in Chapter 2, is based on the use of key pairs. When using a key pair, only one of the keys, referred to as the private key, must be kept (usually) under the control of its owner. The other key, referred as the public key, can be disseminated over publicly available directories for use by any individual who wishes to participate within a security service with the person holding the private key related to that public key. This scheme has been shown to have a high level of efficiency because the keys forming a pair are mathematically related, and it remains computationally infeasible to derive the private key from the knowledge of the public key and the algorithm using them. The dissemination of public keys within PKIs is organized using digital certificates based on the X.509 standard.

In theory, any individual can send the owner of a private key a message encrypted using the corresponding public key (which is available on the related certificate) and only the owner of the associated private key can read the secure message by decrypting it. Similarly, the owner of the private key can establish the integrity and origin of the data he sends to another party by digitally signing it using his private key. Anyone who receives that signed data can use the associated public key (that he can get from the certificate) to validate that it came from the owner of the private key and verify the integrity of the data that have been transmitted across the network.

Since transactions can be no more secure than the system in which they are created and managed, the most important element to protect and authenticate transactions can be made by establishing a way for individuals/customers to locate each other and have confidence that the public keys they use truly belong to the entity (individual or system) with whom they are communicating. A PKI can help in achieving these needs. It is designed to generate digital certificates (or public key certificates) to bind a public key to the identifying information about its owner. It also manages the certificates, certificate statuses, and related elements for the benefit of users, e-business, and enterprises. It also may involve symmetric key cryptography for different purposes including key encryption, secure logs, and protecting archives. Therefore, it can be said that a PKI is used to offer strong authentication technology. In addition, using a combination of secret key and public key cryptography, a PKI enables a large number of basic security services (e.g., data confidentiality, data integrity, non-repudiation), as well as sophisticated services (such as key management and distribution, securing communication protocols over heterogeneous networks).

The basic foundation for PKIs is defined in the ITU-T X.509 recommendation (Housley *et al.*, 2002). Recently, the *PKI Forum*, the *Internet Engineering Task Force* (IETF), and the *Public Key Infrastructure X.509 working group* (or PKIX group) have been the driving forces behind setting up a formal and generic model that is suitable for deploying nation-wide (as well as firm-wide) certificate-based architectures. Several standards and whitepapers have been published to cover the measure issues on PKIs. Among the major documents are: (a) the RFC 2510bis, which describes the Internet X.509 Public Key Infrastructure (PKI) Certificate Management Protocol (Adams *et al.*, 2004); (b) the RFC 2511, which describes the Certificate Request Message Format (CRMF) syntax and semantics (Schaad, 2004); and (c) the path building module, which provides guidance and recommendations to developers building X.509 public key certification paths within their applications (Cooper *et al.*, 2004; Schaad, 2004).

4.2 The PKIX architecture model

Naturally, the Internet is an un-secure medium of communication compared to private networks. On the Internet, transmission of an enterprise's data may be eavesdropped by an adversary equipped with sniffing tools, inducing possible damages to the other communication partner. Public key cryptography is a suitable technique for resolving these issues. A PKI offers the basis of practical use of public key cryptography. PKI has been exploited in many business applications such as secure electronic transactions (SET) and IP security. The PKIX model defines the elements that comprise a PKI including components, documents, and policy instruments. The PKIX architecture model includes the specification of a certificate revocation list. The X.509 v3 certificate format is described in detail, with additional information regarding the format and semantics of Internet name forms in RFC 2832.

4.2.1 Main PKI components

The PKIX model components integrate four major components: the end-entity, public key certificate, certification authority, and repository. Figure 4.1 illustrates our interpretation of this model.

End-entity

End-entities can be considered as the users/consumers of the PKI-related services. The term end-entity is a generic term used to denote subscribers, network devices (such as servers, routers), processes, or any other entity that has applied for and received a digital certificate for use in supporting the security and trust in transactions to be undertaken. An end-entity can also be a relying party (an individual or an organization), who does not hold necessarily a certificate and who may be the recipient of a certificate (during the course of a transaction) and who therefore acts on reliance of the certificate and/or digital signature to be verified using that certificate.

Public key certificate (PKC)

A public key certificate (or just certificate) acts like an official ID card or passport document. It provides a means of identifying end-entities of their identities to public keys. PKCs can be distributed, publicly published, or copied without restriction. They do not contain any confidential information. A public key certificate is a digital document that is associated with an end-entity. Moreover, a PKC is a data structure containing a public key, relevant details about the key owner, and optionally some other information, all digitally signed by a trusted third party, usually called the certificate authority, which certifies that the enclosed

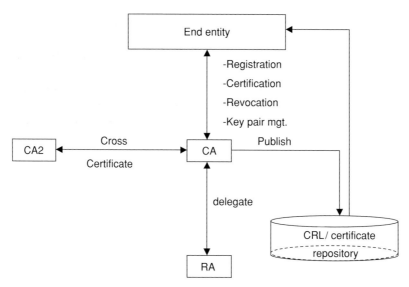

Figure 4.1   PKI model.

public key belongs to the entity listed in the subject field of the certificate. The technical advantage of a certificate is represented by the fact that it is considered impossible to alter any field of the certificate without easy detection of the alteration.

The X.509 v3 certificate is a widely used certificate format within a PKI system (Housley *et al.*, 2002). It is being utilized in the major PKI-enabled applications available in the market place, such as the SSL and the privacy enhanced mail (PEM).

Certification authority (CA)

A certification authority is the issuer of public key certificates within a given PKI. Public key certificates are digitally signed by the issuing CA, which effectively (and legally) binds the subject name to subject public key and the CA's public key that is used to verify the signature on the issued certificates. CAs are also responsible for issuing certificate revocation lists (CRLs), which report on invalidated certificates, unless this has been delegated to a separate entity called a certificate revocation list issuer.

A CA should be involved in a number of administrative and technical tasks such as end-user's registration, end-user's information verification, certificate management, and certificate publication. However, some of the administrative functions may be delegated to optional components, called registration authority (RA). CA's primary operations include certificate issuance, certificate renewal, certificate revocation, and certificate verification. The verification of an end-entity certificate may involve a list of CAs, say $[CA_1, \ldots, CA_n]$ operating as follows:

1. CA_n is the issuer of the end-entity certificate;
2. CA_{k+1} is the issuer of the certificate issued to CA_k to sign certificates;
3. CA_1 is a trusted CA (from the end-entity point of view).

Therefore, the end-entity certificate represents the starting point to validate a given certification path, which represents a list of certificates signed by CAs and delivered for CAs.

Certificate Repository (CR)

A certificate repository is a generic term used to specify any method for storing and retrieving certificate-related information such as the public key certificates issued for end-entities and the CRLs which report on revoked certificates. A repository can be an X.500-based directory with public access facilities via the so-called *Lightweight Directory Access Protocol* (LDAP) or the *File Transfer Protocol* (FTP) so that the certificates can be retrieved by any end-entity for various purposes.

It is possible to offload certain verification functions from the end-entity system to a trusted third party, who will act on its behalf. For example, a specific protocol can be set up at the end-entity site to ask a trusted third party about the revocation status of certificates that the end-entity wishes to rely on. Arguably, the trusted third party could be viewed as a virtual repository since the revocation status and the output verification are derived and returned to the end-entity system in response to a request for PKI-related information.

Certificate revocation list (CRL) issuer

A CRL is a means to notify who wishes to check the status of a given certificate. Typically, a CRL is a signed document that contains reference to certificates, which are decided to be no longer valid. The CRL issuer may be an optional entity to which a CA delegates the verification of information related to revocation, issuance and the publication of CRLs. Usually, the CA that issues a given certificate is also responsible for issuing revocation information associated with those certificates, if any.

A CA may transfer the whole revocation function to another CA. CRLs that are issued by the other CA are referred to as indirect CRLs. Therefore, and for the sake of efficiency and coherence, a certificate should include a field indicating the address of the location where CRLs that might include this certificate are published when it would be revoked.

Registration authority (RA)

A RA is an administrative component to which a CA delegates certain management functions. The RA is often associated with the end-entity registration process. However, it can be responsible for a number of other functions including the following tasks:

- Ensuring the eligibility of applicants to be issued with certificates, while verifying the accuracy and integrity of the required information provided by the applicants.
- Verifying that the end-entity requesting the issuance of a certificate has possession of the private key associated with the public key being provided.
- Generation of key pairs, archiving key pairs and secret keys, and delivering keys to end-entities.

- Verifying revocation requests, relaying requests, and reporting on revoked certificates.
- Conducting the needed interactions with the delegating CA on behalf of the end-entity, in case of key compromise notifications and key recovery requests.

The RAs, however, are not allowed to issue certificates or CRLs. Deployment of RAs can provide two primary advantages. First, RAs can help to reduce overall certification costs. This is especially true in large, geographically distributed companies that require their users to be physically present before specific PKI-related activities are permitted. Second, offloading the administrative functions from a CA allows an organization to operate their CA off-line, which reduces the opportunity that an adversary mounts remote attacks against that CA.

4.2.2   PKI documents

A PKI must be operated in accordance with well-defined policies that define the rules to perform the PKI activities appropriately. The deployment of a PKI system in a company requires the development of a complete and coherent security and related documents to help people's understanding. Four main types of documents are important to PKI business activities. They are: the certificate policy, certificate practice statements, subscriber agreements, and relying party agreements (ISC, 2001). End-entities only have a business relationship with the issuing authority of the certificates they hold or rely on. Subscribers would have this relationship represented in the subscriber agreement.

Certificate policy (CP)

A certificate policy sets forth general requirements that PKI participants must meet in order to operate within a PKI. A CP is also a named set of rules that indicate the applicability of a certificate to a given application. Typically, a CP may be used by an end-entity to help in deciding whether the binding within a certificate is sufficiently trustworthy for a given application, describes the appropriate uses for certificates and the responsibilities of individuals and organizations that can participate in the PKI system.

For example, a CP might indicate that a particular type of certificate may be used to verify digital signatures on e-mails requiring a high level of assurance and security. Another example may set forth rules for certificates used to authenticate the parties involved in a given transaction activity, or may cover a set of applications with common security requirements to establish various assurance classes of certificates (e.g., rudimentary, basic, medium, and high classes of certificates are the certificate classes used in the Canadian PKI).

Certificate practice statement (CPS)

A certificate practice statement defines a comprehensive statement of practices and procedures followed by a single CA or a related set of CAs set out in a CP. A CPS is the means by which providers of digital certification services document and demonstrate their ability to support the requirements of a CP. A CPS is intended to describe what a CA should do from the technical, business, and legal perspectives. While a CP task is establishing requirements, a CPS is oriented towards the expose of practices and procedures. Therefore, a CPS typically contains detailed description of the technologies used and the operating practices followed in order to provide their trust services.

CPs and CPSs do not represent the same concept. They fulfill different objectives. A CP is viewed as a statement of requirements (i.e., what needs to be achieved), whereas a CPS is viewed as the statement of procedures (i.e., those requirements to be satisfied). A CPS may support multiple certificate policies (used to fit the need of different applications). Different trust service providers may use different CPSs to support the same CP. However, it is possible for a trust service provider to have a CPS and not a CP. For example, in an enterprise deploying a private PKI with a single CA, there is no need for a CP. The PKI may have a CPS to reveal to its users what its practices and procedures are.

Subscriber agreements

This is a document representing an agreement between the subscriber applying and receiving a certificate and the issuing authority of the certificate. It focuses on the subscriber's responsibilities, rights, and obligations in using the certificate. It also defines the terms and conditions under which the subscriber may use his certificate and the private key corresponding to the public in the certificate. A subscriber agreement may be as simple as a brief statement of rights and obligations. It also may be adapted for certificates having higher levels of trust assurance. The relationship between a subscriber agreement and a CPS differs based on the business model and the PKI technology.

Relying party agreements

This is typically an agreement between a party that wishes to rely on a certificate and the information contained in it, on one hand, and the issuing authority of the issued certificate, on the other hand. The agreement governs the terms and conditions under which the relying party is permitted to rely upon the certificate. Most common rules in the relying party agreement require that the relying party should check the status of the certificate it wishes to rely on and all certificates in any chain built on that certificate.

4.3   PKIX management functions

The PKIX model identifies a number of management functions that are potentially needed to support the management process to perform the interaction between the PKIX components and handle the lifecycle of certificates. The main functions include: registration, initialization, certificate generation, certificate update, key pair management, certificate revocation, and cross-certification. We describe in what follows the main features and requirements of these functions.

Registration

End-entities must enroll into the PKI by applying to the issuing authority before they can take advantage of the PKI-enabled services. Registration is the first step in the end-entity enrollment process. This step is usually associated with the initial verification of the end-entity's identity and the information he/she provides. The level of assurance associated with the registration process can vary based on the target environment, the intended use of the certificate, and the associated policy. The registration process can be accomplished directly with the CA or through an intermediate RA. It may also be accomplished on-line or off-line depending on the trust level of the required certificate and the security practices implemented by the issuing authority.

Once the required information is provided by the end-entity and verified in compliance with the applicable policies, the end-entity is typically issued one or more shared secrets and other identifying information that will be used for subsequent authentication as the enrollment process continues. The distribution of the shared secrets is typically performed following specific ways and may be based on pre-existing shared secrets.

Initialization

The registration process is followed by the initialization process. This involves initializing the associated trust anchor (or trust point) with the end-entity. In addition, this step is associated with providing the end-entity with its associated key pairs. Key pair generation involves the creation of the private/public key pair associated with an end-entity. Key pair generation can be made prior to the enrollment process or it can be performed in response to it. Key pairs can be generated by the end-entity client system, RA, CA, or some other PKI components such as a hardware security module. However, in the case where the end-entity generates the key pair, the registration process should include the verification that the public key provided by the end-entity is connected to the private key held by the end-entity.

The location of the key pair generation is driven by operational constraints and applicable policies. Moreover, the intended use of the keying material may have an important role in determining where the key pairs should be generated. It is possible that tasks composing the initialization process may occur at different moments and places. However, the task performed by the end-user should not be realized before an explicit certificate request is generated.

Certificate generation

This process occurs after the termination of the initialization process. It involves the issuance of the entity public key certificate by the certification authority. Typically, the generation process organizes the necessary information (including the CA's identity and the revocation address) in a data structure following the X.509 standard and digitally signs it. If the key pair related to the certificate is generated externally to the CA, the public key component must be delivered to the CA in a secure manner. Once generated, the certificate is returned to the end-entity and/or published to a certificate repository.

Certificate update

Certificates are issued with fixed lifetimes (referred to as the validity period of the certificate). The duration of the pre-fixed lifetimes can be of one year or two years (or even longer). On certificate expiration, the key pair used with the certificate may also be required by the end-entity for different reasons. As a result, the certificate is updated (or renewed) and its lifetime is re-fixed. However, it is preferable that a certificate renewal involves the generation of a new key pair, and the issuance of a different public key certificate since it contains a new public key.

Key pair update can occur in advance of the pair's expiration. This will help to ensure that the end-entity is always in possession of a valid certificate. The key pair update may induce a certificate renewal before the associated public key actually expires. It also provides a period of time where the certificate associated with the initial key pair remains unrevoked, meaning that this certificate can be used for a short window of time to verify digital signatures that were created with this key pair. This will help to minimize inappropriate alert messages that would otherwise be generated to the end-entity.

Revocation

Public key certificates are issued with fairly large lifetimes. Nevertheless, the circumstances that existed when the certificate was issued can change to an unacceptable state before the certificate can come to expire normally. Reasons for unacceptability may include private key compromise, change of the information related to the subscriber (e.g., affiliation and name change). Therefore, it may become necessary to revoke the certificate before its expiration date. The revocation request allows an end-entity (or the RA that has initiated the enrolment process) to request revocation of the certificate. Certificate revocation information must be made available by the CA that issued that certificate or by the CRL issuer, to which the CA delegates this function.

X.509 defines a method for publishing the above information via certificate revocation lists (CRLs). The frequency of publication and the type of CRLs used are a function of local policy. Finally, one can notice that end-entities, or third trusted parties operating on their behalf, must check the revocation status of all certificates it wishes to rely on. This will be addressed in what follows.

Key pair management

Since key pairs can be used to support digital signature creation, data encryption, and message decryption, an end-entity may need to rely on the CA for the creation and management of key pairs. When a key pair is used for encryption/decryption, it is important to provide a mechanism to recover the necessary decryption keys when normal access to the keying material is no longer possible, otherwise it will be impossible to recover the encrypted data. Key pair recovery allows end-entities to restore their encryption/decryption key pair from an authorized key backup facility, provided by the CA.

It is also possible that an end-entity's association with an organization can change (e.g., employee resignation, firing, or new appointment), and the organization has a legitimate need to recover data that has been encrypted by that end-entity. It is also possible that access to the keying material may be required in association with legitimate law enforcement needs. Moreover, a CA can provide certification services where key pairs need to be managed at the PKI level. Key pair management includes all functions needed during key life cycle.

Cross-certification

Cross-certification is the action performed by one CA when it issues a certificate to another CA. The basic purpose of a cross-certification is to establish a trust relationship between two CAs, with which the first CA validates the certificates issued by the second CA for a period of time. Cross-certification is provided to establish the proof of certificate paths for one or more applications by allowing the interoperability between two distinct PKI domains or between CAs working within the same PKI domain. While the former is referred to as inter-domain cross-certification, the latter is referred to as intra-domain cross-certification.

Cross-certification may be unilateral or mutual. In the case of mutual cross-certification, a reciprocal relationship is established between the CAs; one CA cross-certifying the other, and vice versa. Unilateral cross-certification simply means that the first CA generates a cross-certificate to the second CA, but the second does not generate a cross-certificate to the first. Typically, a unilateral cross-certificate applies within a strict hierarchy where a higher level CA issues a certificate to a subordinate CA. However, cross-certification adds an important complexity to the process of validating a path certificate. This will be discussed in the following sections.

Additional management functions

Additional functions may be required by a PKI within specific frameworks (e.g., closed groups-based PKI or real-time operational PKIs). The PKIX certificate management protocols (Adams, 2004) provide a comprehensive set of management functions that might be required in today's PKIs. Among the management functions of interest, one can include the prompt announcement of issuance and the CRL broadcast. While the first function provides mechanisms for announcing to the members of a group of subscribers the issuance of a certificate, the second function provides the end-entities, who had accessed a certificate, with any CRL related to that certificate.

4.4 Public key certificates

Parties relying on a public key require confidence that the associated private key is owned by the correct remote subject (individual or system) with which an encryption or digital signature mechanism is engaged. This confidence is obtained through the use of digital certificates, which bind public key values to subjects. The binding is asserted by having a trusted CA digitally sign each certificate. The CA may base this assertion on technical means

Figure 4.2   Certificate format.

(such as proof of possession through a challenge-response protocol) or on the presentation of the private key.

A certificate has a limited valid lifetime, which is indicated in its signed contents. Because a certificate's signature and timeliness can be independently checked by a certificate-using relying party, certificates can be distributed via untrusted communications and server systems, and can be cached in an unsecured storage within certificate-using systems. Therefore, a certificate should contain various fields to make it easier to use. The description of these will be addressed in the following subsection.

4.4.1   Certificate format

A public key certificate is a sequence of three types of attributes: mandatory attributes that should be filled in every certificate, attributes for selected extensions, and additional attributes. Values of this attribute type must be encoded according to the syntax defined by several standards (e.g., RFC2253). Figure 4.2 depicts a certificate that is displayed. Mandatory attributes include the following fields:

- *Certificate version* This field depicts the version of the encoded certificate or of the CRL. Values considered in this field can be 0, 1, 2, or 3. Particularly, if the extension attributes are used in the certificate, the version is set to 2; when extensions are provided, the version field is set to 1.
- *Serial number* The content of this field is an integer assigned by the CA to each certificate. It is unique for each certificate issued by a given CA. Certificates handle serial number values up to 20 octets.
- *Signature algorithm* This field identifies the algorithm used by the CA in signing the certificate.
- *Issuer* This is a string representation of the certificate or CRL issuers. The issuer field should contain a non empty distinguished name. The name can be hierarchical including subfields such as common name, pseudonym, initials, organization, and country name.
- *Validity* The "validity" period in a certificate is the time interval, say [*notBefore, notAfter*], during which the certificate status is maintained by the issuing authority. The field contains a sequence of two timestamps, which define the beginning and end of the certificate's validity period.
- *Subject* This field identifies the end-entity associated with the public key written in the subject public key field. Where it is not empty, the subject field should contain the subject's distinguished name, which should be unique for each certified end-entity.
- *Subject public key info* This field carries the public key of the subject and identifies the algorithm associated with the certified public key.

Attributes for selected extensions allow users/groups to define private extensions to transport information unique to these users. Attributes contain the following list of standard extensions:

- *Authority key identifier extension* This attribute identifies the public key to be used to verify the signature on this certificate or CRL. This extension is used when the issuer has multiple signing keys. The key may be identified based on either the subject key identifier in the issuer's certificate or on the issuer name and serial number.
- *Subject key identifier extension* This attribute identifies the public key being certified. It enables distinct keys used by the same subject to be differentiated. When an end-entity has obtained multiple certificates, especially from multiple CAs, the subject key identifier provides a means to quickly identify the set of certificates containing a particular public key.
- *Key usage extension* This attribute defines the purpose (e.g., encryption, signature, or certificate signing) of the key contained in the certificate.
- *Policy information identifier extension* This attribute contains object identifiers which indicate the policy under which the certificate has been issued and the purposes for which the certificate may be used. In a CA certificate, these terms limit the set of policies for certification paths that include this certificate.
- *Subject alternative name extension* The subject alternative name extension allows additional identities to be associated to the subject of the certificate. Among the defined options, one can mention an IP address, a uniform resource identifier, or multiple name forms.

- *Issuer alternative name extension* The issuer alternative name extension allows additional identities to be bound to the subject of the certificate or CRL.
- *Basic constraints extension* The attribute identifies whether the subject of the certificate is a CA. It also defines the maximum depth of valid certification paths that include this certificate.
- *Extended key usage extension* This attribute determines one or more purposes for which the certified public key may be used, in addition to the basic purposes indicated in the "keyUsage" attribute. If the extension is present, then the certificate should only be used for one of the purposes indicated.
- *CRL distribution points extension* This attribute identifies how the full CRL information for the related certificate can be obtained.

Additional attributes contain the following:

- *Certificate location* This attribute contains a pointer to the directory entry of a certificate. Thus, it is possible to point to the certificate from a white page entry.
- *Certificate holder* This attribute contains a link to the directory entry of the end entity to which this certificate was issued.

4.4.2 CRL format

A CRL is a means of notifying end-entities who wish to verify the status of a certificate that the certificate is revoked. CRLs are issued to mark some certificates unusable, even though they have not yet expired. The circumstances that may lead to certificate revocation in a company may include, but are not limited to, the following cases:

- When a certificate owner is removed from the security domain of an enterprise, due to particular reasons of importance for the enterprise.
- When a certificate holder realizes that his/her private key has been lost or might have been compromised.
- When some information contained in the certificate was no longer correct/valid so that signing with the certificate can be considered inappropriate.
- When security attacks are suspected with the particular certificate or with the associated private key.

CRLs may be used in a large spectrum of generic applications requiring large interoperability. A CRL profile has been defined by RFC 3280 (Housley *et al.*, 2002) to provide a set of information that can be expected in every CRL. Typically, CAs publish CRLs to provide the status information about the certificates that have been issued. Each CRL has a scope, which is the set of certificates that satisfy some specific requirements such as "all certificates issued for a group G that have been revoked for key compromise." A complete CRL contains all unexpired certificates that have been revoked for one of the revocation reasons in the scope of the CRL. A delta CRL, however, only lists those certificates, within its scope, whose revocation status has changed since the issuance of the last delta CRL.

A CRL contains various fields. The first field is a sequence containing the name of the issuer, issue date, issue date of the next list, and the list of revoked certificates. When one or more certificates are revoked, each entry on the revoked certificate list is defined by a sequence of user certificate serial number, revocation date, and optional CRL entry extensions. The second field identifies the algorithm used by the CRL issuer to sign the CRL. The third field contains a digital signature calculated based on the content of the CRL. Optional fields include lists of revoked certificates and CRL extensions. The revoked certificate list is optional to support the case where a CA has not revoked any unexpired certificates that it has generated. Extensions provide methods for associating additional attributes with CRLs. Extensions used within Internet CRLs include:

- *Authority key identifier*: This extension provides a means of identifying the public key corresponding to the private key used to sign a CRL. The identification can be based on the key identifier or on the issuer name and serial number.
- *Issuer alternative name*: This allows additional identities to be associated with the issuer of the CRL.
- *CRL number*: This is a non-critical CRL extension which conveys a monotonically increasing sequence number for a given CRL scope and CRL issuer.
- *Delta CRL indicator*: This is a critical CRL extension that identifies a CRL as being a delta CRL. Delta CRLs are generally smaller than the CRLs they update.
- *Issuing distribution point*: This is a CRL extension that identifies the CRL distribution point and scope for a particular CRL, and indicates whether the CRL covers revocation for end-entity certificates only, CA certificates only, or attribute certificates only.

4.5   Trust hierarchical models

Typically, PKI architectures fall into four configurations: a single CA, a hierarchy of CAs, a mesh of CAs, or bridge certification authorities (Polk and Hastings, 2000). Each of these configurations is determined by a set of fundamental attributes including the number of CAs in the PKI architecture and the trust relationships between the CAs. The most basic PKI architecture contains exactly a single CA that provides all the PKI services for subscribers. Every certification path within a single CA architecture begins and ends with the same certificate. This results in a single user trust point. This configuration is the simplest to deploy; only one CA must be established and all the users understand the applications for which the certificates were issued.

Single certificate authority PKI architecture, however, does not scale easily to support very large or diverse communities of end-entities. The larger the community of users, the more difficult it becomes to support all of the necessary applications. A natural extension is to connect the CA that supports different communities to form a larger and more diverse PKI.

4.5.1   Hierarchical model

Isolated CAs can be combined to form a larger PKI using the so-called *superior–subordinate* relationships, and organized under the form of a tree structure. The root of the tree, called root CA, is the unique CA that is able to issue a self-signed certificate. Subordinate CAs are located at the lower layers of the tree. The leaves represent end-entities that have applied and received certificates. In general, the "root" CA does not issue certificates to users, but only issues certificates to subordinate CAs. Each subordinate CA may issue certificates to users or another level of subordinate CAs. With the hierarchical architecture, all subscribers should trust the same root CA. That is, all end-entities of a hierarchical PKI begin certification paths with the "root" CA's public key.

In a hierarchical PKI, the trust relationship is only specified in one direction. This means that subordinate CAs do not issue certificates to their superior CA. The root CA imposes conditions governing the types of certificates that subordinate CAs are allowed to issue. However, these conditions are not specified in the certificate contents issued by the subordinate CAs; but the related CPSs control them. Hierarchical PKIs have several attractive features due to the simple structure of the PKI architecture and the uni-directionality of the trust relationships. Among these are:

- Hierarchical PKIs are scalable to incorporate a new community of users since the "root" CA establishes a trust relationship with that community's CA.
- Certification paths are easy to build because they are unidirectional and the trust relationships are well controlled. This features results in a deterministic building process of a simple certificate path from a subscriber's certificate back to the trust point.
- Certification paths are relatively short since the longest path is equal to the depth of the tree plus one.
- Users of a hierarchy know implicitly which applications a certificate may be used for, based on the position of the CA within the hierarchy.

Hierarchical PKIs have several drawbacks, which are induced by the reliance of the system on a single trust point, the root CA. The major drawbacks are:

- The compromise of the root CA may result in the compromise of the entire PKI. The nature of a hierarchical PKI is that all trust is concentrated in the root CA and failure of that trust point is highly damaging.
- A second drawback is due to the fact that a general agreement on a single CA may be *politically* impossible. Inter-organizational competition may disqualify such an agreement.
- Finally, the evolution from a set of isolated CAs to a hierarchical PKI may be logistically impractical because all end-entities may need to adjust their trust points.

4.5.2   Mesh PKIs

An alternative to a hierarchical PKI is to organize CAs into a graph using peer-to-peer relationships. A PKI architecture constructed on peer-to-peer relationships is called a *mesh PKI* (as depicted in Figure 4.3). All the CAs in a mesh PKI can be trust points – in general, users will trust the CA that issued their certificate. CAs issue certificates to each other; the pair of certificates describe their bi-directional trust relationship. Since the CAs have peer-to-peer relationships, they cannot impose conditions governing the types of certificates

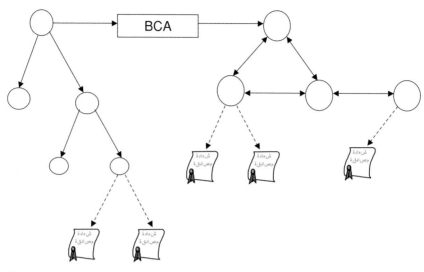

Figure 4.3   A mesh PKI.

other CAs can issue. However, the trust relationship may not be unconditional. If a CA wishes to limit the trust, it must specify these limitations in the certificates issued to its peers.

Mesh PKIs have several interesting features. First, mesh PKIs can easily integrate a new community of end-entities by simply allowing the establishment of peer-to-peer trust relationships with the community's CA (involving one or more CAs participating to the mesh PKI). Second, mesh PKIs are very flexible since there are multiple trust points. Therefore, the compromise of a single CA cannot force the failure of the entire PKI, since the CAs which have issued certificates to the compromised CA should simply revoke them. This will remove the compromised CA from the mesh PKI and leave the other CAs in a valid status. Subscribers of the non-compromised CAs can continue encrypting data and signing messages securely with the remaining end-entities of the mesh PKI.

Recovery from a compromise is simpler in a mesh CA than in a hierarchical PKI because it affects fewer users. A mesh PKI can easily be constructed from a set of isolated CAs because the users do not need to change their trust, provided that the CAs issue certificates to at least one CA within the mesh. This is very advantageous when an organization wants to merge separately its developed PKIs. Nevertheless, mesh PKIs have some unwanted features. These drawbacks are due to the bi-directional trust model they use. Certification path building represents a hard task to perform. Unlike a hierarchy, building a certification path from a subscriber's certificate to a trust point is non-deterministic. This makes path discovery more difficult since there are multiple paths. While some of these are valid, some others may include improper certificates. The building process may get worse since it can be involved in loops of certificates. Another limitation characterizing the mesh PKI is scalability since the maximum length of a certification path may be as high as the number of CAs in the mesh PKI. Moreover, users of a mesh must also determine which applications a certificate may be used for based on the contents of the certificates rather than the CA's location in the PKI. This determination must be performed for every certificate in a certification path, which is a complex process.

4.5.3 Bridge certification authority architecture

The Bridge Certification Authority (BCA) architecture is designed in order to provide a solution for the shortcomings observed with the two previously described basic PKI architectures. It also attempts to connect heterogeneous PKI architectures. Unlike the mesh PKI and hierarchical PKI, the BCA architecture does not issue certificates directly to users. They are intended to be used as a trust point by the CAs operating the PKIs forming the bridge. To integrate a PKI architecture, the BCA needs to establish a bi-directional trust relationship with the root CA, if the PKI architecture is hierarchical, or only with one CA, if the architecture is a mesh PKI. This allows the users from different user communities to have their natural trust points and users to interact with each other through the BCA with a specified level of trust.

- The number of trust relationships required to build a "federating" mesh architecture among the architectures are highly complex to implement.
- The autonomy of some PKIs in the set makes it infeasible to change some particular rules within the running CSPs.

The paradigm of BCA is useful for organizations (e.g., government) where various trust domains are already managed using independent PKI architecture and need to leverage the existing PKIs to support electronic processing with their partners. An application of the BCA concept is the US Federal PKI. However, BCAs are facing technical challenges such as the efficient discovery and validation of certification paths and the interoperability of large PKI directories since the BCAs include some mesh PKI components within its overall structure. Moreover, BCAs need more complex certificate information to constrain trust relationships between PKIs. BCAs face also another challenge characterized by the distribution of certificate directories in a way that is useful to end-entities and their applications to support the collection of certificate status information.

4.6 Certification path processing

Typically, certification path processing is composed of two phases: path construction and path validation. Path construction aims to find one or more certification paths linking a target certification to a trusted certificate (or anchor) that a relying party can trust. Path validation is the process of binding between the subject name and its public key and verifying whether the certificate is legally issued, as represented in the end-entity certificate. It involves making sure that each certificate in the path is valid (i.e., within is validity period), has not been revoked, and has integrity, and that all constraints imposed on the certification are satisfied, including path length constraints, name constraints, and policy constraints.

The path validation process determines the set of certificate policies that are valid for this path, based on the *policy information identifier extension*, which occur in the certificates.

As explained before, this attribute contains the object identifiers, which indicate the policy under which the certificate has been issued and the purpose for which the certificate may be used. The path validation also considers the policy mapping extension occurring in the CA's certificates. This attribute defines set pairs of object identifiers; each pair (*idp, sdp*) indicates that the issuing CA considers its issuer-domain-policy identified by *idp* equivalent to subject-domain-policy identified by *sdp*.

The authority key identifier (AKID) and the subject key identifier (SKID) are certificate extensions that also can be used to facilitate the extension of policies in a certificate path construction process. The AKID is used to distinguish one public key from another when the issuing authority has multiple signing keys. On the other hand, the SKID provides a means to identify the certificate and the public key it contains. This means that, for a given certificate path $\{C_1, C_2, \ldots, C_n\}$ and for all k, the SKID of C_k is the AKID of certificate C_{k+1}.

4.6.1   Path construction

Path construction is a complex process, which so far lacks standardization. It can be achieved through two approaches:

1.   *Forward construction* This approach starts from the target certificate to a recognized trust point.
2.   *Reverse construction* This approach begins from a recognized anchor to the target certificate.

Typically, path construction depends on the PKI architecture associated with the end-entity certificate (Lloyd, 2002). Next, we will consider how path construction is affected by the three abovementioned PKI architectures: hierarchical PKI, mesh PKI and Bridge CA. Formally speaking, the path construction requires obtaining a finite sequence of certificates, say $\{C_1, C_2, \ldots, C_n\}$, which supports that binding occurring in the target certificate (Housley *et al.*, 2002) and satisfies the following conditions:

1.   The subject of certificate C_i is the issuer of certificate C_{i+1}, for all $i \geq n$.
2.   Certificate C_1 is issued by a trusted CA.
3.   Certificate C_n is the certificate to be validated.
4.   Certificate C_i is valid at the time of verification, for all $i \geq n$.

Building certificate paths within a hierarchical architecture is easy. Paths are typically built on a forward direction. A relying party wishing to validate a certificate C_A of a user Us_A starts with the target certificate and works his way to the root CA, which represents a trust point for the relying party. The relying party (or the software acting on behalf) can extract the identity of the CA that issued C_A (as determined by the CA distinguished name in C_A). Let CA_1 be this CA. Since CA_1 has only one certificate C_1 issued for the public key associated with its signing key, the certificate will be either self-signed, meaning that CA_1 is the root, or there is another certification authority CA_2 and a certificate C_2 issued for CA_2 that certifies its public key. This process terminates after a finite number of steps.

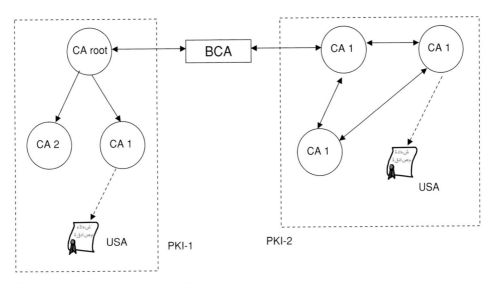

Figure 4.4 Path construction example.

This process determines a certification path $C_n, C_{n-1}, \ldots, C_1, C_A$ and a sequence of certification authority $CA_n, CA_{n-1}, \ldots, CA_1$ such that CA_k is the issuer of C_n and:

$$C A_k = \text{Issuer of } C_k = \text{subject of } C_{k+1}.$$

The length n of the certificate path is smaller than the depth of the hierarchy structure. Moreover, the (AKID, SKID) relation can be used also to build the aforementioned certificate path.

When the PKI architectures involved in the path construction are heterogeneous, the abovementioned attributes can be sufficient. X.509 has defined several other attributes that can be used for this purpose, when cross-certification is present. The attributes include the following:

- The *issuedToThisCA* attribute This is used by a CA's directory entry to store all the certificates issued to this CA.
- The *issuedByThisCA* attribute This is used by a CA's directory entry to store all the certificates issued by the CA to other CAs.

We will use Figure 4.4 to demonstrate how these attributes can be used in a complex PKI structure. For this, we can assume that Us_A, who is the relying party, is receiving a digitally signed message from Us_B. Moreover, we assume that the trust anchor of Us_A is CA_4 and that PKI_1 and PKI_2 represent a hierarchical and a mesh PKI architecture, respectively. A bridge CA (BCA) is introduced between the two PKI domains.

Starting from Us_A's certificate, a first segment of the path can be easily constructed from the target certificate to root CA_{Root} in PKI_1. A second segment can be built to connect CA_{Root} to CA_3 via BCA. This can be done first by discovering, in the CA_{Root}'s directory entries related to the *issuedToThisCA* attribute that there is a certificate issued by BCA to CA_{Root}. Second, another certificate that is issued by BCA to authority CA3 using the *issuedByThisCA* attribute related to BCA is found. Finally, the process attempts to build the third segment to link CA_3 to CA_4. This is accomplished by looking at the *issuedToThisCA* attribute in CA_3's directory entry.

Construction certificates may have various drawbacks, including the occurrence of multiple paths and loops. It is possible that multiple candidate paths can be extracted based on the very general criteria we have so far discussed. In the absence of any other criteria that may be used to benefit one candidate path over the others, the process may lack efficiency since it can be classified as a blind process. On the other hand, looping can occur when a series of cross-certificates leads the process back to a cross-certificate that is already part of a path under construction.

4.6.2   Path validation

A user of a security service requiring knowledge of a public key generally needs to obtain and validate a certificate containing the required public key. If the public key user does not already hold an assured copy of the public key of the CA that signed the certificate, the CA's name, and related information (such as the validity period or name constraints), then it might need an additional certificate to obtain that public key. The primary goal of path validation is to verify the binding between a subject distinguished name or a subject alternative name and the subject public key. This requires obtaining a sequence of certificates that support that binding. The procedure performed to obtain such a sequence of certificates has been shown to provide multiple candidate paths.

A particular certification path may not be appropriate for all applications. Therefore, an application may augment this path validation process to further limit the set of valid paths. The path validation process also determines the set of certificate policies that are valid for this path, based on the certificate policies extension, policy mapping extension, policy constraints extension, and inhibit any-policy extension. To achieve this, the path validation algorithm constructs a valid policy tree. If the set of certificate policies that are valid for this path is not empty, then the result will be a valid policy tree of depth n, otherwise the result will be a null valid policy tree.

The validation process requires the following inputs (Housley *et al.*, 2002): (a) a candidate certificate path of length n; (b) the current data time; (c) the user-initial-policy set, which determines the policies that are acceptable to certificate user; (d) the relying party's trust point information; (e) the initial-policy-mapping-inhibit, which indicates if policy mapping is allowed in the certification path; (f) the initial-explicit-policy, which indicates if the path must be valid for at least one of the certificate policies in the user-initial-policy set; and (g) initial-any-policy-inhibit, which indicates whether the *anyPolicy* object identifier should be processed if it is included in a certificate.

The major actions performed by path validation are the following:

- Basic certificate processing: This contains several actions that are performed on all certificates of the given path including verifying the certificate validity, its non-revocation, and the signature it transports. Furthermore, if the certificate policy extension is present in the certificate, then the processing will update the tree of policies accordingly.

- Wrap-up procedure: This mainly includes the computation of the intersection of the policy tree. The processing terminates by returning a success indication together with the final value of the policy tree, the working public key, and the working public key algorithm.

4.7 Deploying the enterprise's PKI

As PKIs become more widely deployed and cost and expertise become easy to manage, a company may need to pay attention to financial return made possible from PKI-enabled business process. The practical steps that a company can take to deploy its PKI are: (a) assessing the need for certification, and (b) defining the deployment road map. Choosing a PKI solution is another important task for the deployment of PKIs that is handled on the basis of a precise knowledge of PKI solutions. For the sake of simplicity, we will not address this issue in the current section.

4.7.1 Assessing the needs

One of the first steps an enterprise has to consider when comprehensive solutions are to be introduced is to define a feasible set of security policies and assess whether a candidate PKI is able to conform to the security policy. The following sub-steps are essential to this process:

Defining the scope of the enterprise's security policy

During this step, the enterprise should consider the following tasks: (a) the creation of one or more certificate policies and certificate practice statements; (b) the means by which the CA keys will be protected; (c) the process of how a CA key compromise will be handled; (d) the means by which PKI repositories will be protected; (e) the process of how users and CAs will be verified; and (f) finally, the process of how the PKI will be audited.

Defining a trust model

An enterprise has to choose the type of certification path it wants to work with. It has also to decide how it will trust a public key. In a very simple situation, the enterprise issues certificates to its own employees and customers. All certificate holders trust the same CA, so the certification path only requires one certificate. In another situation, the enterprise may need to determine whether it will need to trust certificates issued by other security domains and how the CAs involved in this process are organized.

A trust model is not only important for defining the relationships with the external parties, but also plays a significant role in defining relationships and processes between parties within the enterprise. For example, an enterprise may want to have a root CA, with one layer of subordinate CAs representing different divisions of the enterprise and the trust

PKI enablement of existing applications

Some of the existing applications and services, which are under use, may already support PKI and do not need to be changed. For example, various web browsers support client authentication using public key certificates. The PKI solution must provide a library to support the functions required to build end-entity interfaces. Application developers can use the APIs for the end-entity to access and manipulate the functions provided by the PKI.

4.8   Conclusion

In general, information security systems require three main mechanisms to provide an acceptable level of electronic management and control: enablement, perimeter control, and intrusion detection and response. The public key infrastructure (PKI) discussed in this chapter falls under the first approach. Public key infrastructure is an effective and popular security tool. PKI is a long-term solution that can be used to provide a wide range of security protections. Its outcome can be estimated through the ongoing increase of business applications, which enables enterprises to conduct business electronically and safely. This chapter gave a vendor-neutral introduction of the PKI technology. It explained how security is demonstrated through the use of chains of certificates.

References

Adams, C., S. Farrell, T. Kause, and T. Mononen (2004). Internet X.509 Public Key Infrastructure – Certificate Management Protocol (CMP). IETF, draft-ietf-pkix-rfc2510 bis-09.txt.

Cooper, M., Y. Dzambasow, P. Hesse, S. Joseph, and R. Nicholas (2004). Internet X.509 Public Key Infrastructure: Certification Path Building. IETF, draft-ietf-pkix-certpathbuild-04.txt.

Housley, R., W. Polk, and W. Ford (2002). Internet X.509 Public Key Infrastructure Certificate and Certificate Revocation List (CRL) Profile. IETF, RFC 3280 (available at http://www.faqs.org/rfcs/rfc3280.html).

Information Security Committee (2001). PKI Assessment Guidelines, PAG v0.30, public draft for comment. American Bar Association.

Lloyd, S. (2002). Understanding Certification Path Construction, PKI Forum (available at www.pkiforum.org/pdfs/Understanding_Path_construction-DS2.pdf).

Polk, W. T. and N. E. Hastings (2000). Bridge Certification Authorities: Connecting B2B Public Key Infrastructures. National Institute of Standards and Technology (available at cscr/nist.gov.pki/document/B2B-article.pdf).

J. Schaad (2004). Internet X.509 Public Key Infrastructure Certificate Request Message Format (CRMF). IETF, draft-ietf-pkix-rfc2511bis-07.txt.

5 Biometric-based security systems

Biological features have been investigated extensively by many researchers and practitioners in order to identify the users of information, computer, and communications systems. There are an increasing number of biometric-based identification systems that are being developed and deployed for both civilian and forensic applications. Biometric technology is now a multi-billion dollar industry and there is extensive Federal, industrial, business and academic research funding for this vital technology especially after September 2001.

An automated biometric system uses biological, physiological or behavioral characteristics to automatically authenticate the identity of an individual based on a previous enrollment event. In this context, human identity authentication is the focus. However, generally this should not necessarily be the case.

This chapter aims at reviewing state-of-the-art techniques, methodologies, and applications of biometrics to secure access to e-based systems and computer networks. It will also shed some light on its effectiveness and accuracy of identification as well as trends, concerns, and challenges.

5.1 Introduction

Biometrics deals with the process of identifying persons based on their biological or behavioral characteristics. This area has received recently a great deal of attention due to its ability to give each person unique and accurate characteristics. Moreover, the cost of implementing such technology to identify people has decreased tremendously. Biometrics techniques have been widely accepted by the public due to their strengths and robustness (Obaidat, 1997; Obaidat, 1999).

Identifying the identity of an individual involves solving two major issues: (a) verification, and (b) recognition. Verification is defined as confirming or denying an individual's claimed identity, while recognition (identification) is the issue of establishing a subject's identity from either a set of already existing identities or otherwise. The term positive identification is defined as identifying an object with very high certainty. In today's information-interconnected society, identifying an individual has become a challenge due to the need for accurate identification, verification or authentication as inaccurate verification may entail major losses and risks. Accurate identification of an information system's user could deter e-business and e-government fraud, and crime. In the United States, there are over one million dollars in security benefits that are declared by "double dipping" welfare beneficiaries with fake multiple identities (Woodward, 1997). Credit card companies claim that they lose multimillion dollars per year in credit card fraud. Moreover, millions of cellular phone fraud calls are made by thieves who steal cell phones. Banks report that multimillion dollars are stolen from customers' accounts due to stolen ATM cards and their PINs. Obviously, an accurate method of verifying the right check payee would also decrease the amount of money stolen through fake encashment of checks each year (Obaidat, 1994; Obaidat, 1997; Obaidat, 1999; Langeand, 1997; Clarke, 2001; Woodward, 1997; Joyce, 1990).

Securing access to computer and network systems has become recently an important issue due to the dependence of individuals and organizations on these systems on a daily basis. The reliance of people on e-based systems has increased tremendously in recent years and many businesses rely heavily on the effective operations of their information systems. Many organizations store sensitive information such as trade secrets, credit records, driving records, income tax, bank account transactions, stock market history, classified military data, and alike. There are many other examples of sensitive information that if accessed by unauthorized users may entail loss of money or releasing of confidential and classified information to unwanted parties, competitors, and even enemies.

There are many incidents of information security problems that have been reported recently. Some of these deal with accessing students records in universities, and accessing confidential information in government and military data bases. It is worth mentioning that the pace of technology change of data processing equipment has outstripped abilities to protect information from intentional attacks and breaches (Obaidat, 1997; Obaidat, 1999).

The total number of e-commerce, e- business and e-government users has increased at a phenomenally impressive rate. This is due in part to the decrease of prices of computers, Personal Digital Assistants (PDAs), cell phones as well as decrease of Internet subscription rates and availability of both wired and wireless broadband Internet access at affordable costs.

Attacks on e-based systems can be divided into active and passive attacks (Obaidat, 1997; Obaidat, 1999):

- Active attacks: These attacks involve altering a data stream or creating a false stream. They can be classified into: masquerade, reply, modification of messages, and denial of service. A masquerade happens due to a fake entity. An authentication sequence can be collected and replayed after a valid authentication sequence has taken place. Reply entails the passive gathering of data and its subsequent retransmission to create unapproved access. Finally, modification of a message is basically performed so that some portion of a genuine message is altered, or delayed, recorded in order to create unauthorized or false results.

- Passive attacks: These attacks include eavesdropping and snooping on the transmission of data. Here the intruder seeks to either learn the contents of messages or analyze the traffic. In the first case, the intruder tries to learn the contents of e-mail messages, or a file being transmitted using FTP (File Transfer Protocol), which may contain sensitive information. In the second case, the intruder tries to analyze the traffic in order to observe and learn the frequency and length of an encrypted message that is being sent, which may give more hints and information about the nature and behavior of the communication and parties involved in the session (Obaidat, 1997).

In general, passive attacks are not easy to detect, however, actions can be taken to avoid their occurrence. On the other hand, it is not easy to avert the occurrence of active intrusion. Any security system should have the following main characteristics:

- Integrity This means that the assets of the organization can only be modified by authorized users. There are three major aspects of integrity that are recognized: (a) authorized actions, (b) error detection and correction, and (c) separation and protection of resources.
- Availability This refers to the degree of avoiding any denial of service attempts. Assets should only be accessed by authorized parties. The latter should not be prevented from accessing objects to which they have legal access. Keep in mind that availability applies to both service and data.
- Confidentiality This characteristic refers to the degree of privacy or secrecy. It implies that the e-system should be accessible only to authorized parties where the access can be of the type of read only or both read and write depending on the user's type. Privileges may include viewing, printing, and modification.

The effectiveness of access control in general is based on two concepts: user identification, and protection of the access right of users (Obaidat, 1997; Obaidat, 1999).

5.2 Biometrics techniques

Biometrics is a broad name for the measurements of humans, which is intended to be used to recognize them or verify that they are who they allege to be. Most biometric schemes have been designed in the last few decades, and the methods have already been employed in various applications and environments. It is a common term that includes a wide range of metrics of human physiology and behavior. Some of these technologies measure relatively steady features of the body such as DNA, features of the retina and ear-lobes, fingerprints, and thumb geometry. Others concentrate on the dynamic aspects of human behavior such as the method of establishing a hand-written signature, and the procedure of keying a password on a keyboard of a computer, PDA or a cell phone. Below are the major categories on which biometric technologies can be based (Clarke, 2005):

1. **Appearance** These are supported by still images such as the descriptions used in passports including color of the eyes and skin, height, visible physical markings, gender, race, facial hair, and wearing of glasses.
2. **Natural physiography** These include thumbprint, fingerprint sets, handprints, retinal scans, earlobe capillary patterns, hand geometry, and DNA-patterns.
3. **Bio-dynamics** Among these are the manner in which one's signature is written, statistically analyzed voice characteristics, keystroke dynamics, particularly login-id and password.
4. **Social behavior** This category includes habituated body-signals, general voice characteristics, style of speech, and visible handicaps. They are usually supported by images.

5. **Imposed physical characteristics** In this case, we use bar-codes and other kinds of brands, and embedded micro-chips and transponders.

Biometrics is expected to become a vital component of the identification technology since:

- Prices of biometrics sensors have declined.
- The underlying technology has become more mature.
- The public has become aware of its strengths and robustness.

The desirable characteristics in a good biometric are: (a) Universality, (b) Uniqueness, (c) Permanence, and (d) Collectability. A brief description of each of these characteristics is given below.

- **Universality** This means that every person should possess this characteristic. In practice, this may not be the case; otherwise, the population of non-universality must be small. For example, not every individual has a right index, therefore, a biometric device based mainly on this cannot be considered universal.
- **Uniqueness** This means that no two individuals have the same characteristic. In other words, each person should have a different version of the biometric. For example, fingerprints are considered a unique biometric.
- **Permanence** The characteristic does not vary in time, that is, it is time invariant. At best this is a good approximation. Level of permanence has a great impact on the system design and long term operation of biometrics.
- **Collectability** This means that the characteristic can be quantitatively measured. Practically, biometric collection must be: (a) non-disturbing, (b) dependable (reliable) and robust, and (c) rewarding and cost effective.
- **Performance** This refers to the reachable identification accuracy, the resource requirements to achieve an acceptable identification accuracy and the environmental factors that affect the identification credibility.
- **Acceptability** This specifies to what extent people are willing to accept the biometrics system for identification purposes.
- **Circumvention** This refers to how easy it is to trick the biometric system by fake techniques.

Of course, we should look into other assessment aspects especially those related to technology such as maturity, scalability and usability. Since sensors are often used for biometrics technology, characteristics of these devices such as physical sensing principle, power consumption, robustness as well as sensor size and weight are of great interest and concern to researchers and developers working in this field.

Figure 5.1 shows a general block diagram of a typical biometric system, while Tables 5.1 and 5.2 show the major advantages and disadvantages of key physiological and behavioral-based biometrics techniques. It is worth mentioning that some convenient and widely accepted biometrics schemes are not necessarily the most accurate ones.

A brief description of the major biometrics techniques is given below.

Table 5.1 *General physiological characteristics with their main features when applied to biometric authentication.*

Physiological Characteristics	Advantages	Disadvantages
Fingerprint	(a) Easy to use (b) A mature technology (c) Advanced sensing and algorithm technology	(a) Quality may vary within a population (b) When applied to wireless networks, the lack of supervision is noteworthy (c) Accessibility issues due to criminological history
Face	(a) Incorporation of cameras into many wireless devices (b) Contactless achievement of templates (c) Universal characteristic	(a) Wireless use sets further difficulty for imaging techniques (b) Weak accuracy (c) Imaging conditions are essential
Iris	(a) Patterns are unique and stable (b) Contactless achievement of templates (c) Higher accuracy	(a) Requires imaging conditions (b) Due to high cost and some operation difficulties, it is potentially less acceptable for wireless environment

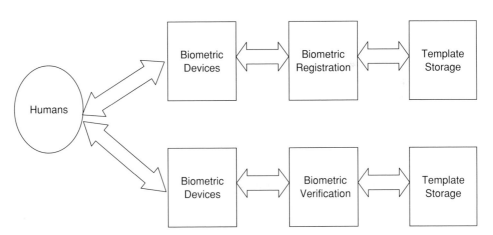

Figure 5.1   A functional diagram of a typical biometric system.

Voice

Voice is considered a popular and common characteristic of an individual that has been used for centuries to identify persons with good accuracy. However, it is not expected to extract features from a voice signal that can effectively identify/recognize an individual from a huge database of identities. Extensive research efforts have been exerted at developing algorithms to extract features from voice signals in order to help in identifying speakers and speech (Obaidat, 1998, 1999).

Table 5.2 *Universal behavioral characteristics with their characteristics for automated biometric authentication.*

Behavioral Characteristics	Advantages	Disadvantages
Signature	(a) Acceptable due to long history and popular acceptance	(a) Performance is arguable since the individual may decide to change his signature
	(b) Does not depend on native/natural language	(b) Usual variability of signatures
	(c) In addition to static features, dynamic features such as acceleration, velocity, and trajectory profiles of the signature are also employed which helps to enhance performance	
Voice	(a) Universally available in communication systems and networks	(a) It is not widely accepted in public places due to various impairments such as noise and interference
	(b) Interactive authentication protocol is possible	(b) Environment and state/mode of the subject affect accuracy and performance
	(c) Needs hardware, which is available, for example, in cell phones	
Keystroke dynamics	(a) Available in computer environment	(a) Accuracy depends to a certain degree on the keyboard and mode of the subject

In the context of speaker identification, we differentiate between the text-dependent scheme and authentication and text-independent scheme that does not rely on a fixed phrase. In the voice-to-print authentication scheme a complex technology transforms voice into text. A more challenging task is to devise schemes that do not depend on natural languages of the speakers. This biometric scheme has the most promise for expansion since it does not require any new hardware; almost all personal computers or laptops are already equipped with a microphone. Nevertheless, reduced quality and surrounding noise can affect the authentication accuracy. Moreover, the enrollment process is usually more complex than with other biometrics. This latter characteristic has led to the perception that a voice-based authentication scheme is not easy to use. Hence, voice verification software requires enhancement. It is expected that in the coming few years a voice-based scheme will be used along with a finger-scan scheme to provide a robust biometric-based security scheme to identify computer and network systems' users. This is due to the fact that many people regard finger scanning as a high accuracy authentication form, and if voice biometrics is used as well then it will replace or enhance passwords, account names or PINs (Liu, 2001).

Fingerprints

This scheme relies on the unique features found on the fingertips of the system's user. Fingertip patterns offer a variety of methods of fingerprint authentication. A number of them imitate the established police method of matching details; others use straight pattern-matching devices. Some verification schemes can detect when a live finger is presented while others cannot. It is worth mentioning that there are more fingerprint devices available than for any other biometric technology.

A fingerprint is made of a series of edges (ridges) and furrows (grooves) on the surface of the finger. The uniqueness of the fingerprint can be determined mainly by the pattern of ridges and furrows and the minutiae points. The latter are basically local ridge features that arise at either a ridge junction or a ridge ending.

Fingerprint identification technique has been in use by law enforcement agencies for over a hundred years and has become pretty much the de facto international standard for positively identifying individuals. The first fingerprint recognition system was used in law enforcement about four decades ago.

Fingerprints patterns are based on DNA. This means that identical twins will have nearly identical fingerprints. Nevertheless, state-of-the-art systems have the capability to distinguish between them (Jain, 1999).

In general, fingerprint authentication can be a good choice for in-house systems, where users can be given enough clarification and guidance, and where the system operates in a restricted environment.

Face

This method analyzes the unique shape, pattern and positioning of facial features. Basically, it requires a digital camera to develop a facial image of the user for verification. This scheme has received substantial interest. It is interesting to mention that the casino industry has capitalized on this technique to produce a facial database of scam artists for quick recognition by security personnel.

In general, face recognition is a complex scheme and mainly software-based. Artificial Intelligence (AI) is used to simulate human interpretation of faces. It is a non-intrusive biometric technique, which gives it an advantage over other schemes. Moreover, it is based on characteristics and features which are exposed to the public constantly. Therefore, capturing face images, as opposed to scanning fingers, will not violate individual privacy. This is why this scheme is becoming popular in airports. Moreover, this technique is good for providing simple and convenient ways to verify users in low to medium security applications. The chief approaches that are usually taken for face recognition are (Weng, 1999):

1. Transformation technique This approach consists of representing a face image with a set of orthonormal basis vectors.
2. Attribute presentation-based technique This relies on extracting biometric features such as nose, eyes, etc., from the face. Furthermore, the geometric properties of these extracted features are also used and analyzed.

It is worth noting that these days mobile phones are equipped with cameras, which makes this scheme possible even for wireless systems.

Retina

This technique entails analyzing the layer of blood vessels positioned at the back of the eye. It employs a low intensity light source through an optical coupler in order to scan the unique patterns of the retina. Retinal verification is considered an automatic method that provides accurate identification of the person. These retinal scans can be quite accurate but do require the user to look into a receptacle and focus on a given point. This is not particularly suitable if the user wears glasses or is concerned about having close contact with the detecting tool. Due to these reasons, retinal scanning is not widely accepted by all users despite the fact that this technology works fine.

The first practical working retina identification system was built in 1981. The system consists of a camera using infrared light and a personal computer. The latter was used to analyze the reflected light waveforms. The main task of the analyzing algorithms is to extract features and perform a correlation operation. In general, any effective retina-based identification system should consist of the following main functions:

1. Enrollment: Here, a person's reference eye signature is constructed and a PIN number and text is linked with it.
2. Verification/Authentication: In this part, an individual previously registered alleges an identity by inputting a PIN number. The retina identification system scans the ID subject's eye and compares it with the reference eye signature linked with the entered PIN. If there is a match, then access is granted.
3. Recognition: The system scans the ID subject's eye and compares it with the available database to find the correct reference signature. If a match is found, then access is granted.

The three major subsystems of a retina identification system are:

1. Imaging and signal processing subsystem A camera is needed to convert a circular scan of the retina into a digital signal.
2. Matching subsystem This verifies or identifies the acquired eye pattern with a stored template.
3. Representation This block compares the retina signature with the corresponding stored identification database information.

Despite the many advantages for this scheme, it has some limitations. Among these are (Hill, 1999):

1. Perceived health concerns: Although low light level is used for retina identification, there is a perception that retina identification can hurt the retina.
2. Sever astigmatism: Since people with eyeglasses should remove them in order to have proper identification, individuals with severe astigmatism may have problem aligning the dots in the camera align target. For such subjects, what they see can be quite different from dots, which can result in ambiguous feedback during the alignment phase of the process. This causes the eye pupil to be outside the accepted part of the scan. Clearly, this part of the scan will be an invalid one.

3. Cost of sensor: This scheme is considered more expensive than other biometric schemes such as speech-based ones.

4. Ergonomics: This technology requires bringing the eye to the identification system or the identification system to the eye, which makes it a more difficult approach than some other biometric schemes.

5. Outdoor vs. indoor environment: It is known that small pupils can minimize the false rejection error rate. This is due to the fact that light must pass the eye pupil twice; once incoming and once departing the eye. The arrival light can be reduced by as much as four times when the ID subjects' pupils are small. Moreover, outdoor surroundings are less conducive to dependable retina identification systems performance than indoor surroundings due to local light circumstances.

Hand geometry

This scheme is concerned with measuring the physical individuality of the user's hand and fingers from a 3-D viewpoint. It measures the shape, width, and length of fingers and knuckles. It provides a good balance of performance characteristics and is relatively easy to use. Moreover, it is flexible enough to accommodate a wide spectrum of applications and has the potential to provide good verification accuracies (Liu, 2001; Cavoukian, 2005; Young, 1981; Zunkel, 1999).

In this method, the user places his/her hand on a special reader, aligning fingers with particular positioned guides. In order to capture the top view and side view, a camera is used to scan these areas. Such a set up provides length and width information in order to compute the thickness profile. It is known that each hand has unique features including finger length, width and thickness, curvatures as well as relative location of these features. In general, the scanner employs a CCD camera, inferred LEDs with special mirrors and reflectors. Among this scheme's advantages are speed, and reliable verification.

This technique is being used in Ontario, Canada at nuclear power generation stations as well as by the United States Immigration and Naturalization Services (INS). At the University of Georgia, this technology is being used to control access to its student cafeteria. Basically, when students visit the cafeteria, they swipe their ID cards using a reader and have their hands verified before being able to enter the dining area. Athletic clubs in the United States use this technique to allow members access to the club without a key or ID using a hand scanner (Young, 1981; Zunkel, 1999). Obviously, this scheme has great potential for e-commerce, e-government and e-business.

Iris

This scheme has some similarities to finger scan. An iris image is usually captured using a non-contact imaging procedure, using a CCD camera. In order to capture the image, cooperation from the subject is expected. Of course, this cooperation is needed in order to register the image of the iris in the middle imaging area and make sure that the iris is at the correct distance from the focal plane of the CCD camera (Rhodes, 2003).

The human iris consists of elastic connective tissue called the trabecular meshwork whose premature morphogenesis is completed during the eighth month of conception. It is composed of pectinate ligaments sticking into a knotted mesh revealing striations, ciliary processes, crypts, rings, wrinkles, a halo, and other features. In the first year of life, a layer of chromatophore cells frequently changes the color of the iris; however, the recent clinical data indicate that the trabecular pattern itself is steady throughout an individual's lifetime. Due to the fact that the iris is protected at the back of the corona and the aqueous humor, it is resistant to the environment with the exception of its pupillary response to light. Pupillary movement, in the absence of light changes and related expandable/elastic deformation in the iris surface, offer one check against photographic or other simulacra of a living iris in highly secured applications. It was found that even visibly dark-eyed individuals reveal plenty of unique iris details when imaged using infrared light (Adler, 1965; Daugman, 1993; Daugman, 1999).

Implementation of this scheme for authentication purposes at the desktop can be performed similarly to finger scan. Simply, by storing and matching the biometric template on a smart card and employing the biometric to unlock smart card confidential information. Here no physical contact is required to capture the image. The iris image is usually captured by a CCD camera with the subject about one meter or less from the camera. Once captured, the iris image is parsed to create a template. Typically, an iris can yield about 173 independent points. Such information is more than five times the information derived from a fingerprint. Moreover, it is more than needed to differentiate a single person among the entire population of the world (Adler, 1965).

It is worth mentioning that the human iris is stable throughout one's life and is not susceptible to injury or wear. Moreover, contact lenses or eye glasses do not interfere with the use of biometric identification systems. This approach has been utilized in some ATM machines as an alternative to passwords or PINs by the Bank of United Texas in the United States, and the Royal Bank of Canada (Cavoukian, 2005). Other possible applications include telecommunications and Internet security, nuclear power station control, prison control, computer login authentication, e-commerce and e-business security, e-government, and different government applications.

Signature

This method analyzes the way a user signs his/her name. Signature is considered as a series of activities, which contain unique biometric features. Examples on these features include personal pace, acceleration, and pressure flow. It is worth noting that unlike electronic signature capture that treats signature as a graphic image, signature recognition technique determines how the signature is signed. In this scheme, the subject signs his/her name on a digitized graphics tablet or Personal Digital Assistant (PDA). Then the identification system analyzes signature dynamics such as velocity, stroke order, stroke count and pressure. Moreover, the system can follow each person's biological signature fluctuations over time. The collected signature dynamics data are encrypted and condensed into a special template (Rhodes, 2003).

It was found that signing features such as pressure and velocity are as essential as the completed signature's static form. Signature authentication has the advantage of being known to people more than other existing biometrics schemes as people are familiar with signatures as a means of transaction-related identity authentication and most people would see nothing uncommon in broadening this to include biometrics. In general, signature authentication devices are quite precise in operation and clearly are ideal for applications where a signature is an accepted identifier. Despite the fact it is an easy scheme to implement, this technique is not a very popular biometric scheme (Liu, 2001).

Keystroke dynamics

Since the start of the twentieth century, psychologists and mathematicians have tested and experimented with human actions and behaviors. Psychologists have established that human actions are expected in the performance of recurring, and normal routine tasks (Obaidat, 1997; Obaidat, 1999; Umphress, 1985; Bryan, 1973; Obaidat, 1993a; Obaidat, 1993b; Bleha, 1991; Bleha, 1993). It was found since 1895 that telegraph operators have unique patterns of keying messages over telegraph lines (Bryan, 1973). Moreover, an operator often knows who is typing on the keyboard and sending information simply by listening to the characteristic configuration of dots and dashes.

Nowadays, the telegraph keys have been replaced by other input/output peripherals such as the keyboard and mouse. It has been established through empirical studies that keyboard characteristics are rich in cognitive traits and possess promise as an individual identifier. A person sitting near a typist or a computer user is usually able to recognize the typist by keystroke patterns (Obaidat, 1993a; Obaidat, 1993b; Bleha, 1991; Bleha, 1993; Obaidat, 1994).

It has been demonstrated that the human hand and its environment make written signatures difficult to forge. In Joyce and Gupta (1990), it has been shown that the same neurophysiological factors that make a written signature inimitable are also exhibited in an individual typing pattern. When a computer user types on the keyboard of a computer, he/she leaves a digital signature in the form of keystroke latencies (elapsed time between keystrokes and hold/dwell times). Human nature says that a person does not just sit before a computer and overwhelm the keyboard with a nonstop stream of data entry. Instead, the individual types for a time, stops to gather thoughts and feelings, pauses again to take a rest, resumes typing, and so on. In devising a scheme for identity authentication, a general baseline must be recognized for finding out which keystrokes describe the individual's key pattern and which do not. Physiologists have analyzed the human interface with information systems and devised several paradigms describing the interface to computers. One of the well-liked paradigms is the keystroke-level model developed by Card *et al.* (1980). Their model explains the human–machine interaction during a session at a computer terminal. It was meant to evaluate alternative designs for highly interactive programs. The keystroke level model summarizes the terminal session using the expression below:

$$T_t = T_a + T_e,$$

where T_t corresponds to the duration of the terminal session; T_a corresponds the time needed to evaluate the task, construct mental illustration of the functions to be achieved, and select a scheme for answering the problem; and T_e stands for the time needed to carry out all functions making up the task. Keep in mind that T_a varies according to the degree of the deemed task, familiarity of the user, and knowledge of the functions to be carried out. Obviously, this term is not quantifiable; therefore, T_a cannot be used to characterize an individual. In contrast, T_e represents actions that can be expressed as:

$$T_e = T_k + T_m,$$

where T_k is the time to enter information and T_m is the time required for mental preparation. Shaffer (1973) has found that when a typist is inputting data, the brain behaves as a buffer, which then yields the text onto the keys of the keyboard. Cooper (1983) found that the normal pause points are between words as well as within words that are longer than 6–8 characters.

Despite the fact that handwriting and typing are individual manual skills, they both have measurable uniqueness (Obaidat, 1993a; Obaidat, 1993b; Bleha, 1991; Bleha, 1993; Obaidat, 1994). Umphress and Williams (1985) have carried out an experiment for keystroke characterization by using two sets of inputs for user detection, namely, a reference profile and a test profile. It was found that a high degree of correlation could be achieved if the same person typed both the reference and test profiles. Several medium confidence levels were assigned in cases where the typists of the profile differed. Nevertheless, in most cases test profiles had low scores when the typist was not the same person who typed the reference profile.

Obaidat and his colleagues (Obaidat, 1997; Obaidat, 1999; Obaidat, 1993a; Obaidat, 1993b; Bleha, 1991; Bleha, 1993; Obaidat, 1994) explained a method of recognizing a user based on the typing technique of the user. The inter-character time intervals measured as the user types a well-known sequence of characters were used with conventional pattern recognition methods to classify the users, with good verification results. By necessitating the character sequence to be typed twice, and by using the shortest measurements of each test, better results were achieved than if the user typed the sequence only once. The minimum-distance classifier provided the best classification accuracy. To achieve the least classification error, their analysis considered the effect of the dimensionality reduction, and the number of classes in the verification system. The measurement vector was obtained by calculating the real-time durations between the characters input in the password. Obaidat and Macchiarolo (1994) and Obaidat and Sadoun (1997) employed some traditional neural network paradigms along with classical pattern recognition techniques for the classification/identification of computer users using as feature the interkey times of the keystroke dynamics.

Obaidat and Sadoun (1997) authenticated computer users using hold times of keystroke dynamics as features to verify system users. The subjects in the test were requested to key in their login user ID during an eight-week period. A computer program gathered key hit and key release times on a desk top PC to the nearest 0.1 ms. The system measures the time durations between the moment every key button is hit to the moment it is released. This process was performed for each letter of the user ID and for each participant. A scan code is spawned for both the hit and release of any key. The system produced good results in terms of low rejection rate of authorized users and low acceptance rate of unauthorized users. Actually, when some neural network paradigms were used for the classification process, the two error rates (error rate of accepting an unauthorized user and error rate of rejecting an authorized user) were zero.

Clearly, keystroke dynamics are fertile with individual traits and they can be used to extract features that can be used to authenticate access to information systems and networks including e-based systems. The keystroke dynamics of a computer user's login string offer a characteristic pattern that can be used for verification of the user's identity.

5.3   Accuracy of biometric techniques

We reviewed in the previous section the major biometric techniques that can be used for securing access to e-based systems. Each of these schemes has its strong points and weak points.

E-commerce and e-government developers are looking carefully at the use of biometrics and smart cards to more accurately verify a trading party's or a citizen's identity. For instance, many banks are attracted to this combination in order to better validate customers and ensure nonrepudiation of online trading, buying, and banking. Point-of-sales (POS) system retailers are working on the cardholder authentication methods that would join smart cards and biometrics to substitute signature verification process. It is expected that adding a smart-card-based biometric verification scheme to a POS credit card payment will reduce fraud significantly (Liu, 2001).

Some organizations are using biometrics to get secure services over the telephone through voice verification. Voice verification is currently deployed in many parts of the world by Home Shopping Networks. It has been found that a risk management method can help identify the necessity and use of biometrics for security.

It has been reported to the United States Congress that about two percent of the American population does not have a legible fingerprint and hence cannot be signed up into a finger-print biometric-based security system. The report suggested a system utilizing dual bio-metric measures (indicators) for authenticating individuals. This approach is often called a multimodal biometric system and has been shown to improve identification accuracy and minimize vulnerability to spoofing while increasing population coverage. The basics of multimodal biometrics rely on a combination of various biometric mode data and the feature extraction, match score, or decision level. Feature level combination brings together feature vectors at the demonstration level to offer higher-ranking dimensional data points when creating the match result. Match grade level fusion brings together the individual scores from various matchers. Determination level fusion combines accept or reject verdicts of different systems (NIST, 2000; Snelik, 2003; Ross, 2001).

One example of multiple biometrics is combining facial and iris recognition. Moreover, multiple biometrics could also entail multiple instances of a sole biometric, such as one, two, or ten fingerprints, two hands, and two eyes, etc. It is possible to have a system that integrates fingerprint and facial recognition technologies in order to provide better robust and accurate authentication/identification. There are some commercially available

Table 5.3 *A summarized comparison of major biometric schemes.*

Biometric Scheme	Universality	Uniqueness	Robustness	Collectability	Performance	Acceptability
Face	H	L	M	H	L	H
Fingerprint	M	H	H	M	H	M
Hand Geometry	M	M	M	H	M	M
Keystrokes	L	L	L	M	L	M
Iris	H	H	H	M	H	L
Retina Scan	H	H	M	L	H	L
Signature	L	L	L	H	L	H
Voice	M	L	L	M	L	H

systems that combine face, lip movement, and speaker recognition to control access to physical structures and small office computer networks. Experimental studies have found out that the identities established by systems that use more multi-biometrics have much better results and performance, especially when applied to large target populations (Rhodes, 2003).

Table 5.3 shows a summarized comparison between the major biometric schemes in terms of universality, uniqueness, robustness, collectability, performance and acceptability (Jain, 1999), where H stands for high, M stands for medium and L stands for low.

The match between application and biometric scheme is decided based on: (a) nature and characteristics of the application, (b) requirements of the application, (c) aspects of the biometric technology, and (d) cost. When dealing with biometric systems, the application is described in terms of: (a) need for verification, (b) whether the system is attended, unattended, or semi attended, (c) whether the users adapted to or are familiar with the biometric system or not, (d) nature of application in terms of secret or not, (e) whether the subjects were cooperative or not, (f) storage constraints, (g) degree of constraints on performance requirements, and (h) acceptable types of biometrics to the users, which are usually related to local culture, customs, privacy concerns, religion, hygiene standards of the society (Jain, 1999).

The main applications of biometric-based identification include: (a) banking security fund transfers, ATM security, e-commerce, (b) physical access control in airports, government and corporate offices, (c) information system security such as databases using login tools, (d) e-government applications including passport, national ID, and driving license issuing and renewal processes, immigration and naturalization services, voter registration, government benefit distribution, (f) customs, (g) customer loyalty and preference in e-commerce and e-business, and (h) cellular bandwidth access control.

Biometrics is considered a young technology, and its performance has become accepted recently. The desired good characteristics include: robustness, acceptability by users and community, uniqueness, ease of collection of data, and of course performance in terms of low acceptance error rate and low rejection error rate.

It is important to analyze the performance metrics to find out the strengths and weaknesses of each biometric scheme. The three major performance metrics of interest are: (a) false match rate (FMR), (b) false non-match rate (FNMR), also called false rejection, and (c) failure to enroll rate (FTER). A false match happens when a system incorrectly matches an identity, and FMR is the probability of subjects being incorrectly matched. In verification and positive identification systems, unauthorized people can be granted access to systems as the result of incorrect matches. In a negative verification system, the result of a false match may be to deny access to authorized individuals. For instance, if a new applicant to a public benefits program is wrongly matched with a person previously registered in that program under another identity, the applicant may be deprived of access to benefits.

A wrong non-match happens when a system denies a valid user from accessing the facility, and FNMR is the probability of authorized users being erroneously not matched.

There are several biometric issues that are not easy to quantify; however, they can be assessed qualitatively. These are: (a) security, which consists of schemes and protocols used to protect an application against some threats from adversaries, (b) usability, which explains the easiness to use the system, keeping in mind that usability and convenience may often play a conflicting position in security of systems, (c) privacy, which is usually related to data confidentiality, specifically, biometric traits and associated data are considered very personal, and (d) acceptability, which is a basic aspect for meeting the requirements of any biometric-based security system (Bencheikh, 2005).

In authentication and positive recognition systems, subjects may be not permitted access to some facility or resource as the result of a failure in the system. In negative verification systems, the outcome of a false mismatch may be that a user is given access to system resources to which he should be not be allowed. Forged matches may happen due to the fact that there is a high level of relationship between two persons' characteristics. Fake non-matches happen due to the fact that there is not enough strong resemblance between an individual's enrollment and test patterns, which could be due to a number of conditions. For instance, a user's biometric data may have changed as a result of aging or injury. If biometric verification systems were ideal, both error rates would be nil. Nevertheless, because biometric systems cannot verify users with 100% accuracy, a compromise exists between the two. False match (false identification) and non-match rates (false rejection) are inversely related. Thus, they must be evaluated in tandem, and satisfactory risk degrees must be balanced with the drawbacks of inconvenience to users. Obaidat and Sadoun (1997, 1999) have suggested that if the classification does not match, several things could happen:

1. The user is denied access. This is the highest security level.
2. The user is granted access, after providing a higher-level password.
3. The user is granted only limited access.
4. The user is granted access, but a "warning" is signaled to the administrator, and the user's actions are intercepted for later analysis. This is considered the lowest security level. In general, there is a tradeoff to consider with any security system; the risk of security breach balanced with the user inconvenience.

5.4   Issues and challenges

There are many issues and challenges in using biometric-based identification systems. Some of these issues are known open problems in the allied scientific area such as pattern recognition, neural networks and computer vision, while others require methodical cross-disciplinary efforts. Innovative engineering designs have the potential to resolve most of these open problems.

Whereas biometric technology is presently obtainable and employed in a variety of applications, questions continue regarding the technical and operational efficiency of biometric technologies in large-scale environments. A risk management approach has the potential to help describe the need and use for biometrics for information security. Moreover, a determination to use biometrics should take into consideration the costs and returns of such systems and their prospective outcome on ease and privacy. In order to have an effective risk management policy, certain aspects should be identified and clarified. These include:

1. Identification of assets being protected: Here there is a need to identify assets that should be sheltered and the consequence of their possible loss.
2. Identifying the opponents: It is important to identify and characterize the danger to these possessions. The target and capacity of an adversary are the main criteria for creating the level of threat to these assets.
3. Way of exposing: This entails identifying and characterizing weaknesses that would permit known threats to be a reality.
4. Priorities: It is important to assess risk and determine priorities for protecting assets. Risk assessment checks the likely for the loss or harm to an asset. Risk degrees are identified by evaluating the impact of the loss or damage, threats to the asset, and vulnerabilities.
5. Actions to be taken: It is important to identify what actions are to be taken. The pros and cons of these countermeasures must be weighed in opposition to their drawbacks.

Determining the performance of a biometric system is essential. When testing a biometric system, we basically perform this in three modes: technology, scenario, and operational evaluations. Technology assessment aims at comparing the performance of schemes from a specific technology, for instance, person identification based on iris or fingerprints. Scenario assessment investigates the performance of the entire biometric system within a replicated application modeling a real case. Operational evaluation tests the performance of the whole system in a specific application with a certain population. It is worth mentioning that reliable performance estimates are obtained only by well-defined test and control conditions (Bencheikh, 2005).

Another important issue is deviation of characteristics and mannerisms. There is no particular biometric scheme that can be employed for the whole population of users in all cases. Clearly, this makes the process of automatic authentication of all users with a sole biometric scheme unfeasible. Therefore, other and proper schemes of identification for those incapable of employing the biometric should be an essential part of any effective recognition system. Moreover, both the behavioral and physiological features/characteristics may change with time. For instance, fingerprints can be stained by scratches, contact to chemicals, or entrenched dirt; voice can vary due to colds; speed of keying in a keyboard may change due to illness and signature may change as a person ages. In addition, general reasons like anxiety, health, environmental circumstances, and time limitations can affect the communication between the subject and the system, which may lead to discrepancies. It is worth mentioning here that behavioral characteristics are more vulnerable to deviations than physiological traits. A subject can vary his voice, deliberately or not, much more readily than he changes his fingerprints. Comparative investigation studies on different biometric-based identification systems have shown an inferior degree of accuracy in behavioral-based biometrics than with physical-based systems (Obaidat, 1997; Cavoukian, 2005).

The attitude of users towards biometric systems can also impact the overall performance of the biometric-based authentication system. The first time an individual uses the system will be different from the fifth or tenth time. Each of the five or ten biometric measures will be distinctive. However, if a helpful user turns out to be more familiar with a biometric system, the merit of captured data will get better. It is known that some schemes are more acceptable to users than others. For example, one investigation study has found that iris and retina scans are most unacceptable. Hand geometry and fingerprint geometry were found to be more acceptable, while face, voice and signature-based schemes were found the most acceptable (Cavoukian, 2005; Jain, 1997). People's acceptance of biometric systems may have to deal with culture and customs as they may be perceived as an intrusive tool. Speed of registration and inconvenience are also a concern. For instance, a subject may be apprehensive about a retina scan since it entails a bright beam of light into the eye and needs accurate alignment while the scan is completed. Clearly, this may cause uneasiness on the part of subjects. Some biometric schemes can be performed in a more ordinary and relaxed way, such as dynamic signature authentication.

The level of uniqueness of a biometric system in identifying users depends mainly on the kind and extent of the application. In applications with small population, the uniqueness of the biometric characteristic is not an issue. It may be enough for the feature to only be unlikely to be identical. However, in systems with large population, systems based on non-unique characteristics may be more likely to have a false acceptance rate due to resemblance.

The size of the collected and used data is another issue that affects the accuracy of identification. It is estimated that the amount of data collected in bytes/second is about 750 bytes/s for fingerprints, 265 bytes/s for iris scan, 35 bytes/s for retina scan, and 9 bytes/s for hand geometry. This is why in e-government and law enforcement applications fingerprints are preferred (Cavoukian, 2005).

The accuracy rates reported by manufacturers of biometric-based authentication systems may be different from what customers may find in reality. This is due to the fact that manufacturers test their systems under certain conditions and environments that may not apply to the real applications. However, with the increased interest in biometric systems, there is a call to have independent organizations that test the systems and report results.

The degree of security of a biometric-based security system varies depending on the type of biometric characteristics employed and the method of applying them. Of course, there are strong and weak biometrics from the viewpoint of security. In general, biometrics are considered secure since they are based on the assumption that only the individual who has exactly the identical iris, face, voice, fingerprints, or whatever, can impersonate another individual. Because this is considered impossible, such biometric systems are considered highly secure. Nevertheless, the reality indicates that some biometric schemes can be spoofed or circumvented easier than others. In terms of impersonations, there are two types of concerns: (a) active imposer acceptance where an imposer presents a customized, fake biometric sample, purposely in order to connect it to another individual who is an authorized enrollee and he is incorrectly verified by the system as being the authorized enrollee, (b) passive imposer acceptance where an imposer presents his own biometric trial and states the identity of another individual either purposely or unintentionally, and he is wrongly recognized by the system (Cavoukian, 2005). Despite that there are some reported cases of spoofing, it should be stated that it is very hard, if not impossible, for an individual to fake an actual biometric trait. Nevertheless, it is easy for an individual to copy a digital image of a subject's biometric. Therefore, customers and users in general should be alert that as biometrics are becoming more popular and more online applications are spreading, such as e-government, e-services and e-commerce, a considerable security issue could arise. Once a biometric is digitized, it can be compromised and seized without appropriate security controls. Keep in mind that an identity stealing in this context does not require the genuine biometric characteristic to mimic an individual. Only the digital image of that person's biometric is needed. If not, the holder of the biometric has to be observed when the live scan is retrieved, or there is some extra verification measure used, otherwise the system will not be able to make out the difference.

The biometric schemes that became widely accepted are the ones that are easy to use, and perceived as the least intrusive to the person's privacy such as fingerscan, hand geometry and facial recognition schemes. There is no proof that biometric collection has harmfully affected or hurt the subject. Furthermore, there is no evidence that any of the commercially available biometric-based security systems has presented health dangers or left skin marks (Cavoukian, 2005).

5.5   Concluding remarks

Biometric technology aims at identifying an individual based on his unique physiological and behavioral characteristics. There are numerous applications of biometric-based security systems and new ones are emerging. Many challenges in using this promising technology such as privacy, accuracy, variation in characteristics and trait, and user attitude have to be addressed before the technology can be widely accepted worldwide. Legislation should be put in place in order to protect the privacy rights of individuals as well as endorsing the use of biometrics or legitimate their uses. It is essential to keep in mind that efficient security cannot be accomplished by depending on technology alone. Technology and users should work together as part of a whole security process. Any weakness in any of these areas weakens the success of the security process. Before starting the process of designing, developing and implementing biometric-based security systems, several aspects should be considered. Among the major ones are: (a) the manner in which the technology will be used, (b) a comprehensive cost–benefit analysis must be performed to find out whether the benefits obtained from a system outweigh the costs, and (c) an analysis must be performed to find out the best compromise between the improved security, which the use of biometrics would offer, and the effect on the individual's privacy and convenience. Finally, this technology has great promises and we should see more and more of its applications for securing of e-based systems, especially in e-government, e-commerce and e-business, and computer networks.

References

Adler, F. H. (1965). *Physiology of the Eye: Clinical Application*, 4th edn, C. V. Mosby.

Bencheikh, R. and L. Vasiu (2005). Hybrid authentication systems. Proc. of the 2005 International Workshop in Wireless Security Technologies, pp. 130–7.

Bleha, S. and M. S. Obaidat (1991). Dimensionality reduction and feature extraction applications in identifying computer users. *IEEE Transactions Systems, Man, and Cybernetics*, Vol. **21**, No. 2.

Bleha, S. and M. S. Obaidat (1993). Computer user verification using the perceptron. *IEEE Transactions Systems, Man, and Cybernetics*, Vol. **23**, No. 3, 900–2.

Bryan, W. L. and N. Harter (1973). Studies in the physiology and psychology of the telegraphic language. In *The Psychology of Skill: Three Studies*. E. H. Gardener and J. K. Gardner (eds.), NY Time Co., pp. 35–44.

Card, S., T. Moran, and A. Newell (1980). The keystroke level model for user performance time with interactive systems. *Communications of ACM*, Vol. **23**, 396–410.

Cavoukian, A. (2005). Consumer Biometric Applications. Information and Privacy Commissioner of Ontario, Canada. http://www.ipc.on.ca/docs/cons-bio.pdf

Clarke, R. (2001). Biometrics and privacy. Notes available at: http://www.anu.edu.au/people/Roger.Clarke/DV/Biometrics.html

Cooper, W. E. (1983). *Cognitive Aspects of Skilled Typewriting*. Springer-Verlag.

Daugman, J. G. (1993). High confidence visual recognition of persons by a test of statistical independence. *IEEE Trans. Pattern Recognition Machine Intelligence*, Vol. **15**, No. 11, 1148–61.

Daugman, J. G. (1999). Recognizing persons by their iris patterns. In *Biometrics: Personal Identification in Networked Society*, A. Jain, R. Bolle, and S. Pankanti (eds.), Kluwer, pp. 103–21.

Hill, R. (1999). Retina identification. In *Biometrics: Personal Identification in Networked Society*, A. Jain, R. Bolle, and S. Pankanti (eds.), Kluwer, pp. 123–41.

Jain, A., L. Hong, S. Pankanti, and R. Bolle (1997). An identity-authentication system using fingerprints, *Proc. of IEEE*, Vol. **85**, No. 9, 1366.

Jain, A., R. Bolle, and S. Pankanti (1999). Introduction to biometrics. In *Biometrics: Personal Identification in Networked Society*, A. Jain, R. Bolle, and S. Pankanti (eds.), Kluwer, pp. 1–43.

Joyce, R. and G. Gupta (1990). Identity authentication based on keystroke latencies. *Communications of ACM*, Vol. **33**, No. 2, 168–76.

Langeand, L. and G. Leopold (1997). Digital identification: it is now at our fingertips, *EEtimes*, March 24, Vol. **946** (http://techweb.cmp.com/eet/823/).

Liu, S. and M. Silverman (2001). A practical guide to biometric security technology. *IEEE Computer Magazine*, January/February 2001, 27–32.

NIST (2000). Report to the United States Congress, Summary of NIST Standards for Biometric Accuracy, Tamper Resistance and Interoperability.

Obaidat, M. S. (1993a). A methodology for improving computer access security. *Computers & Security*, Vol. **12**, 657–62.

Obaidat, M. S. and D. T. Macchairolo (1993b). An on-line neural network system for computer access security. *IEEE Transactions Industrial Electronics*, Vol. **40**, No. 2, 235–41.

Obaidat, M. S. and D. T. Macchairolo (1994). A multilayer neural network system for computer access security. *IEEE Transactions Systems, Man and Cybernetics*, Vol. **24**, No. 5, 806–13.

Obaidat, M. S. and B. Sadoun (1997). Verification of computer users using keystroke dynamics. *IEEE Transactions on Systems, Man and Cybernetics*, Vol. **27**, No. 2, 261–9.

Obaidat, M. S. and B. Sadoun (1999). Keystroke dynamics based authentication. In *Biometrics: Personal Identification in Networked Society*, A. Jain, R. Bolle, and S. Pankanti (eds.), Kluwer, pp. 213–30.

Obaidat, M. S., A. Brodzik and B. Sadoun (1998). A performance evaluation study of four Wavelet algorithms for pitch period estimation of speech signals. *Information Sciences Journal*, Vol. **112**, No. (1–4), 213–21.

Obaidat, M. S., C. Lee, B. Sadoun and D. Nelson (1999). Estimation of pitch period of speech signals using a new dyadic Wavelet algorithm. *Information Sciences Journal*, Vol. **119**, No. 1, 21–39.

Rhodes, K. A. (2003). *Information Security: Challenges in Using Biometrics*. United States General Accounting Office. http://www.gao.gov/new.items/d031137t.pdf

Ross, A. and A. Jain (2001). Information Fusion in Biometrics. In *Proceedings of AVBPA*, Halmstad, Sweden, June 2001, Springer, pp. 354–9.

Shaffer, L. H. (1973). Latency mechanisms in transcription. In *Attention and Performance*, Vol. IV, S. Kornblum (ed.), Academic Press.

Snelik, R., M. Indovina, J. Yen, and A. Mink (2003). Multimodal biometrics: issues in design and testing. In *Proc. of the 2003 International Conference on Multimodal Interfaces* (IMCI 2003), Vancouver, Canada, ACM.

Umphress, D. and G. Williams (1985). Identity verification through keyboard characteristics. *International Journal Man-Machine Studies*, Vol. **23**, 263–73.

Woodward, J. D. (1997). Biometrics: privacy' foe or privacy's friend? *Proceedings of the IEEE-Special Issue on Automated Biometrics*, Vol. **85**, 1480–92.

Weng, J. and D. Swets (1999). Face recognition. In *Biometrics: Personal Identification in Networked Society*, A. Jain, R. Bolle, and S. Pankanti (eds.), Kluwer, pp. 65–86.

Young, J. R. and R. W. Hammon (1981). *Automatic Palmprint Verification Study*. Rome Air Development Center, Report No. RAD-TR-81-161, Griffith AF Base, New York.

Zunkel, R. L. (1999). Hand geometry based verification. In *Biometrics: Personal Identification in Networked Society*, A. Jain, R. Bolle, and S. Pankanti (eds.), Kluwer, pp. 1–43.

6 Trust management in communication networks

Trust management is a major component in the security of e-services. Issues in trust management include: (a) expressing security policies and security credentials; (b) ascertaining whether a given set of credentials conforms to the relevant policies; and (c) delegating trust to third parties under relevant conditions. Various trust management systems have been developed to support security of networked applications. Unfortunately, these systems address only limited issues of the trust management activity, and often provide their services in a way that is appropriate to only special applications. In this chapter, we present a comprehensive approach to trust management, consider the major techniques and functionalities of a trust management system, and describe three well-known trust management systems.

6.1 Introduction

Recent advances in Internet computing, paired with the increase in network resources and end-node processing capabilities, have led to the growing need of organizations and administrations to use large Intranets to connect their offices, branches, and information systems. They also pushed for the development of e-services for the need of their customers. All the emerging applications and e-services have different notions of the concept of resource. They share one thing in common: the need to grant or restrict access to their resources according to the security policy appropriate to that e-service.

Resources handled by e-services are of different types. While a clinical information system considers that a resource is a patient's record, a banking payment system considers accounts and money as the major resources to manage (Guemara-ElFatmi *et al.*, 2004). Similarly, a secure network layer protocol (such as IPsec) would associate resources such as *"security associations."* While different resources can be considered as the aggregation of more basic resources, it is often convenient to refer to these aggregations as single resources, authorizing relevant security operations. Therefore, a generic security mechanism for access-granting or execution-authorizing should be able to cope with the number and types of resources.

Traditionally, access control systems and authorization systems fail to provide efficient and robust tools for handling the security at the scale and need of the emerging e-services. For example, access control lists (ACLs), which have been widely used in distributed systems to describe which access rights a user (or principal) has on a resource, are inappropriate for at least two reasons: (a) changes in the e-service security policy often require reconfiguring or even reconstructing the ACL components or functions; and (b) ACL-based mechanisms enforce limited policies. This often prevents differentiation between principals based on their profiles, the specific need to protect the resource to be accessed, or the level of trust that can be given to the individual locally. Besides, when applied in a distributed environment, ACL-based approaches present a number of shortcomings including the protection of the operations that modify the ACLs and the security level (or inadequacy) of the authentication mechanisms used in conjunction with ACLs (Blaze *et al.*, 1999a).

On the other hand, traditional authorization mechanisms that use the "binary" authorization model may face various drawbacks, since they grant access based on the fact that the requesting principal presents a certificate issued by a trusted certification authority. Granting within this model grants a full access and use, or simply rejects the request. Typically, this functionality is implemented in network communication. For example, security protocols such as the transport layer security protocol (TLS) support authentication of both clients and servers using public keys. Nevertheless, since communication protocols are implemented by untrusted codes, users may find security methods based on environments using these protocols untrusted. For example, to authenticate the client to the server, TLS protocol needs access to the user's private key. This represents a serious limitation since the client may not want to give the key to an untrusted code.

Several solutions have been proposed in the literature to overcome the abovementioned shortcomings. Most of the proposed solutions are based on a centralized security architecture, which limits their utility in distributed environments where centralized authorities do not exist. Trust management has been introduced by Blaze *et al.* (1996) to provide a general-purpose, application-independent mechanism for checking proofs of compliance with policies. Several trust management systems have been proposed to address authorization in decentralized environments. These systems include but are not limited to: simple public key certificate (SPKI) (Ellison, 1999), PolicyMaker (Blaze *et al.*, 1996), keynote (Blaze *et al.*, 1999b), and delegation logic (Li *et al.*, 2000). These systems are based on the notion of delegation, whereby one entity delegates part of its authority to other entities.

Basically, trust management uses a set of unified mechanisms for specifying both security policies and security credentials. The credentials are defined as signed statements (or digital certificates) about what principals (or users) are allowed to do. Even though they are called certificates, these credentials are fundamentally different from traditional CA-issued certificates. The access grants are permitted directly to the public keys of principals, no matter the name is globally or locally verified. This is why trust management systems are sometimes called key-oriented PKIs. The unified mechanisms are also designed to separate the mechanism from policy. In addition, by using the same mechanisms for verifying digital credentials and trust policies, a trust management can be used by many different applications. This constitutes a major difference with the ACL concept, whose structure and semantics usually reflect the needs of one particular application. Therefore, a single-branch enterprise may use a unique trust manager system to control the access to all its applications, while allowing different trust policies.

6.2 Trust definition

Trust is a complex concept that is hard to define rigorously. However, a large spectrum of definitions of trust has been provided in the literature, many of which are dependent on the context on which interactions occur. An example of such definitions is given by McKnight and Chervany (1996) stating that: *"Trust is the extent to which one party is willing to depend on something or somebody in a given situation with a feeling of relative security, even though negative consequences are possible."*

Following this definition, one can say that different entities (individuals or system components) may have different kinds and levels of trust in the same target entity (server or applications). It is also worthwhile to notice that trust is correlated to the purpose and nature of the relationship. For example, an employee may be trusted by an enterprise to work with a set of documents that are classified confidential, but not higher than that. The same employee might not be trusted to deal with financial transactions of any amount, while another employee can be trusted by the same enterprise to sign financial transfers up to a certain amount.

Trust can be expressed with three basic dimensions: The first dimension represents the trust origin (or *trustor*); the second dimension represents the trust relationship (or *trust purpose*); and the third dimension represents the target of the trust (or *trustee*). In addition, a trust measure can be added to each trust relationship. The measure could have different forms. It could be Boolean (i.e., fully trusted or not trusted at all); discrete (e.g., strong distrust, weak distrust, weak trust, and strong trust); or continuous (i.e., based on a continuous function such as probability or a belief function of trustworthiness). Another important component can characterize a trust relationship, the time parameter, since the trust measure (or trust of the trustor in the trustee) regarding a specific purpose might change with time.

It is obvious that trust is not implicitly transitive. Nor is it symmetric. However, transitivity is recommended for a lot of trust relationships, meaning essentially that, for a given trust relationship, if entity Us_A trusts entity Us_B with a trust value α_{AB} and entity Us_B trusts entity Us_C with a trust value α_{BC} then Us_A trusts Us_C with a trust value α_{AC}. Trust value α_{AC} is determined depending on the nature of the trust measure and the purpose of the trust. If the trust measure is Boolean, then the conjunction $\alpha_{AC} = \alpha_{AB} \wedge \alpha_{BC}$ might be used. In the case where the trust measure is a probability function, α_{AC} may be equal to the product $\alpha_{AB} \times \alpha_{BC}$. When the nature is more complicated, α_{AC} may be more difficult to compute.

6.2.1   A model for trust

There are a number of situations where trust relationships need to be established to allow access to resources or authorize the execution of a requested action. But because principals may be involved in different security domains, it is often infeasible to have a central entity in charge of authorizing every possible principal. In order to scale authorization and reduce the induced activity, delegation is required, where local sites can be assigned the authority to authorize principals based on local security policies.

In addition, once entities have deployed e-services, resource usage and authorization process need to be monitored to ensure that resource usage is not being misused and that rules have not been hacked. Typically, examples of resource misuse include but are not limited to services performing distributed measurement to a target node, services sending a large number of ICMP packets, and services performing experiments that investigate thoroughly a large number of IP addresses (Chun, 2004). Examples of attacks against authorization include attempting to modify rules and replaying requests. When anomalous resource usage occurs, one needs a mechanism to rapidly identify the responsible entity, the individual who authorized that entity, and so on along a chain of trust. Resolving the problem means being able to localize where the anomalous resource use is occurring and to suspend the associated service, if needed. Detecting anomalous resource usage is complex since it should be made real-time. This strengthens the necessity for mechanisms which proactively identify anomalous resource usage and perform appropriate actions in an automated way.

A trust model therefore should provide various tools for delegation, authorization, access, and monitoring. Typically, a trust model can be represented by a three-layer architecture having a first layer to express and verify trust relationships in a flexible and scalable way, a second layer to monitor the use of trust relationships, and a layer to manage and evaluate trust relationships based on criteria such as historical traces of past behavior.

The first layer is responsible with authentication, authorization, and policy definition. In order for a system to scale, trust management in a distributed environment must be decentralized to allow trust to be flexibly delegated between principals. At the same time, while delegation of trust is essential for scaling, trust also needs to be fully accountable to allow delegation to be trusted along paths of delegated trust. The first layer provides the mechanisms to express, delegate, and verify trust relationships and perform authentication and authorization using public key cryptography and assuming fully accountable path of trusts. The second layer in the trust architecture provides two major services. First, it provides a continuous monitoring on how the trust relationships are being used by the principals of the system. Second it provides periodic logging of monitoring of data to create useful information for tracing delegated trust. Monitoring and logging allow trust relationships to be continuously inspected so that, if a misuse occurs, the appropriate log is immediately retrieved in order to find: (a) the chain of trust that asserted that principal's identity and (b) the chain of trust that authorized the principal to perform the trusted action.

The third layer in the model architecture aims to perform automatic detection of anomalous resource usage and to perform appropriate actions in response. Local anomaly detection allows the system to flag resource usage on individual nodes that is deemed suspicious or likely to cause problems. The third layer can implement monitoring and logging primitives to observe resource usage information for each principal on each node. It then applies rules that map resource usage information over time to anomalous events along with alerting level. For each type of anomalous event and each alerting level, there is a set of actions that can be applied.

6.2.2   Delegation of trust

A principal can request the execution of any action that he has to execute or has been delegated the right to execute, provided that the local host authorizes it. The principal can also delegate this right to other principals, if he has been authorized to subsequently delegate. Therefore, a delegation itself can be defined as the right that can be delegated, in the sense that a principal could be given the ability to execute some actions, but not to further delegate them, or the ability to delegate some actions but not the ability to execute them. Hence, the concept of delegation leads to the concept of a chain delegation satisfying the property that, if any delegation link, which gives the ability to execute an action, becomes invalid, the execution is refused.

A simple specification of delegation including conditions, which put constraints on the execution of requested actions, should at least present fields such as: the issue time, which defines when the statement is issued; the start time, which determines when the delegation becomes valid; and the end time, which states when the delegation becomes invalid. The delegation specification also contains information about the delegator principal, the delegate principal, and the list of delegated actions. In addition, for each action, it describes the conditions on the delegation and a Boolean value stating whether the delegatee has the permission to redelegate the right to execute the actions.

In order for a delegation to be valid, two actions need to be successfully performed. The first action considers the construction of a valid chain of trust delegation. This action is often based on the techniques developed for the certificate chain construction described in the previous chapter. The second action considers the compliance of the delegation content with the local rules that manage the access to resources and the execution of actions at the local site. The chain of trust delegation is therefore based on the use of signature schemes and delegation issuers (or authorities). Different types of delegations have been defined in the literature including the following list of categories (Kagal *et al.*, 2001):

- *Time bound delegation* This is a delegation that is valid only for certain periods. Typical periods of validation vary from several hours to several weeks.
- *Group delegation* This type of delegation can be used to delegate rights to a group who satisfy a given set of conditions.
- *Action restricted delegation* This category requires the delegatee to satisfy certain conditions before the requested action(s) can be authorized.
- *Redelegatable delegation* With this type of delegation, a field is provided to inform whether the right can be delegated along with the right to redelegate.
- *Strictly redelegatable delegation* This type of delegation allows a right to be re-delegatable without giving the delegatee the right to actually request the execution of the action(s).

A typical format of a delegation is given by the following:

> *Delegate[IssueTime, StartTime, BndTime,*
> *From: delegator,*
> *To: Y,*
> *canDo(Y, Action, Condition-on-action),*
> *(Condition-on-delegation, CC)]*

where delegate[-] contains the information that delegation is issued at *IssueTime* and that it is valid for the period [*StartTime, BndTime*]. The fields *From[-]* and *To[-]* contain the delegator and delegatee identities, respectively. The *canDo[-]* field states that requester Y (or delegatee) is authorized to execute *Action*, only if *Condition-on-action* is satisfied. Finally, *CC* is an attribute whose value defines whether the delegation is redelegatable or not.

Illustrative examples

The following delegations are examples of strictly redelegatable delegation and action restricted delegation, respectively. The first example states that employee *Smithson* is given the right to further delegate the action, denoted by *access*, but not the permission to execute the action himself (Kagal *et al.*, 2001). The second delegation says that only employees of enterprise Cie and named John can perform *Action*. Therefore, all employees have been delegated the right.

1. *Delegate(IssueTime, StartTime, BndTime,*
        From: Smithson,
        canDo(Y, access, name(Y) ≠ Smithson)),
        (Condition-on-delegation= true)
2. *Delegate(IssueTime, StartTime, BndTime,*
        From: John,
        canDo(Y, Action, name(Y)=John and employee(Y, Cie)),
        (Condition-on-delegation= false)

The delegation in the first example can be redelegated, while it cannot be redelegated in the second.

6.3   Digital credentials

Digital credentials represent an efficient mechanism to specify, certify, and convey identity and profile information along with means of verifying their worthiness. Credentials are signed by administration's key and contain the various conditions under which a specific user (as identified by his key in the digital credential) is allowed to access a resource. Entities who wish to gain access to some resource first need to acquire a digital credential from one of the credential issuers. The user then provides this credential to the relevant service (or relying party). Figure 6.1 depicts a generic model for digital credentials.

The traditional X.509 certificates (described in Chapter 4) represent a limited type of digital credentials. They are valid for a predefined period of time. Their content can be used for authentication and authorization purposes. Moreover, they must be revoked whenever their content is out-of-date or has been compromised. Several other forms of digital credentials are used. In the following subsections we will consider two types of digital credentials, the active digital credential and the SPKI certificate.

6.3.1   Active credentials

Active digital credentials are mechanisms to extend traditional static credentials by providing means for dynamically updating their content along with the assessment of their

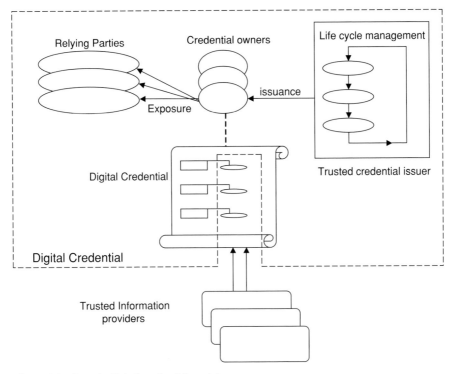

Figure 6.1 Generic digital credential model.

trustworthiness. Their purpose is to present identity and profile data of the credential owner. They are meant to satisfy three requirements: (a) provide an up-to-date content of digital credentials; (b) support up-to-date evaluation of the trustworthiness and validity of digital credentials; and (c) reduce the management complexity of digital credentials. In contrast with traditional digital certificates, which have static content and a predefined validity period, active credentials provide certified mechanisms to dynamically retrieve, compute, and update their content and assess their current level of trustworthiness and validity (Mont and Brown, 2002). Assessment concerns the values of credential attributes, the validity and trustworthiness of these attributes, and the validity and trustworthiness of the whole digital credential.

A typical architecture of an active digital credential contains a list of attributes along with embedded functions; see Figure 6.2. A credential attribute is characterized by a set of properties whose values can be determined dynamically by executing embedded functions. The properties of an attribute include the values of the attribute, its default value, and its level of trustworthiness. Attributes represent information such as name, public key, address, credit-rating, etc. The functions embedded in an active digital credential are digitally signed by a credential issuer and certified as trusted methods to compute property values.

An example of an active digital credential is a digital credit card credential. The list of associated attributes contains, but is not limited to, a credit card number, a validity period, a credit limit, and a credit rate. An associated function can dynamically compute the level

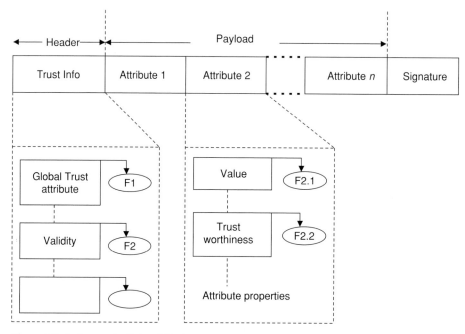

Figure 6.2 Active credential architecture.

of risk and trustworthiness associated to the credential owner, for example by interacting with a certified credit rating company (Mont and Brown, 2002). Levels of trust could have any value in a numeric range; such as [0, 1], or values from a predefined lattice (such as the basic lattice defined by {no-trust, low-trust, medium trust, high-trust}).

Active credentials involve four main actors: (a) the credential issuer, which issues active digital credentials, manages their life cycles, and certifies the functions which dynamically retrieve the content of a credential; (b) the trusted information provider, which supplies up-to-date information about the identity and profiles under well-defined constraints based on a trust relationship with credential issuers; (c) the credential owner, who specifies which information provider must be used to retrieve identity and profile information; and (d) the relying party that supplies services upon receipt of credentials. Access might be granted to the active credential owner depending on the dynamic evaluation of the content of these credentials. An active digital credential can be seen as a contract between the credential owner, credential issuers, and information providers. As the owners can impose the criteria for the disclosure of their data, active credentials enable them to retain control over their data.

The immediate assessment of the attribute values allows active credentials to remain valid and trustworthy even if part of their attributes is not anymore. The advantage is to reduce the number of revoked credentials and limit the need for short credential lifetimes (particularly when the objective of the credential is to provide the identity compared to supplying authorization rights). Active credentials however present a drawback since they depend on the availability of external systems and services. When those entities are not available, the credential content cannot be retrieved. To mitigate this shortcoming, default values for attributes properties and local elaborations of information can be introduced.

Like digital certificates and traditional digital credentials, active digital credentials need proper infrastructure.

6.3.2   SPKI certificates

Simple Public Key Infrastructure (Ellison, 1999) represents an alternative to the public key infrastructure that eliminates the notion of global name and reduces the complexity due to X.509 standard. SPKIs incorporate the notion of local name and view certification authorities as being associated with principals, and public keys. For the sake of uniformity, we consider that there are three types of certificates in SPKI: naming certificates, which provide for local names, authorization certificates, which grant a specific authorization from the issuer to a subject, and revocation lists certificates, which report on invalid naming certificates. The naming certificate and the authorization certificate can be considered as digital credentials. While the former identifies an individual by a public key and contains information guaranteed by the issuer, the latter identifies the public key to which the right to execute a set of actions is delegated.

Let us consider that for a given principal there is a local name space, and that a local name's meaning may vary from one principal to another. We define a fully qualified name (Ellison *et al.*, 1999b) as a compound expression of the form:

$$(namek n_1 \ldots n_s)$$

where k is a public key and $n_1 \ldots n_s$ are local names. Naming, authorization, and revocation certificates may have the following forms of signed messages:

⟨**Cert** (**Issuer** (**Name** k n)) (**Subject** p), Val ⟩;
⟨**Cert** (**Issuer** (**Name** k)) (**Subject** p), Act, Del, Val ⟩; and
⟨**Crl** (**Issuer** (**Name** k)) (**Revoked** $c_1 \ldots c_s$), Val ⟩;

where k is a public key that represents the issuer (who should have signed the certificate), n is a local name of the issuer, p is a fully-qualified name of the owner of the digital certificate, *Val* is an attribute describing validity constraints on the certificates, *Act* stands for a set of valid actions that the issuer delegates the right of execution to subject p, *Del* is an optional delegation attribute, which indicates the conditions under which the certificate is valid (or the conditions under which the subject is authorized to propagate the authority of performing actions in *Act* to other individuals, and $c_1 \ldots c_s$ represent certificates (or hash values that identify the certificates) that the issuer declares as invalid.

An action description is represented by a pair of the form:

$$\langle (\text{Action name}) \text{ in-parameters, } (\text{Env\_state, } \pi) \rangle$$

where name is the name of the action, in-parameters is a set of variables for which input values should be provided in order to be executed, and π is a predicate (or condition to satisfy) involving the local environment that is in charge of executing the action and the parameters occurring in in-parameters. Predicate π represents an optional set of conditions that the system in charge of executing the action should check before execution. The delegation *Del* is basically a statement to be satisfied when the authorization is processed. This may vary from a static statement reduced to a bit, which controls the transfer of the authority of delegation, to a dynamic predicate involving time and delegation criteria. In the easiest case, the value 1 of the delegation bit states that the owner of the authorization certificate can issue a new authorization certification for the actions he is allowed to execute.

The validation Attribute *Val* contains the period validity, which is typically described as an interval of time, say [*Start_time, End_val*], during which the certificate is valid. It also may include different constraints, depending on the type of the certificate. For example, it can contain for the *authorization certificate* a predicate, denoted by *now(t)*, stating that the authorization should become invalid at the time where it is used to execute an action occurring in it where the parameter *t* represents the issuing time.

Like the X.509 standard, various operations can be performed on SPKI-based digital credentials including generation, revocation, and on-line checks. While generation and revocation of naming certificates are typically realized by a trusted credential authority, the online checks are performed locally based on local trust policies. Authorization certificates, however, can be issued by individuals who are authorized to delegate rights on the execution of actions. The generation process is initiated by the user and managed by the certificate signer, who should be in charge of the appropriate verification, especially when some decisions or/and information are to be included in the authorization, delegation and validation fields of the requested certificate. The generation process can be decomposed into three sub-processes:

User interfacing process This is an interactive tool responsible for the collection of the needed information. It also delivers in a secured manner generated certificates and verifies the delivery acknowledgment process.

Checking process This process is responsible for the verification and the validation of the credential content. While the verification checks trustworthiness of the information to be included in the credential, the validation checks whether the content of attributes Act, Val and Del comply with the statements present in the parent certificates and needed for the generation.

Propagation process This process is responsible for updating all archives, directories and services about the generation. It addresses the publication of the naming certificates.

The revoking process of certificates is an operation that can be achieved when triggered by the certificate owner or by the certificate signer. This process can be decomposed into various sub-processes based on the certification practice within the system where the certificates are utilized. The publication of the revocation certificates is part of the revocation process.

Among the major on-line checks, four types can be considered: validity, authorization, delegation, and status checks. All on-line checks can have the following format:

$$(\textbf{Check}\,(\text{Checker check-type})\,(k\,p)\,(\text{Act Cred}\langle\text{opt}\rangle))$$

where Checker represents the specification of one or more uniform resource identifiers that can be contacted to request the check, and check-type defines the type of the check (e.g., validation or delegation), the pair (k, p) represents the subject requesting the on-line check, Act is the action or set of actions, and Cred is digital credential or authorization certificate for the check (if any). Finally, ⟨opt⟩ represents an optional constraint that the requestor may need to check on the executing environment before authorizing the action(s). Its field also contains parameter values to be used in the on-line checks.

On-line checks may involve constructing different credential chains. In addition to the construction of naming certificate chains, which is similar to the X.509 infrastructures, another chain is important to construct: the authorization chain. Typically this is done by finding a sequence $Cert_1, Cert_2, \ldots, Cert_n$ that satisfies the following statements:

- $Cert_1$ is a valid naming certification
- $Cert_k$ is a valid authorization certificate issued by the subject of $Cert_{k-1}, k > 1$
- The delegation field in $Cert_k$ is true, for $k < n$
- The Act content in $Cert_k$ is compliant with the Act content in $Cert_{k-1}$

Informally speaking, an authorization chain supports the proof needed to check that an authorization certificate is valid. It states that all issued certificates involved in the issuance of the credential provided with the request should be valid and should authorize the required action and allow the delegation.

6.4 Authorization and access control systems

Authentication, access control and authorization are the typical means to achieve confidentiality, integrity and availability services. Authentication denotes the act of proving the authenticity of any object or piece of information. This definition includes the fact that authentication denotes the act of proving the authenticity of the identity of the communicating individual (or message originator) as it has traditionally be understood in the literature. Authentication protocols are used to enhance the collection of evidence available to the communicating parties so that one or more of them can believe that a given piece of information is authentic.

6.4.1 Access control systems

An increasing number of organizations have networked their resources with more restrictive access control policies and used various protection mechanisms to enforce their policies. Access control includes the means and methods by which active entities (e.g., individuals, processes and threads) are limited in their ability to manipulate objects in a computer or opening connections, receiving messages, or accessing services available on the network. The purpose of access control is to maintain confidentiality, integrity and availability by making it impossible for unauthorized parties to read information, modify content, or consume resources. *Capabilities* represent a useful means to represent access control information (Nikander and Metso, 2000). A capability is a security token associated with a subject that lists a number of permissions, which define objects and the actions a principal may perform.

Access control management systems appropriate for the scale and complexity of today's networks need several requirements including: (a) the systems must be able to support policy requirements of the large number of applications in use today; (b) the systems must be able to adapt to different management structures; (c) the systems should be able to handle large numbers of users, applications, and policy enforcement points; and (d) the systems should be efficient and not impose significant overheads on existing services. Therefore, policy should be expressed in a way that is easy to distribute the enforcement points.

One way of expressing low-level policy is in the form of public key credentials. An administrator can issue signed statements that contain privileges of users; enforcing points can verify the validity of these credentials and enforce the policy encoded therein. An additional benefit is that they need not be protected when transmitted over the network. Three schemes can be used to distribute access policies. First, the policies can be pushed directly to the enforcement points. Second, the policies can be pulled by the enforcement points from a policy repository as needed. Third, the policies can be distributed to the user systems, which are responsible for delivering them to the enforcement points.

Several shortcomings can be observed with the abovementioned approaches. While the first scheme requires all policy information be stored locally at an enforcement point, which may present problems for embedded systems or routers, the second requires that the enforcement point performs additional processing and overhead when evaluating a security request. Finally, the third approach requires modification of the user application. One advantage of choosing to use credentials as a means for distributing policy is the fact that in a credential-based access control system, adding a new user or granting more privileges to an existing user is simply a matter of issuing a credential. The inverse operation (removing a user or revoking issued privilege) means notifying entities that might try to use the relevant credential that is no longer valid, even though the credential itself has not expired (Keromytis and Smith, 2002). Potential reasons for the revocation include theft or loss of the administrator key used to sign the credential.

6.4.2   Authorization systems

Authorization is the activity of determining whether an entity has the right or the authority to perform a certain set of actions on another entity such as resources, processes, or applications. An entity might be granted *authorization privileges* depending on its attributes, profiles, and the nature of the action(s). Authorization mechanisms check whether the entity can exercise its privileges by verifying whether it satisfies pre-established local authorization conditions. These conditions can be expressed by high-level policies, called authorization policies, based not only on user and profile privileges, but also on service dependent information and external data related to the executing environment. Therefore, the authorization question in trust management can be translated as follow:

"Does the set of credentials prove that the request complies with the local security policy?"

When a third party receives information and has no guarantee of the validity of the information that is presented to it, trust issues arise and *digital credentials* can be used to cope with aspects of trust; authorization has to deal with them. Digital credentials represent a powerful way to describe both identity and attributes associated to people and services. They can be used programmatically by authorization mechanisms to make decisions involving trust issues. For this, the certificate authorities underwriting the digital credentials must provide ways to measure and judge trust. Nevertheless, the ultimate decision on trusting digital credentials and their contents has to be taken through definition of proper trust policy.

Binding the authorization information directly to several attributes including public keys has apparent security advantages. The proof whether a set of actions is allowed can be provided based on this credential. This approach also enables the delegation of issuing credentials. On the other hand, if the delegation network is large, the original intent of the service provider might be lost. Another problem is that malicious entities in the delegation network can over control the whole authorization mechanism. The authorization attributes are assigned by credential issuers. Any principal may issue credentials. The role of delegation is to enable finding the right issuer, who has reliable information about the key owner. The right to issue credentials can be delegated. In this way we do not bind the authorization to keys, but the authorization decision can still be made relying on credentials and policies.

6.4.3   Trust policy

A policy is the combination of rules and services, where rules define the criteria for resource access and usage. Each policy rule is comprised of a set of conditions and a corresponding set of actions. The conditions define when the policy rule is applicable. Once a policy rule is activated, one or more actions contained by the policy rule may be executed. Policies can contain other policies allowing one to build complex policies by aggregating simpler policies. Traditional examples of policies for access control are authorization policies, which are designed to protect target entities and are enforced by these entities. Typically, authorization policy enforcement is the responsibility of an enforcement point or agent, which should intercept actions and perform checks on whether the access is allowed.

Access control management, which provides the rules and procedures for governing the decision in the behavior of the propagation of authority towards legitimate users who have been successfully authenticated, involves various tasks including handling delegation and propagation of authority and management of occurring events. Often, access control management is achieved by means of three major types of trust policies (Damianou *et al.*, 2001):

- Delegation policies: These policies define rules which allow the subject of an authorization (grantor) to delegate some or all of his access rights to a new entity (grantee). When a delegator issues a delegation, he mainly adds new rules to the delegation policy, since he adds at least a rule saying that the delegatee is allowed to request the access to a resource or an action. The delegator can also add rules imposing constraints on the requested resources, if this can be supported by the accompanying digital credentials.
- Negative authorization polices: Such policies are useful for situations where the delegator does not trust the target to enforce a policy. The rules consist of policies that are enforced at the subject site and apply to the requested actions.
- Obligation policies: These policies contain event-triggered rules that accomplish management tasks on a set of target entities or on the subject itself. Obligations are useful

for automating the reaction of a system when security violations or reductions on the quality of service are observed. They differ from the aforementioned policies by the fact that they are concerned with managing access control tasks.

Policy rules occurring in trust policies also can be classified into different categories including the following:

- *Usage rules*: These are rules which describe how to allocate available services on the network and control the selection and configuration of entities based on specific usage data.
- *Security rules*: These rules identify client, permit or deny access to resources, and select and apply appropriate authentication mechanisms.
- *Event rules*: These rules define what alarming actions can be performed on the occurrence of events that are considered crucial for the security of the system or the quality of the service provision.

Each policy rule is interpreted in the context of a specific request for accessing and using resources. It is assumed that taking pre-determined actions are based on whether a given input policy conditions are satisfied.

Illustrative example

The following rules give a practical meaning for the abovementioned classification of policies. Their description is kept informal for the sake of simplicity.

Rule 1 (Limiting number of accesses) If a digital credential is presented for the nth time with a request to access a local resource, and n is a fixed limit of access, then the request should be rejected. This rule is an event that can be included into a negative authorization policy.

Rule 2 (Refrained credential issuance) A digital credential should not be submitted with a request to access a local resource, if the target site is located in a particular class of IP addresses. Such a rule can be included in an obligation policy located at the subject host.

Rule 3 (File access rule) A usage rule to control access to a given resource classified as "secret" can accept digital credentials that contain a particular signature.

Trust policies are particularly useful for specifying trust intentions. Analysis of specifications is made easy by the availability of a framework explaining how trust intentions are communicated between entities, and therefore controlling the propagation of trust. A suitable trust policy specification language, such as the one described in (Damianou *et al.*, 2001) is required to allow explicit policies. However, publishing policies to allow subjects to interoperate makes the rules vulnerable to attacks. Thus, there is a high need to provide analysis capabilities and protect the integrity of the policies themselves. The Extensible Markup Language (XML) is the universal format for structured documents and data on the Web. It can be considered a good candidate to support textual descriptions of trust policies that may be easy to generate and read and allow viewing policy information with common web browsers. However, it is limited for exchanging policies between systems.

Clearly, any trust management system should contain a compliance checker to provide a proof that a requested transaction complies with the implemented trust polices. The basic questions that must be answered before the design of the compliance checker include:

1. How should the concept "*proof of compliance*" be defined, analyzed and implemented?
2. Which language (or notation) should policies and credentials be expressed in?
3. How should responsibility be divided between the trust compliance checker, the other entities of the trust management system, and the requestors?
4. How credential discovery is performed to complete a proof of compliance?

6.5 Trust management systems

In this section, we survey several well-known trust-management systems and explain how answers to the questions explored in the previous sections are provided. A trust-management system has five basic components:

- A language for describing "*actions*," which are operations with security consequences that are to be controlled by the system.
- A mechanism for identifying "*principals*," which are entities that can be authorized to perform actions.
- A language for specifying application "*policies*," which govern the actions that principals are authorized to perform.
- A language for specifying "*credentials*," which allow principals to delegate authorization to other principals.
- A "*compliance checker*," which provides a service to applications for determining how an action requested by principals should be handled, given a policy and a set of credentials.

6.5.1 *PolicyMaker*

PolicyMaker and KeyNote are trust-management solutions developed at the AT&T Research Laboratories (Blaze *et al.*, 1996 and 1999). They have been developed to address the authorization problem directly rather than handling it via authentication and access control. PolicyMaker provides an application-independent definition of the proof of compliance for matching up credentials, requests and policies. It is typically a query engine that evaluates whether a proposed action is consistent to a local policy. The inputs of the PolicyMaker interpreter are the local policy, the received credentials, and a specification of the actions that a public key wants to execute. The interpreter's response to a query can be either yes or no or a list of restrictions that would make the actions executable.

PolicyMaker considers that a policy is a trust assertion that is made by the local system and is unconditionally trusted by the system. A credential is a signed assertion made by other entitties and the signature must be verified before the credentials can be used. Credentials and policies are written in an assertion language such as Java or Safe-TCL. The syntax of an assertion is of the following form:

Source ASSERTION AuthorityStruct WHERE Filter

where Source is the originator (or source) of the assertion, AuthorityStruct defines the public key(s) to whom the assertion is applicable and Filter is the predicate (or condition) that the actions must satisfy for the assertion to hold. The following policy gives an example of assertion specifying that any doctor who is not a plastic surgeon should not be trusted to give a check-up (Grandison, 2000):

policy
ASSERTS doctor key
WHERE filter that allows check-up if the field is not plastic surgery
A Query to policymaker has the following format:
key$_1$, key$_2$, . . . , key$_n$, REQUESTS ActionString

The above can be interpreted as a requester represented by n keys, key$_1$, key$_2$, . . . , and key$_n$ has requested the execution of the actions listed in ActionString. The processing of the actions in ActionString and the signature verification are left entirely up to the calling application. This means that any signature scheme can be used, provided that the application uses the appropriate procedure to perform signature verification. Finally, PolicyMaker engine uses the credentials presented to it to prove that the requested action complies with the policy.

KeyNote was developed to overcome various weaknesses in PolicyMaker. KeyNote has the same principles and assertions as PolicyMaker. It however includes two additional features. First, the compliance engine is extended to include signature verification and use of a specific assertion language. Second, the KeyNote engine is passed with a list of credentials policies, the public keys of the requester, and a list of attribute-value pairs. This is generated by the application and contains all the information relevant to the request. Thus, this information accurately reflects the application's security requirements.

Similar to PolicyMaker, assertions in KeyNote can either be policies or credentials. An assertion however has the following form:

$$Authorizer: X, Licensees: Y, Comments: C, Conditions: H, Signature: S$$

where *Authorizer* field identifies policies that are locally trusted and so do not need a signature, *Licensees* field specifies the principal(s) to which authority is given, *Comments* field includes several comments about the request, *Conditions* field describes the requirements needed to allow the execution of actions, and the *Signature* authenticates the assertion.

6.5.2   Referee

Referee is an acronym for *Rule-controlled Environment For Evaluation of Rules, and Everything Else*. It is an environment for evaluating compliance with policies (rules), but the evaluation process itself may involve dangerous actions and hence is under policy control.

Referee uses the term trust management to refer to the problem of deciding whether requested actions, supported by credentials, comply with a set of policies. It considers a policy that sets rules about which credentials must be present in order to allow the execution of an action. Policies also determine which statements must be made about a credential before it is safe to run it. Finally, policies determine whether and how credentials are to be authenticated before the statements they create are declared to be trustworthy. For example, some kinds of statements may be used only if the credentials were signed using a particular cryptographic key.

The "Everything Else" refers to credentials, whose evaluation also needs to be under policy control. In placing everything under policy control, Referee differs from PolicyMaker, and the trust management system. In particular, PolicyMaker does not permit policies to control credential-fetching or signature-verification; it assumes that the requesting entity has collected all the relevant credentials and verified all digital signatures before calling the trust management system.

Three primitive data types are used in Referee: tri-values, statement-lists, and programs. A tri-value can be true, false, or unknown. A statement-list is a collection of assertions expressed in a particular format. Each program takes an initial statement list as an input and may also take additional arguments. A program may invoke another program during the course of its execution. Referee considers that a policy governing the execution of a particular action is a program that returns true or false answers depending on whether the available statements are sufficient to infer compliance, non-compliance with a policy, or returns unknown if no inference can be made. Finally, a credential is represented by a program that examines the initial statements passed to it and derives additional statements. This generalizes the usual notion of a credential as directly supplying statements.

Referee allows policies and credentials to return pairs of tri-values and statement lists, which specify the returned justification that can be expressed as a list of statements along with a tri-value answer. For example, a policy may reject downloading of code and provide statements, which indicate whether the code is known to be malicious or the local machine is currently experiencing a heavy load to allow a downloading action. It is also useful for credential programs to indicate whether the execution was successful (a tri-value) in addition to returning a list of statements.

Referee uses a rule-based trust policy language, where rules are evaluated top down and the returned value of the last rule, in a policy, is the policy's returned tri-value and statement-list.

6.6   Trust-management applications

Applications of trust management have appeared in the literature and have addressed several areas in information and communication technology. A particular emphasis is given to clinical information and medical document distribution, communication traffic filtering, and electronic payment. We discuss in this section these applications.

6.6.1   Clinical information systems

Clinical information systems provide a fully integrated electronic patient record. They include traditional clerical information about appointments and admissions; results from specialties in areas such as pathology, radiology, and endoscopy; drug treatments;

procedures; and problem lists. They also generate and store plans for nursing care, clinical correspondence, and dictated notes from ward rounds. Rules can be applied to the electronic patient record to maximize his/her privacy, but professionals in healthcare information technology have been unwilling to adopt these principles on the basis that they would be expensive to implement and unwieldy, hard to monitor and maintain.

Nine principles of data security have been provided by Anderson (Anderson, 1996) to define the core of trust security:

Principle 1 (Access control): Each identifiable clinical record shall be marked with an access control list (ACL) naming the people or groups of people who may read it and append data to it. The system shall prevent anyone not on the list from accessing the record in any way.

Principle 2 (Record opening): A clinician may open a record with himself (or herself) and the patient on the ACL. When a patient has been referred, the clinician may open a record with himself or herself, the patient, and the referring clinician(s) on the access control list.

Principle 3 (Control): One of the clinicians on the access control list must be marked as being responsible. Only this clinician may change the access control list, and he/she may add only other healthcare professionals to it.

Principle 4 (Consent and notification): The responsible clinician must notify the patient of the names on his/her record's access control list when it is opened, of all subsequent additions, and whenever responsibility is transferred. The patient's consent must also be obtained, except in emergency or in the case of statutory exemptions.

Principle 5 (Persistence): No one shall have the ability to delete clinical information until the appropriate time has expired.

Principle 6 (Attribution): All accesses to clinical records shall be marked on the record with the name of the person accessing the record as well as the date and time. An audit trail must be kept of all deletions.

Principle 7 (Information flow): Information derived from record A may be appended to record B only if B's access control list is contained in A's.

Principle 8 (Aggregation control): Effective measures should exist to prevent the aggregation of personal health information. In particular, patients must receive special notification if it is proposed to add a person to their access control list who already has access to personal health information on a large number of people.

Principle 9 (Trusted computing base): Computer systems that handle personal health information shall have a subsystem that enforces the above principles in an effective way. Its effectiveness shall be evaluated by independent experts.

The first three principles listed by Anderson identify the need for a clinical information system to limit a user's access to the records of his/her own patients and no others. Anderson proposes that this is done through ACLs that identify which individual users are responsible for a patient. Different solutions can be provided based on the generic trust management systems such the SPKI system, PolicyMaker, and KeyNote. All these solutions may differ on the language they use to describe credentials and the policy deduced from the aforementioned principles, but they all share the following three-step process to provide a clinical information trust management system:

1. Define the form of credentials, the resources involved in the systems, principals, and actions to be performed.
2. Derive from the nine principles the set of rules that constitute the trust policy. The policy should be complete, non conflicting, and minimal (if possible).
3. Decide where the enforcement points should be deployed.

Typical design decisions include the definition of resources, access control lists, and actions performed. Resources represent the storage related to the patient clinical information. Actions are represented by the read, write and modify operation provided on these patient records. ACLs can be signed documents or protected files that can be created, accessed, modified, and appended. Digital credentials are assumed to identify clinicians, record controllers, and patients. Signed messages include: requests, system notifications, patient comments, and control messages. They can be useful to append all revocation information concerning the names included in it to an ACL. Finally, requests can include at least the following actions:

- Generate an access list for a record
- Open a record
- Make a clinician in charge of record
- Alter/Add health care information to a record
- Notify the patient of the names on the ACL linked to his/her record
- Delete clinician information
- Append to a record the information derived from another record
- Issue and publish a revocation list of names.

Trust policy rules are carefully deduced and verified using the above principles. Using Principle 1 for example, one can show a clinician can submit a request to access a record provided that she/he is named on the record ACL. An access request, if accepted, would trigger the execution of two tasks: record display and sending a signed notification to the patient owning the record

6.6.2   E-payment systems

Efficient electronic payment systems represent an essential prerequisite for electronic commerce. The design of such systems poses many security-related problems. Among these problems, one can mention the protection of the clients' privacy and the security of transactions. In many payment systems, this protection relies exclusively on administrative and legal measures. By using cryptographic tools, it is possible to design electronic payment systems that allow participants to remain anonymous during a transaction.

We define in this subsection an anonymous payment system, which identifies virtual accounts using an account identifier accompanied by a pin number, assuming that it can authorize anonymous payments. Actions in this payment system are of four types: deposit, payment, transfer, and balance check. A deposit action is realized by debiting another account. The payment action allows the account holder to pay bills or to authorize a third party to be paid, while a transfer action withdraws an amount of money to be transferred.

The balance check retrieves the current state of the amount of money available in the virtual account. Balance checks are allowed to be performed only by the account owner.

Our trust management model is assumed to satisfy the following principles as stated in (Boudriga and Obaidat, 2003):

- Principle 1 (*Owner identification*): Each account owner has in his/her hands a naming certificate, which identifies the account and establishes a unique link between the owner, the account, and any authorization certificate issued by the owner.
- Principle 2 (*Account visibility*): All information related to an account can only be accessed by the account owner and the employees authorized by the bank. Each access should be stored.
- Principle 3 (*Authorization definition*): Each account owner can issue two types of authorization certificates: payment certificates, which authorize the payment to a third party (a merchant, for example), and new authorization certificates that are re-issued by the owner for his/her need or for the purposes of security (e.g., change of personal key).
- Principle 4 (*Payment description*): A payment certificate authorizes the bank to perform only one payment and cannot be transferred to another party.
- Principle 5 (*Reissuance description*): The re-issuance of an authorization certificate invalidates, and not revokes, the last authorization certificate that was used before the issuance.
- Principle 6 (*Pass-phrase usage*): A pass-phrase is introduced (occasionally) by the account holder during the process of generation of the authorization certificate, while pin number is not. The pass-phrases are allowed to be modified through the process of certificate re-generation. The pin number and the pass-phrase are submitted by the account owner through a secured connection.

Credentials in this trust management system as defined in Boudriga and Obaidat (2003) can be described in a SPKI-based language where the naming certificate identifies account owners and banking employees and should be generated only by the bank. Authorizations are issued by account owners and cannot be delegated (Principal 4 requirement). Credential discovery and authorization certificate are easy to implement since they have very limited length. Therefore, an account in our model is represented by a naming certificate of the form:

$$\langle\text{cert (issuer (name } k \ n)) \text{ (subject } p), \textit{Val} \rangle,$$

where k and n are the public key and the name of the bank, which are in charge of managing the account and signing the naming certificates. The expression p has the form: $k' \ n'$ where k' is the public key of the account owner, n' is the account identifier (including a pin number), and $\textit{Val}$ is reduced to an interval of one year duration, for example. An authorization certificate has the form:

$$\langle\text{cert (issuer (name } k' \ n \ h(n_1))) \text{ (subject } p) \ \textit{Act Del Val} \rangle$$

where k' is the public key identifying the issuer of the certificate, subject p identifies the beneficiary of the unique action in *Act*, *Val* contains an interval of time or the predicate

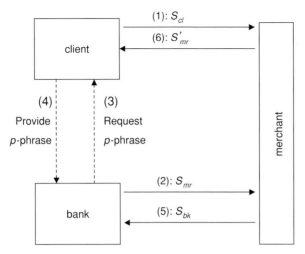

Figure 6.3 Payment protocol.

$now(t)$, and Del is equal to 0 if the certificate is a payment certificate. This means that the propagation of delegation is not allowed in our trust management system. Finally, we assume that field Val may contain other predicates such as a predicate, called *excess-aut* (u), which is true when the amount of money to be paid exceeds the available amount in the account. It is, however, assumed to be lower than an authorized excess u.

Now, let us consider the different messages used in the generic payment protocol. Figure 6.3 depicts the payment action and the role of the system entities involved in the payment of a transaction. The requests' formats are derived from the requirements for implicit guarantees and extend those used in Bellare (2000). We first simplify our notation by introducing the following. Let S_{cl}, S_{mr}, S_{bk}, and S'_{mr} be the following signed data structures:

$$S_{cl} = Sig(cl, mr, bk, pay(a), trx),$$
$$S_{mr} = Sig(mr, bk, cl, pay(a), trx),$$
$$S_{bk} = Sig(bk, mr, cl, pay(a), trx, ack),$$
$$S'_{mr} = Sig(mr, cl, bk, pay(a), trx, bill),$$

where cl defines a client (as represented by its payment certificate), mr represents the merchant, bk is the payment system (or bank), a is the amount of payment, trx contains the transaction information including at least the transaction identifier trx_id, and transaction time trx_t. Transaction acknowledgment ack is a binary value stating whether the authorization is approved. With S_{cl}, customer cl authorizes a payment with an amount a to merchant mr, who then adds his/her own signature S_{mr} to form an authorization request for the payment. S_{bk} constitutes the system's authorization response, after checking the conformance of request mr, creates a payment receipt S'_{mr} (including the appropriate bill), and sends it together with S_{bk} to cl. The bk system may initiate a communication with cl to request the pass-phrase p needed to complete the payment (principles 3 and 4).

6.6.3 Distribute firewalls

IPSec is a network-layer protocol that allows protecting packets. It is based on the concept of packet encapsulation. A packet is encrypted and encapsulated as a payload in another network packet, making the encryption to any intermediate node that participates in the routing of the protected packet. IPsec offers four measures of security services: authentication of packet sources, integrity of transmitted messages, confidentiality of transmitted messages and anti-replay attacks. Outgoing packets are therefore encrypted, authenticated and encapsulated, and incoming packets are verified, decrypted and decapsulated upon receipt. The key distribution process supporting IPSec is based on the concept of *security associations* (SA), which contains the encryption and authentication keys. SAs are managed at each end node, while each protected packet transports a unique SA identifier.

When an incoming packet arrives at the destination, the receiving node applies two types of filtering. The first type applies the traditional filtering as performed by network firewalls. The second takes place if the packet is encapsulated under IPSec. In this case, the appropriate SA is accessed and several rules that provide decisions for whether the SA contains correct information, and whether the resulting packet (after decryption) should be accepted (this case occurs when a packet attempts an access to a non permitted port). A node sending an outgoing packet makes similar decisions. Therefore, IPSec security policy should not be limited to packet filtering rules, but should contain rules that control when the SAs are created, how they are used, and what packet filters are associated with them. Some other types may be needed to be included in the new version of IPSec.

Policy management for SAs should consider several issues including:

- SAs have duration and limited size for SA's identifying values (2^{32} different values);
- SA creation involves different resources, packet exchanges, and operation of policy verification;
- Negotiating SAs may generate parameters that might affect low level filtering.

The problems of controlling IPsec security associations and related filtering activity are easy to formulate as a trust management task since a check for compliance with the local policy is involved where actions represent the packet filtering rules required to allow two nodes to conform to each other's policies.

For the sake of simplicity we consider in what follows the SA creation process.

A KeyNote-based solution has been developed in the literature. Typically the developed framework considers that:

- Each node has its own KeyNote specified policy governing SA management. This policy describes the classes of packets under consideration and all types of SAs, and defines under what conditions a node can initiate SA creation with other nodes.
- When two nodes find that they require an SA, they each send proposed packet filter rules, along with credentials that support the proposal. Any delegation structure between these credentials might include trusted third-parties.
- KeyNote interpreters are queried whether the proposed packet filters comply with the local policy. If this is the case, the SA containing the specified filters is created.

The solution provided by Ionnides *et al*. (2000) considers credential discovery and acquisition. Although nodes may be expected to manage locally policies and credentials that are directly referred to them, they may not know about intermediate credentials that may be required by the nodes with which they want to communicate securely. To do this, a policy query protocol to acquire or update credentials relevant to a specific key exchange is needed.

6.7 Concluding remarks

In this chapter, we presented the main aspects of trust management in the security of any e-based system or network. We showed that the main issues in a trust management system include: expressing security policies and security credentials, ascertaining whether a given set of credentials conforms to the relevant policies and delegating trust to third parties under relevant conditions. We discussed digital credentials, access control schemes and trust policies. We reviewed the main trust management systems that have been developed to support security of networked applications. We also presented a comprehensive approach to trust management, and described the three most well known trust management systems. The main applications of trust management systems such as clinical information systems, e-payment systems and distributed firewalls were investigated and discussed. The chapter includes examples and cases studies on the applications and techniques of trust management systems.

References

Anderson, R. (1994). Liability and Computer Security: nine principles. In *Proceedings of the Third European Symposium on Research in Computer Security (ESORICS 94)*, Springer-Verlag, pp. 231–45.

Bellare, M., J. Garay, R. Hauser, A. Herzberg, H. Krawczyk, M. Steiner, G. Tsudik, E. Van Herreweghen, and M. Waidner (2000). Design, implementation and deployment of the iKP secure electronic payment system. *IEEE Journal on Selected Areas in Communication*. Vol. **18**(4), 1–20.

Blaze, M., J. Feigenbaum, and J. Lacy (1996). Decentralized trust management. In *Proceedings of the 17th Symp. on Security and Privacy*, IEEE Computer Press, pp. 164–73.

Blaze, M., J. Feigenbaum, J. Ioannidis, and A. D. Keromytis (1999a). The Role of Trust Management in Distributed Systems Security. In *Secure Internet Programming*, 1603 LNCS, Springer-Verlag, pp. 185–210.

Blaze, M., J. Feigenbaum, J. Ioannidis, and A. Keromytis (1999b). The KeyNote Trust-Management System Version 2, RFC 2704.

Boudriga, N. and M. S. Obaidat (2003). SPKI-based trust management systems in communication networks. Proc. 2003 Intern. Symp. On Perform. Evaluation of Comp. and Telecomm. Systems (SPECST'03), pp. 719–726, Montreal, Canada, July 20–24, 2003.

Chun, B. N. and A. C. Bavier (2004). Decentralized trust management and accountability in federated systems, HICSS 2004, Jan. 5–8, 2004, Hawaii.

Damianou, N., N. Dulay, E. Lupu, and M. Salomon (2001). The Ponder specification language, in *Proc. Workshop on Policies for Distributed Systems and Networks (Policy 2001)*, HP Labs Bristol UK, 29–31 Jan 2001, Springer-Verlag LNCS 1995, pp. 18–39.

Ellison, C. (1999a). *SPKI certification theory*. IETF, RFC 2692.

Ellison, C., B. Franz, B. Lampson, R. L. Rivest, B. M. Thomas, and T. Ylonen (1999b). *SPKI requirements*, IETF, RFC 2693.

Grandison, T. and M. Sloman (2000). A survey of trust in Internet applications. *IEEE Communications Surveys and Tutorials*, Vol. **3**, No. 4, 2–16.

Guemara-ElFatmi, S., N. Boudriga, and M. S. Obaidat (2004). Relational-based calculus for trust management in networked services, *Computer Communications Journal*, **27**, 1206–19.

Ioannidis, S., A. D. Keromytis, S. M. Bellovin, and J. M. Smith (2000). Implementing a distributed firewall. In *Proceedings of the 7th ACM Conference on Computer and Communications Security*, ACM, pp. 190–9.

Kagal, L., T. Finin, and Y. Peng (2001). A Delegation Based Model for Distributed Trust. Proceedings of the IJCAI-01 Workshop on Autonomy, Delegation, and Control: Interacting with Autonomous Agents, pp 73–80, Seattle, August 6, 2001.

Keromytis, A. D. and J. M. Smith (2002). *Requirements for Scalable Access Control and Security Management Architecture*, Technical report CUCS-013–02.

Li, N., B. N. Grosof, and J. Feigenbaum (2000). A practical implementable and tractable delegation logic, *Proc. IEEE Symp. on Sec. and Priv.*, IEEE Comp. Press, pp. 27–42.

McKnight, D. H. and N. L. Chervany (1996). *The Meanings of Trust*, Technical Report no:94–04, Carlson School of Management, University of Minnesota.

Mont, M. C. and R. Brown (2002). *Active Digital Credentials: Provision and Profile Information*, HP-Laboratories, Technical Report HPL-2002–50.

Nikander, P. and L. Metso (2000). Policy and Trust in Open Multi-Operator Networks. *SMARTNET 2000*: 419–436.

Part III

E-security applications

Introduction to Part III

The Internet is growing to be the major means through which the services can be delivered electronically to businesses and customers. Today, the current developments in service provision through communication networks are moving from tightly joined systems towards services of loosely linked and dynamically bound components. The evolution in this category of applications is represented by a new paradigm, the so-called e-service, for which project developers and service providers are encouraging the definition of techniques, design of methods, and construction of tools as well as infrastructures for supporting the design, development, and operation of e-services and e-government systems. The development of e-services can increase the business opportunity for making available value-added e-services by: composing existing e-services (which may be supplied by different providers), customizing and updating e-services, or creating them from formal specifications.

Service composition appears to be an important issue that can provide a competitive advantage for organizations since they can reduce the needed effort and increase the efficiency of e-services to build. Recently, many e-services have been made publicly accessible and therefore are offered in an unsecured manner. However, some among these e-services will need to use encrypted communications and authentication services. To provide security mechanisms for the operation of e-services at a low level granularity, it is important to define: (a) how the e-services authenticate customers in a reasonable fashion; (b) how e-services' standards address the problem of securing the assets in offering e-services; and (c) how cryptographic elements are managed and distributed.

Among the large set of e-services, e-commerce and e-government are addressed in a very different way since e-commerce has been developed during the early days and e-government attempts to respond to a specific need of authentication, availability, and scalability. E-commerce is dramatically changing the way that goods (tangible and intangible) and services are produced, delivered, sold, and purchased. Due to this development, trading on the Web becomes a requirement for enterprises in order to succeed. In addition, e-government is able to reduce the expenses of the enterprises and facilitates their relation with the administration, enabling them to interact securely in places and at times that are convenient to them.

The purpose of Chapter 7 is to examine the e-service paradigm, discuss the technical features it depicts, and study the security challenges it brings forward. It also describes well-established e-services and shows how they are composed, discovered, and delivered.

Chapter 8 provides key support to service providers wishing to provide e-government services in a trusted manner. It lays the foundations for enabling secure services that will really transform the way citizens and businesses interact with government.

Chapter 9 discusses the e-commerce requirements and defines the major techniques used to provide and protect e-commerce. A special interest is given to the SSL, TLS and SET protocols. Electronic payment and m-commerce are also addressed.

Chapter 10 reviews and investigates the security of wireless local area networks (WLANs). The major techniques and their advantages and drawbacks are presented. Moreover, the chief issues related to WLANs security are discussed.

7 E-services security (Web-services security)

The Internet is growing to become the major means through which the services can be delivered electronically to businesses and customers. System vendors and service providers are pushing toward the definition of protocols, languages, and tools that support the improvement, use, and operation of electronic services. The main goal of the e-service paradigm is to provide the opportunity for defining value-added composite services by integrating other basic or composite services. However, security issues need to be addressed within such an open environment.

7.1   Introduction

The notion of service is getting increasingly valuable to many fields of communications and information technology. Nowadays, the current development in service provision through communication networks are moving from tightly joined systems towards services of loosely linked and dynamically related components. The major evolution in this category of applications is a new paradigm, called *e-service*, for which project developers and service providers are pushing for the definition of techniques, methods, and tools as well as infrastructures for supporting the design, development, and operation of e-services. Also, standards bodies are urging to specify protocols and languages to help the deployment in e-services.

E-services are self-contained and modular applications. They can be accessed via Internet and can provide a set of useful functionalities to businesses and individuals. Particularly, recent approaches to e-business typically view an e-service as an abstraction of a business process, in the sense that it represents an activity executed within an organization on behalf of a customer or another organization. Among the attractive characteristics of e-services, one can bring up: (1) The ability to discover e-services whose locations are unknown to the requester. The services can be eventually offered by different providers while they meet the terms of requesters' demand. (2) The negotiation of service contracts by allowing appropriate quality of service requirements, security constraints, and accounting rules. (3) Having the e-services delivered where and when the requesters need them (Tsalgatidou and Pilioura, 2002). Generally, an e-service platform authorizes the e-service providers to register service descriptions and advertise them in ad-hoc web directories. They also give the e-service providers the capability to monitor service provisions.

The development of e-services can increase the business opportunity for making available value-added e-services by integrating/composing existing e-services (which may be supplied by different providers), customizing and updating e-services, or creating them from specifications. E-services behave like business processes with well-defined life-cycles, meanwhile service composition appears to be an important issue that can provide a reasonable advantage for organizations since they can reduce the needed effort and increase the efficiency of e-services to develop. For example, service composition can be used to provide a travel agency with an e-service integrating at least three e-services: an e-service for car rental, an e-service for hotel registration, and an e-service for flight tickets purchase. It is clear that the aforementioned services can be combined because they may bring together independent services from hotel chains, airlines companies, and car rental organizations.

Composite services are typically similar to workflows in the sense that they specify complex services by composing basic and composite services, specify dependencies between services, and provide input and output parameters to those services. However, they differ from workflows in different features; they involve distributed actors on the Web and consider security as a major issue. Among the basic notions characterizing the activities related to e-service development and operation are: message transportation, dynamic binding, encapsulation, and user authentication. In addition to this, quality of service requirements and security protection issues need to be addressed to provide quality assurance, privacy, and auditing of services.

For a company, e-services offer the opportunity to electronically outsource non-vital functions without supporting the traditional coordination overhead. E-services can be viewed typically as web-based applications that meet service needs by easily taking together distributed and specialized resources to make executable a large spectrum of transactions, which may have a real-time nature (Tiwana and Ramesh, 2001) and involve various types of activities. They also support primary business processes (e.g., supply chain management and workflow management) with flexibility and cost effectiveness.

Today, a large spectrum of e-services is made publicly accessible and therefore may be offered in an unsecured manner. Nonetheless, a large number of e-services will need to use encrypted communications with authentication. To provide efficient security mechanisms operating at a granularity level comparable to the composition granularity, it is important to describe:

1. How do the e-service providers authenticate customers and build user's security privileges?
2. How do e-services standards address the problem of securing the principal assets occurring by e-service offerings?
3. How are key management and security performed within e-services?
4. How are traceability and non-repudiation guarantees achieved?

The purpose of this chapter is to examine the e-service paradigm, the technical features, and the security challenges it brings forward. Some important e-services are discussed in this chapter in order to show the usefulness of the paradigm.

7.2   E-service basic concepts and roles

E-services are self-contained and modular applications that can be described, published, located, and invoked over a communication network (e.g., the Internet). E-services enable application developers to respond to a specific need described by a customer or an organization, by using appropriate services published on the web, rather than developing the related code from specification, while offering potential benefits as compared to traditional applications. Furthermore, the e-service paradigm simplifies business applications and interoperation of the applications deployed on a public network. Additionally, it significantly serves end-user needs by enabling them, using browser-based interfaces, to choose, configure, and compose the e-services they like.

E-service activities that are available in a service-based environment integrate the following tools and techniques:

- *E-service description tools and languages* Description languages of e-services allow the specification of service properties and interfaces by allowing the description of what the e-service can provide, where it is located, and how it is invoked? Some other properties such as service content, commercial conditions, and pricing information can be included in the description to be provided. Advanced semantic aspects can be integrated into the description languages as well.
- *E-service creation tools* Creation tools offer the creation of efficient platforms to design, implement, test, and protect e-services. The tools include, but are not limited to, traditional software development tools and service composition tools, which provide the opportunity for defining value-added composite services using basic services and composite e-services that can be looked for on the Internet, if they are available.
- *E-service publishing/unpublishing procedures* Publishing procedure is in charge of the activity that makes an e-service known and available on the Internet (or with an Internet service provider) for potential (or dedicated) customers. Specific registries can be distributed across Web sites and be used to contain business information, service information, and binding information. The unpublishing process is responsible for the removal of an e-service in the case where the e-service is no longer needed or it has to be updated to satisfy new requirements before it is republished.
- *E-Service discovery tools* Discovery tools allow e-service potential customers to describe their needs at a high conceptual level (e.g., by expressing a set of characteristics and specifying the values of parameters that they are willing to accept from the e-service) and make sure that their needs are conforming to what an e-service available on the Internet (or on an intranet accessible by the customer) can provide among the list of candidate service offers conforming to the described needs.
- *E-service binding and invocation tools* Binding activity refers to when and how the interoperation between e-services is established. It uses different types of information such as quality of service requirements, message ordering, and delivery time constraints. Invocation allows a potential customer to request the execution of a service by simple messaging service. It may provide a QoS framework that supports reliable delivery and security constraints.

An e-service classification can be provided using the nature of business activity that the e-service supports and the nature of participants, identifying mainly four classes of e-services:

1. Business-to-Customer (B2C) e-services This class considers a company providing service to an individual directly.
2. Business-to-Business e-services (B2B) This category includes business models that can be used by an organization to provide a service to another peer organization.
3. Government-to-Business (G2B) e-services This category includes services provided by governmental agencies to business organizations.
4. Government-to-Consumer (G2C) e-services This class represents services provided by governmental agencies to citizens.

Let us now notice that customer-to-customer e-services are not directly considered in this chapter, since their models involve unavoidably a third party that will act as a link provider, trusted party, and mediator.

The type of end-product that the e-service supports can be used to present another classification. Three classes can be considered for this dimension (Tiwana and Ramesh, 2001):

1. The physical product processing e-service, which is concerned with the design, assembly, delivery, and tracking of physical goods. Selling books on the Internet represents a traditional example.
2. Digital product delivering e-service, which processes digital goods that exist typically in electronic form (e.g., software package and online music).
3. Pure service delivery applications, which provide true services that have not a tangible form and do not deliver a tangible product to the service consumer (e.g., the e-voting service and the online tax filing and payment).

In addition to the aforementioned tools, procedures and classification, there are various other activities that need to be addressed effectively in an e-service environment in order to take full advantage of the Internet. Such activities include e-service composition, management and monitoring, billing, and security. Moreover, three roles can be distinguished in the provision of e-services and the security services they may require while involved in e-service usage. They are: the service provider, the service requester, and the service broker. We only describe the typical use of these roles as follows.

E-service provider

An e-service provider (ESP) is the entity that provides applications for specific needs as services. The service providers publish, unpublish (i.e., remove) and update e-services. The service provider is the owner of the service, while, from the architectural point of view, the service provider is the platform that provides the implementation of the service. From a business point of view, the ESP is the owner of the service, while from the architectural point of view the ESP is the platform the service is deployed onto.

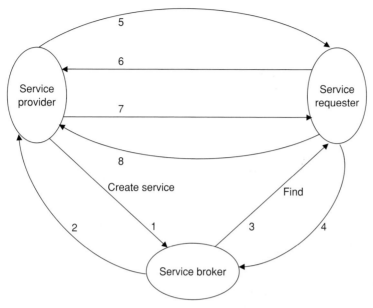

Figure 7.1 E-service model.

E-service requester

A requester is the entity (i.e., an individual, organization, or application acting on behalf of a customer) that has a need that can be fulfilled by a service available on the communication network. From the business point of view, a service requester is the business that requires certain functions to be accomplished. From the architectural perspective, a service requester could be the networked application that is looking for and invoking a service.

E-service broker

An e-service broker is the entity that acts as a mediator between requestors and providers to provide a searchable repository of e-service descriptions where service providers publish their services and service requesters find services and obtain the binding information of these services. The broker should satisfy features such as having the capability to enable universal service to service interaction, negotiation, and selection of services.

Since the service provider, service requester, and service broker interact with each other, they should use standard technologies for service descriptions, communication and data formats. In this perspective, we are currently witnessing the rapid development of a set of interrelated standards that are defining the e-services infrastructure along with development tools. Figure 7.1 depicts the relationships between these actors. Steps 1 and 2 in Figure 7.1 provide the initialization of an update of service registrations. The provider can do these actions through the exchange of messages with the service broker. Steps 3 and 4 allow searching the registries for service registrations. This activity takes place between the service requester and service broker. The registry can be used to locate the appropriate file (steps 5 and 6). The service can be accessed using steps 7 and 8.

The deployment of e-services provides a large spectrum of advantages. Among the major benefits, one can mention the following:

Interoperability

E-services support interoperability by providing standard description languages for sent messages and remote procedure calls such as the simple object access protocol (SOAP, 2000), interface definition language such as the web services description language (WSDL, 2000), and protocols for collaboration and negotiation. These objects represent an important requirement for shared understanding between a service provider, a service broker, and a service requester. By defining precisely what is completely required for interoperability, cooperating e-services can be truly e-service platform and language independent.

Integration

E-services that make possible real-time integration and collaboration can be interrelated dynamically at low cost. A service requester describes the capabilities of the service required and asks the e-service broker infrastructure to find a list of appropriate service offers. Once a list of e-services with the required capabilities is found, an e-service is selected and the information retrieved from the service broker is used to bind to it.

Moreover, by permitting legacy applications to be involved in broker directories and exposed as e-service offers, e-services architecture easily facilitates new interoperability between these applications. In addition, security, middleware and communications technologies can be included to participate in an e-service as environmental prerequisites. Directory technologies, such as LDAP, can be integrated to operate as a service broker.

Encapsulation

An e-service can reduce its intrinsic complexity by using the concept of encapsulation since what is important is the type of behavior an e-service provides and not how it is implemented. A service description document is the mechanism to describe the behavior encapsulated by a service. Encapsulation is a key issue to:

*   Coping with complexity System complexity is reduced when application designers do not have to worry about implementation details of the services they are invoking.
*   Flexibility and scalability Substitution of different implementations of the same type, or multiple equivalent services, is possible at runtime.

7.3   Examples of e-services

Nowadays, a large spectrum of e-services can be observed on the Internet. Examples of e-services, for instance, can be found in incorporating situations, where an enterprise must integrate different IT systems and business processes. A service-oriented architecture would greatly facilitate a faultless integration between these systems. Another type of example can be found in the combination of the product applications industry with pervasive computing, when largely mainframe-based product applications can be exposed as services and made available for use by various remote systems in a service-oriented environment. Within such environments, new services can be created and dynamically published and discovered without disrupting the existing environment.

To illustrate the use of e-services, we consider, in this section, four important e-services that we will be using in the sequel to analyze the main concepts in e-service provisioning and security. They are: (a) the *real-time stock e-service*; (b) the *anonymous payment e-service*; (c) the *postal e-service*; and (d) the *travel industry e-service*.

Real-time stock e-service

This is an e-service that can be used in an application offering real-time stock market information and processing (Tsalgatidou and Pilioura, 2002). The e-service can be accessed through an ad-hoc portal (for example) offering/integrating a series of different services including, but not limited to, the following:

- *A real-time view of the stock market* This service provides information on the real-time status of a given set of stocks.
- *A stock quote service* This service enables investors to retrieve a quote in a specific currency, given the ticket-symbol of any publicly traded stock.
- *A portfolio management service* This service allows an investor to track the performance of his/her shares and perform appropriate transactions.
- *A tool for account balance check* This service checks the investor's account in order to make sure that he/she has the necessary credit/funds for the completion of a transaction.
- *A tool for currency conversion* This tool can be accessed by an investor needing to get the actual conversion between specific currencies.
- *A news service* This service provides the headlines of the latest financial news related to stock market and investment opportunities.

Implementing the real-time stock e-service within a portal can be achieved using three approaches: (a) developing the system from a complete specification, (b) customizing an existing e-service available on the Internet, and (c) composing e-services available with other vendors or business partners. A specific method can choose to develop the news service from scratch, while it assumes that the money conversion service can be acquired, and the portfolio management service be composed. The security issues to be addressed would mainly consider that sending messages should be made confidential, when they contain sensitive information, and be signed when the formation is used as a basis for transactional decisions.

Anonymous payment e-service

Similar to the payment system described in Chapter 6, we consider that an anonymous payment e-service identifies virtual accounts using an account number and a pin number

(Guemara-ElFatmi and Boudriga, 2004). Actions in this system are deposit, transfer, payment, and balance check. Our model for the payment e-service is assumed to satisfy the following rules:

- Each account owner has in his/her hands a naming certificate, which identifies the account and establishes a unique link between the owner and the virtual account.
- Each account owner can issue (by digital signature) payment orders, which authorize the payment of, or the transfer of money to, a third party's account (e.g., merchant). A payment order authorizes to perform only one payment and cannot be transferred.
- An account owner can submit a transaction to deposit an amount of money based on a valid payment order signed by another account owner.
- A user presenting a valid payment can obtain opening a virtual account with an initial deposit amount equal to the amount written on the signed order. Therefore, the requester should be able to operate on the new virtual account (the order may for example include a public key for this purpose).

The design of the anonymous payment e-service can be based on the integration of the following five e-services:

Account balance e-service This e-service allows checking the client's account in order to make sure that he/she has the necessary amount for the completion of a payment order.

Naming certificate generation e-service This service is invoked by a potential client to start the process of creating a virtual account and generate the appropriate naming certificate.

Payment order generation e-service This service is invoked by an account holder when he/she needs to generate a payment order for the benefit of a third party.

Certificates revocation service This service is used to invalidate a naming certificate issued and to close the related account.

LDAP-based verification e-service This service is invoked when there is a need to validate a certificate or an order submitted with a transaction to be performed.

The security needs of the anonymous payment e-service require mainly the LDAP be secured and that the infrastructure used for certificate and order generation be operated following a well-defined security policy. An e-service can be composed to provide an anonymous payment service using certificate generation and revocation services, a certificate verification service, an online banking system, and an anonymous payment management.

Postal e-services

Various traditional postal services can be extended to provide a large spectrum of applications on the Internet in a way that conforms to the primary objectives of providing packet transportation and delivery, stamp production and selling, and money transfer and payment. A postal e-service can be built using the following non-exhaustive list of services:

A service for tracking packet statuses Such service can be accessed by a citizen needing to send a packet or get accurate information about the status of a postal packet he/she has requested the delivery (e.g., tracing transportation and delivery of the packet).

A service for the provision of universal e-mail addresses Such service can be used to provide an individual with an authenticating e-mail address and cryptographic elements that he/she may use to sign electronic letters, order transfer, or transfer money. The service can be involved in non-repudiation and time-stamping services.

A service for money transfer Such tool can be used to send C2C amounts of money. It also can be accessed by a customer needing to get the actual conversion between specific currencies.

A service for stamp factoring industry This service can be used to design and produce postage stamps. The service can be accessed to purchase stamps and collect relevant information about stamp industry and history.

Postal e-services are complex to build since they should cope with the scale of a country, comply with constraining quality of service, and include a large spectrum of citizen oriented services. Composition of e-services is appropriate since various postal services are outsourced or provided.

Travel agency e-service

A composite e-service can be developed to allow a traveler to design a trip while performing all needed reservations including flights, hotels, cars, and tours. The e-service can integrate a set of interoperating e-services including the following services:

Flight reservation composite service This service retrieves useful information about the different flights going to a given destination and allows a real choice of flights to take. It also can provide flight reservation and ticket purchase capabilities.

Hotel reservation service This service allows collecting information about hotels and performing hotel selection and reservation based on a search among reservation services available at chains of hotels and based on a set of constraints that can be composed by the passenger.

Car rental service This service offers to perform car rental and provides users with useful information on the type of car to rent, pickup and return point, and the tourist sites to visit.

Weather information service This service provides requesters with the real-time status of the weather all around the world. It also provides information for forecasting the weather conditions.

Cultural information service This service can provide useful general information related to the culture, history, and social activities available at the place to visit. The e-service can provide reservation and purchase of tickets to attend concerts, tours, and other cultural events.

7.4   Basic technologies for e-services

The development of e-services has generated the design of various technologies including service description languages, message passing protocols, composition technologies, and broker servers. We attempt in this section to describe the major initiatives developed in this regard.

Two major initiatives have been made available for the development of e-services: the UDDI/SOAP/WSDL and ebXML initiatives. The ebXML scheme follows a top-down approach identifying requirements for successfully conducting e-business over the Internet and then working to implement specifications to meet these requirements. On the other hand, the UDDI/SOAP/WSDL approach, however, follows a bottom-up approach implementing specifications that meet individual requirements such as service description, discovery, and invocation.

7.4.1   The UDDI/SOAP/WSDL initiative

The UDDI/SOAP/WSDL initiative (Snell *et al.*, 2002; Cerami, 2002) is constructed on the basis of three standards: the web service description language (WSDL); the simple object access protocol (SOAP); and the universal, description, discovery, and integration (UDDI). UDDI is a specification that defines a service registry of existing e-services and allows a business to publish a description of the e-services it makes available to the registry. A service requester can form a request using SOAP messages to the e-service registry to discover a service provider offering the requested services. Therefore, it allows checking whether a given business entity (or provider) offers a particular e-service, finds a provider based on a given type of e-service, and locates information about how a provider has published a service in order to understand the technical details required to interact with the service. The UDDI specifications use an XML scheme for the communication of messages and the description of UDDI APIs specifications. The UDDI APIs contain messages for interacting with UDDI registries. Inquiry APIs are provided to locate businesses and services.

The UDDI model identifies four major data structures using the XML scheme. These are: (a) the *business entities*, which provide information about businesses including their names, the services they offer, and contact information; (b) the *business services* (or simply e-services), which give more details about each offered e-service; (c) the *binding templates*, which describe the access information (i.e., a technical entry point for the service and providing how and where the service is accessed; and (d) the service type definitions (called *tModels*), which describe what particular standard a service is using. A business entity contains a finite set of business services the business wishes to publish, while a tModel provides the ability to describe compliance with a specification or a shared design. Finally, let us note that UDDI provides two major mechanisms: (a) a mechanism for service providers to register themselves and the services they offer with the UDDI, and (b) a mechanism for service requesters to search for the available services from UDDI. The latter specification, described in (Bellwood *et al.*, 2002), provides a platform-independent way for describing and discovering services.

UDDI can be used to construct, view, update, and delete e-service registrations. It also can be used to search the registry for e-service registrations. UDDI is not, however, a service description language. Services need to be described efficiently by a service description language such as WDSL (WSDL, 2000), and also require a remote invocation mechanism like SOAP (SOAP, 2000). A UDDI business registry is itself a SOAP web service and the UDDI APIs are also based on SOAP. Each e-service, described by a WSDL document, defines services as a collection of network ports and messages. *Messages* are abstract descriptions of the data being exchanged, and *port types* are abstract collections of operations. The WSDL is a general purpose XML language for describing the interface, protocol bindings and the deployment details of an e-service. A WSDL document describes how to invoke a service. A description document uses the following seven elements to define an e-service.

- *Type* This is a container for data type definitions. The WSDL type element contains the complex data types used within the message and defined by the service provider.
- *Message* This is an abstract typed definition of the data being communicated.
- *Operation* An abstract description of an action supported by the service. It contains input and output parameters of the service that are defined in the WSDL message element.
- *Port type* An abstract set of operations supported by one or more ports.
- *Binding* A concrete protocol and data format specification for a particular port type. It describes the protocol, data format, security and other attributes for a particular service interface.
- *Port* A single endpoint defined as a combination of a binding and a network address (or URL).
- *Service* It contains a collection of ports. A port associates an endpoint with a binding from a service interface definition.

SOAP is a standard for sending messages and making remote procedure calls over the Internet. It uses XML to encode data and HTTP as the transport protocol. It also can use other transport protocols such as TCP/IP, SMTP, and FTP. SOAP defines two types of messages, request and response, in order to allow service requesters to request a remote procedure and permit service providers to respond to such request. A SOAP message has two parts, a header that is different from the transport header and a payload, which is equal to the XML payload. SOAP payload request has three fields: the *envelope*, which defines the various namespaces that are used by the rest of the SOAP message; the *header*, which is an optional element for transporting auxiliary information for authentication; and the *body*, which is the main payload of the message. A SOAP response is returned as an XML document within a standard http reply. The XML document is structured just like the request except that the body of a response contains the encoded method result.

SOAP is an easy way to invoke remote procedure calls on another machine in a request/response manner. The client application forms a request that is represented by a correct SOAP XML document and sends it to the other machine over HTTP. The machine processes the SOAP request, validates it, invokes the requested method with the supplied parameters (as transported by the request), constitutes a valid SOAP document and sends it back over HTTP to the requester.

Reviewing the UDDI/SOAP/WSDL initiative, one can say that service registration, discovery, and invocation are implemented by SOAP calls, in a way such that:

1. Service provider sends a UDDI SOAP request, together with WSDL service description of the e-service to UDDI registry operator. Acknowledgement of service successful registration returns a SOAP message from the registry.
2. After a service is registered in UDDI registry, service requestor can discover the service from the registry. For this, the requestor sends UDDI a SOAP request. The registry then locates the service and returns its WSDL description.

3. Finally, the requestor can invoke the service using the binding details in the service description.

7.4.2   ebXML Initiative

The ebXML initiative defines a framework allowing businesses to find peer businesses and conduct business based on XML messages within a set of standard of business processes that are ruled by mutually negotiated agreements. The ebXML initiative integrates four major standards: (a) the ebXML Technical Architecture Specification (ebTA, 2001), which defines the overall technical architecture for ebXML, (b) the ebXML Technical Architecture Risk Assessment Technical Report, which defines the security mechanisms necessary to negate anticipated, selected threats, (c) the ebXML Collaboration Protocol Profile and Agreement Specification (ebCPP, 2001), which defines how one party can discover and/or agree upon the information the party needs to know about another party prior to sending them a message that complies with this specification, and (d) the ebXML Registry/Repository Services Specification (ebRS, 2001), which defines a registry service for the ebXML environment.

A typical example, which is depicted below, describes six activities involving two enterprises called *A* and *B*.

1. A standard mechanism for describing a *Business Process* and its associated information model.
2. A mechanism for registering and storing *Business Processes* and *Information Meta Models* so they can be shared and reused.
3. Discovery of information about each participant including the *Business Process* they support, the *Business Service Interfaces* they offer in support of the *Business Process*, the *Business Messages* that are exchanged between their respective *Business Service Interfaces* and the technical configuration of the supported transport, security, and encoding protocols.
4. A mechanism for registering the aforementioned information so that it may be discovered and retrieved.
5. A mechanism for describing the execution of a mutually agreed business arrangement, which can be derived from the information provided by each participant.
6. A standardized business *Messaging Service* framework that enables interoperable, secure and reliable exchange of messages between *Trading Partners*.
7. A mechanism for configuration of the respective messaging services to engage in agreed on Business Process in accordance with constraints defined in the business arrangement.

During Activity 1, enterprise *A* learns that an ebXML Registry is accessible on the Internet. After reviewing the content of the discovered registry, enterprise *A* decides to build its own ebXML conforming application during Activity 2. Enterprise *A* then uses Activity 3 to submit its own business profile information to the ebXML registry. The business profile submitted by enterprise *A* describes the company's ebXML capabilities and constraints, as well as its supported business scenario, which are XML versions of the business processes and associated information bundles. After receiving verification that the format and usage of a business scenario is correct, an acknowledgement is sent to company A. Enterprise B is involved during activities 4, 5 and 6. During activity 4, B discovers the business scenarios supported by enterprise A in the ebXML registry. In step 5, enterprise B sends a request to enterprise A indicating that it would like to engage in a business scenario with enterprise A using ebXML. Before engaging in the scenario, enterprise B submits a proposed business arrangement directly to enterprise A's ebXML compliant software interface. The business arrangement contains mutually agreed upon business scenarios, information pertaining to the messaging requirements for transactions to take place, and security related requirements.

The description of business process definition permits an organization to express its business process so that is understandable by other organizations. It identifies such objects as the overall business process, document flow, legal and security aspects. This enables the integration of business processes within a company or between companies. The information models define reusable components that can be used in a standard manner within a business context.

The Collaborative Protocol Profile (CPP) contains typical information about the service provider including, but not limited to, contact information, industry classification, supported business processes, interface requirements, and messaging requirements. CPP also may contain security details and error handling. The ebXML collaborative protocol agreement is created during an engagement between two parties and includes information about the service and the process requirements by which all parties involved in the ebXML initiative support transactions. The running of a process is made up of transactions. A transaction is an atomic unit of work between parties, which is simply an exchange of document, such as a purchase order. A transaction can succeed or fail. If it succeeds, it is legally binding. If it is unsuccessful, its sub-effects are compensated (or rolled back). Transaction semantics can specify a number of parameters including those required for the security and reliability of service.

The ebXML messaging service enables cooperative, secure, and reliable exchange of payload between two companies or between companies and the registry. The service allows any application level protocol to be used (e.g., HTTP and FTP). Cryptographic techniques can be used to implement strong security. Secure protocols such as https can be used to implement confidentiality. Moreover, digital signature can be applied to messages to guarantee authenticity. The ebXML registry is similar to UDDI for discovering services. A registry stores information about items in the repository that are created, updated, and deleted through requests made to the registry.

7.5   Technical challenges and security

Technical challenges occur initially during the analysis of requirements and the description of user needs and security requirements. Moreover, challenges can occur in almost all phases of the e-service life cycle, where security is a major issue. Typically, an e-service life-cycle is a six-phase process, represented by the description, publishing, invocation, integration, and management of e-service. The major technical challenges that have attracted the attention of many researchers and e-service developers are: description,

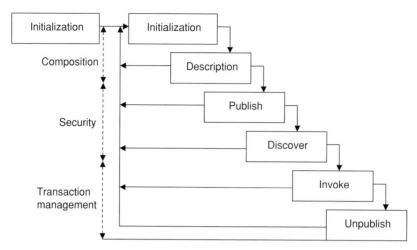

Figure 7.2 E-service life-cycle.

discovery, brokering, composition, publishing, reliability, management, accountability, testing, and traceability of services. Figure 7.2 depicts the life-cycle activity of an e-service and main technical challenges. Starting with the creation challenges, testing is an important feature to check (specially, when security should be provided). Description challenges address the specification of service based on user needs. They address also the syntax and semantic definition of system functionality, reliability, capabilities and security guarantees.

There is a set of basic security requirements that are important to have for trustable e-service applications. These features include authentication, data integrity, data confidentiality, non repudiation, confidence in the reliability and validity of an identity (or trust), user anonymity, user location traceability, transaction traceability, privacy, and security dependability. Advanced e-services (e.g., e-medicare and grid services) may have more restrictive requirements in addition to the abovementioned requirements or may also impose additional restrictions. This puts more emphasis on the technical challenges.

Therefore, security is a major challenge for publishing, discovery, and invocation. Discovery presents an efficient support to allow customers to specify at a high level of description their needs and requirements and attempt to match the resulting specification against registered services. Important issues in security that are related to discovery address protecting exchanged messages, protecting registries, securing the activity of brokers. Finally, one can notice that transaction management involves monitoring service states, security levels, and transaction protections. All activities either basic or value-added, when applied either to a simple or a composite service, are expected to expose their functionality at two different management levels: (a) at a syntactic level, where implementation aspects are addressed and (b) at the semantic level, where the conceptual aspects of services are facilitated.

Major security challenges, however, are encountered when addressing brokering, reliability, monitoring, transaction protection. We briefly define these activities and highlight the role of security and complexity of the related technical challenges. We discuss in the following sections how security requirments are addressed when composing e-services, registering services, and exchanging messaging. One good reference is Mehta *et al.* (2000) that gives an overview of the security in e-services.

Brokering

Brokering is the activity of mediating between requestors and providers in order to match customers' desires and providers' offerings. Brokering is a more complete activity than discovery since the broker must be able to facilitate service-to-service interactions, nego-tiations, and selection of services based on the optimality criteria. This activity is mainly dependent on the way the services are modeled, described, and protected. If the service specification contains the QoS required (including security requirements), then the broker can construct a query that specifies which data are asked for and the QoS requirements on acceptable replies. The broker can in turn access the available registries to find out which e-services are able to provide the required QoS. The broker then orders the replies and selects the optimal offer. Different contributions in the area of service quality and automatic ser-vice selection via service brokering are available in the literature such as in Georgakopoulos *et al.* (1999) and Scannapieco *et al.* (2002).

Brokering discusses the appropriateness of e-service classification and finds the optimal solution among services presented by suppliers having the same or similar objectives. Secu-rity problems related to brokering address the protection of message and the specification. In order to secure a message, a broker should consider two types of threats: (a) the message can be modified or read by a malicious adversary and (b) an adversary may send messages to a service that, while well-formed, lacks appropriate security claims to secure transaction processing. Message integrity is provided by using signature to ensure that messages are transmitted without modifications. The integrity mechanisms may be designed to support multiple signatures and should be extensible to support various signature formats.

Reliability

Reliability of an e-service is the capability of the service or its sub-services to perform its required functions and interoperate with brokers and requesters under stated conditions for a specified period of time. Stated conditions may include privacy protection, intrusion tolerance and service availability. Based on the protection mechanisms they implement, e-service providers can provide more reliable e-services than other providers. Therefore, the availability of mechanisms allowing measuring reliability and controlling the security system state are highly needed. Transmitting reliability information helps to improve the reliability of composite e-services. It is also needed to provide reactive measures to improve reliability and recover an e-service locally when the e-service provider system goes off-line temporarily in situations such as temporary outages, caused by backup or maintenance actions or even a damage due to security attacks.

Furthermore, it should be required to specify what happens when a service goes off-line during the execution of a transaction involving the e-service. Moreover, e-service providers may have to implement appropriate processes for disaster recovery and the migration of all their business partners to new e-services. Finally, since a provided (or requested) QoS typically could be expressed using a set of parameters including throughput, response time, and cost, an interesting contribution to e-service reliability can include the development of models, techniques, and mechanisms to allow each service to have a set of QoS metrics that gather information on the expected QoS of the e-service execution. The satisfaction of the QoS requirements, however, may affect the level of security protections since they have a nature opposed to performance-related parameters.

Monitoring

Monitoring takes place during e-service management (including invocation and operation). Once a service requester and an e-service provider have been linked together in the provision or the use of an e-service, the e-service execution needs to be continuously monitored. This is a difficult task to realize because a large set of sensors and profiles is needed to be implemented. In fact, modifications are required to be incorporated in real-time and without affecting operations at the requester's site. This becomes a challenging issue for at least three reasons: (a) the e-service may be running on a system that is not under control of the requester, (b) the e-service requester cannot interoperate (or simply communicate) with the running operating system, and (c) the security countermeasures are hard to comply with.

E-service monitoring becomes more complex to perform in the case of composite e-services. The properties of a composite e-service are dependent on the properties of its component services and the managers of these components. Composing e-services may need to be coordinated in a constraining way. One solution to this is to outsource the management of some e-services to other e-service providers. Monitoring may have to address the quality of service provision and security requirements to be observed while the invocation task is processed. In this case, the monitoring process should integrate the appropriate tools for handling events and related metrics.

Compositions

Composition refers to the combination of simple e-services (possibly offered by different providers and accessible at different sites) into value-added services that satisfy requester needs. A composite service is similar to a workflow in the sense that it represents a complex activity specified by composing other basic or complex activities (Casati *et al.*, 2000b). This is why several approaches have addressed service composition using the workflow paradigm. Issues in e-service composition consider how to find appropriate services for combination and how to combine. Composite services, nonetheless, have several characteristics that make them different from workflows. The main characteristics are listed below:

- A component e-service is typically more complex than a workflow activity. While the workflow is modeled as a single function, an e-service can offer several operations to be invoked.
- Composite e-services need to be more adaptive and more dynamic to cope with the dynamic features of e-service environment and the continuous need to add new services.

- E-services consider security as a key issue since services do not reside within a protected single site and the sensitivity of resources is different from one site to another.

In order to meet the requirements imposed by the need for efficiency and reliability of composite e-services, several models and techniques have been developed for the description, analysis, and optimization of composite e-services. Some of these models were workflow-based. Composition of services is addressed by eFlow (Casati *et al.*, 2000b) and by the service model CMI (Schuster *et al.*, 2000). eFlow supports the dynamic composition of simple e-services. It allows configurable composite services, where different requesters invoke the desired subset of features that this composite service can offer. The service model of CMI provides another solution that also enables dynamic composition of e-services at a semantic level by separating the service interfaces from service implementation and defining the service interfaces in terms of service state machines and its input/output parameters.

7.6 Message protection mechanisms

It is probable that HTTP over SSL protocol provides a basic security for e-services. However, various services may need a high level of granularity that is not provided with that service. Therefore, it is important to define how an e-service authenticates users and how it learns about their security privileges. Available standards such as XML encryption, XML signature, and OASIS security services seem to be able to be integrated into e-service systems. Particularly, ebXML can handle digital signature and digital encryption.

7.6.1 Security needs

Because e-services may contain sensitive information with respect to providers and customers, the security of e-services is an important issue for service provider and service requester. The request, payment and delivery of a service, for example, usually involve information exchange between the customer and the provider. Thus the confidentiality of this information becomes an issue. The security assessment for e-service systems can be organized with the following goals: (a) to look at causes of e-service failure and (b) to investigate how far the reliability of e-service monitoring and access control could compromise the e-service system if they allow hackers to take advantage of a security hole.

The more typical way of performing a risk analysis is to identify the assets and their values, discover the threats related to these assets, and contain their vulnerabilities. Then the analysis has to suggest the more economic protection (Brooke, 2000). Several methodologies were proposed to this end. Failure modes and analysis of effect attempt to look at individual components and functions of the system and investigate their possible modes of failure. They consider the possible causes for each failure mode, while estimating their possible consequences. The effects of failure are determined for each component and then evaluated for the complete system. Countermeasures can be suggested after this task.

Two types of messages, denoted by *request* and *response*, are used to allow service requesters to ask for a remote service and authorize e-service providers to respond to requests. Each message consists of two parts: a header and a payload. The header may carry auxiliary information for authentication and transactions. The payload of the message contains, in the case of request for remote services, the method name, arguments and e-service target address. The response method is structured just like the request, except that payload should contain the encoded content result.

E-service security must provide secure message exchange and support a large spectrum of security models including X.509, SPKI, and Kerberos. More precisely, message exchange security should provide support for multiple formats, multiple trust domains, multiple signature formats, and multiple encryption technologies. It is also helpful that the secure exchange takes into consideration how keys are derived, how trust is established, how security policy is exchanged, how authentication is performed, and how non-repudiation is provided. In the following subsections, two major contributions are discussed: the SOAP message security and the ebXML message security.

7.6.2   SOAP message security

The message protection provided by SOAP message security takes care of the protection of messages against threats aiming at modifying flowing messages and sending well-formed messages to an e-service that lacks appropriate security of the claims they contain. SOAP message security provides solution for the following concerns: (a) message newness to protect against replay attacks and limit delays, (b) man in the middle attacks, (c) integrity of security elements used to provide confidentiality, integrity, and authentication, and (d) proper use of digital signature and encryption.

An abstract message security model is built in terms of security tokens combined with digital signatures to protect and authenticate SOAP messages. A security token represents a list of claims (i.e., information included in the token by an entity including name, identity, group, security keys, or privileges). A signed security token is a security token that is cryptographically signed by a trusted authority. X.509 digital certificates and the Kerberos tickets constitute typical examples of signed security tokens. Therefore, signed security tokens can be used to guarantee bindings between authentication elements and the principal identity (i.e., key owner's identity). In the absence of approval by a trusted authority or a third party, a security token provides no guarantees of the claims it contains. In this case, it is the responsibility of the token recipient to accept or reject the claims made in the token.

Message headers can be used as a mechanism for conveying security information and can allow for security tokens to be directly inserted into the header. Security tokens have different types and are attached to message headers. Three types of security tokens can be considered: user *name tokens*, binary *security tokens*, and *XML tokens*. User name tokens are introduced as a way to provide a username and optional password information. The additional security property induced by the use of password is made by appending two optional elements, a nonce or a timestamp. A password-digest is computed by the following formula, where the two elements are assumed available. However, if only one of these elements is present, it must be included alone in the digest:

$$Password\_Digest = SHA\,(nonce + timestamp + password).$$

This helps hiding the password and offers a basis for preventing replay attacks.

Signatures are naturally used to verify message origin and message integrity. They are also used by the message producers to demonstrate knowledge of the key used to confirm the claims made in the token and bind their identities to the message they produce. The following example illustrates the form of a secure SOAP message where the body of a message only is signed. The example shows where to place the used security token:

```
<SOAP envelope
< secured message header
    <Timestamp creation time>
    <security token>
    <signature
        < signature-information >
        < signature-value >
        < key-information >>
< message body > >
```

In this example, we assume that the message producer uses a security token and a signature. The token contains a data transporting a symmetric key, which is assumed to be properly authenticated by the recipient. The message producer uses the symmetric key with a hash function to sign the message body, while the recipient uses the same key and hash function to validate the signature. The field *signature-information*, in this example, specifies what is being signed in the message and the type of standards being used to normalize the data to be signed. It includes indications on the signature method, the digest method, the elements that are signed and how to digest them. The field *signature-value* specifies the signature value of the normalized form of the data as defined in Nadalin *et al.* (2004). Finally, the *key-information* field provides information as to where to find the security token associated with this signature. This may indicate, for example, that the token can be accessed and retrieved from a specified URL.

To carry out its requirements, SOAP message security necessitates that a message recipient should be able to reject messages with missing necessary claims or whose claims have unacceptable values. In addition, since it is important for the addressee of a message to be able to determine the freshness of a message and protect it against replay attacks, time stamping can be used to provide such services. The specification offered in Nadalin *et al.* (2004) defines and illustrates the use of time references in terms of what is defined in the XML schema. The *timestamp* module provides a mechanism for expressing the creation and expiration times of the security elements in a message. It assumes that all times should be written in *UTC format* as specified by the XML schema type. The expiration time, however, is relative to the requester's clock. In order to evaluate the expiration time, a recipient needs to recognize that requesters' clocks may not be synchronized to its clock. The recipient therefore, should make an assessment on the level of trust to be placed in the requester's clock and estimate the degree of clock alteration. The aforementioned example places in the *timestamp* field the creation data of the secured message.

Finally, let us notice an important feature. There are many situations where an error can occur while processing the security information associated with a sent message. This may happen in the case of an unsupported type of security token, invalid security token, invalid signature, decryption failure, or unreachability of the referenced token archive. To this end, a SOAP fault mechanism can be made available so that the errors can be reported. The SOAP fault mechanism can help detect the cause associated with the reported error, if any, provided that a file is available for this task.

7.7   Securing registry services

An e-service registry offers a set of services that facilitate sharing of information about e-service offerings between interested parties. E-service registries use mechanisms to guarantee that any alteration of the content submitted by an entity can be detected. These mechanisms must support explicit identification of the responsible entity for any registry content. To this end, registry clients may have to sign any content before submission, otherwise the e-service registry should be able to reject the content. The signature ensures that any alteration (e.g., modification or tampering) of the content's authenticity can be discovered using its association with the requesting entity. A registry client, denoted by Cl_A, wishing to check whether a given client Cl_B has really published a given content, that is available at the registry service, can rely on the following components of the response he/she gets from the registry service, which is implementing a signature-based solution: (a) the payload of the response containing the content that has been published by Cl_B, (b) the public key for validating Cl_B's payload signature, using appropriate information so that Cl_A can retrieve that key, and (c) the signature element containing the header signature of the response (as made by the registration authority).

The aforementioned type of protection has been implemented by various methods. Next, we briefly describe how the ebXML integrates this approach and provides security mechanisms for the ebXML registry service (ebRS, 2002). For lack of space in this chapter and the similarity of methods, we do not consider the other aforementioned schemes.

7.7.1   *ebXML registry security*

The ebMXL registry service specifies the ebXML scheme definition and provides the security solutions for the access control of ebXML registries. It provides an authorization mechanism and proposes another mechanism for the confidentiality of registry content. It also develops a message header signature, presents a payload signature, and discusses a key distribution scheme. In the following, we review the major features of these solutions.

Access control in ebXML registries is provided by creating a default access control policy that grants the default permissions to registry customers. Three policies are defined and the related permissions are set up: (a) the content owners are authorized to access all the content they own, (b) the registry administrators are granted the access to all contents, and (c) the registry guests (or unauthorized users) can only access all read-only content.

In addition, the access control assumes that: (a) any user can publish content, provided that he/she is a registered user, (b) any user can access the content without entering an authentication process, and (c) at content submission, the submitting entity is assigned the default content owner role as authenticated by the credentials in the submission message.

The signature process accompanying the ebXML registry takes care of two issues: the payload signature and the header signature. The payload signature is packaged with payload forming a multipart message encoded in MIME, where the first body part contains an XML signature. Message headers are signed to provide data integrity while the message is flowing through the network. This is made through a hash digest computation of payloads. Header signature requires that the algorithm be identified using an algorithm attribute. Signature in the header is not included in the signature calculation.

To verify a signature, the receiver of the signature requires the public key associated with the signer's private key. To this end, the participants may use a key-information field or distribute the public keys using a trusted third party (or certification authority). Moreover, the ebXML registry service has to assume that: (a) the registration authority and the registry client have their digital certificates, (b) the registry client should register its certificate with the registration authority, and (c) a registry client obtains the registration authority's certificate and stores it in its own protected archive.

In conclusion, one can say that the ebXML registry is able to authenticate the identity of the principal associated with a client request by verifying the message header signature with the principal's certificate, which may be included in the message itself or provided by the registration authority (through unspecified ways). The authentication of a payload is achieved using the signature associated with the payload. It is made on a per message basis, since all messages can be considered independent and there is no need to establish a connection to send a request. The verification includes the identification of the privileges that the principal associated with the message has with respect to specific contents in the registry.

7.7.2 Service-side protection of registries

It is obvious that securing an e-service registry requires a significant effort. In fact, protecting a registry server from attacks implies considering issues including security policy, user management, security auditing, system configuration, and log management. An overview of these issues is discussed in the following. The use of a security policy helps considerably to prevent against configuration inconsistencies and reacts efficiently to security incidents. A security policy is a progressive document that aims to protect registry related systems, delimit acceptable uses, elaborate personnel training plans for security policy enforcement, and enforce security measures. It should be customized to the organization's need to cope with its business activity, and resource sensitivity. Registry security policies need to be reviewed on a regular basis, and also on the occurrence of situations where new vulnerabilities can be a serious concern. These situations may require changing security rules or developing new plans.

To prevent damages and reduce the risks associated with their use, requesters should be managed effectively. Default configurations of e-services and registries often include default accounts with known passwords, as well as active guest accounts. These faults may be used by the intruders' community. Assigning full trust to a registry server user, who has been authenticated, may cause major damages since the unauthorized user can, for example, take control of an active session while its legitimate user is not in command. To reduce this threat, the registry operating system can be configured to lock any user session after an idle period. Such a solution seems to be completely insufficient in highly sensitive contexts.

From the instant where a server risk mitigation plan is enforced and a security policy including procedures and configuration settings is established on a registry, a security audit needs to be continuously conducted on the registry since it remains necessary to check whether the security policy is being enforced correctly. The auditing aims to check that no breaches in the registry configurations have been introduced during system operation and that the server is sufficiently protected. Depending on the activities affecting the server, a security audit needs to be performed on a regular basis and whenever undesired conditions are observed. To check whether security measures set up to protect a registry server have actually protected the server, the security audit should attempt to identify whether the server has been targeted by intruders. To this end, the set of available logs (which include network logs as an intruder may have cleaned the registry log content before leaving the machine) have to be examined, statistical analysis performed, anomalies detected, and alerts correlated in order to identify attack scenarios and decide the appropriate countermeasures.

Log files represent the most common source for detecting suspicious behaviors and intrusion attempts. The efficiency of these files is strictly affected by any failure or attack on the related data collection mechanisms and any weakness in protecting their output. Log protection is very important in the sense that intruders may access these files or the tools that manage them, in order to remove, alter signs of malicious activity, or even add erroneous information to these files. Log file protection can be altered due to bad configuration of access permissions, insecure transfer to remote hosts, and storage in public areas. Moreover, logging data locally is easy to configure; it allows instantaneous access but it is less secure as the log content may be lost whenever the registry is compromised. On the other hand, remote log storage is better protected, but requires strengthening the communication security medium using, for instance, a separate subnet path or an encryption mechanism.

Another important challenge that registry security administrators have to face is the guarantee of registries' availability. Registry overload should be efficiently managed to avoid being a victim of QoS deterioration, where a client response time increases over an acceptable level and the target reaches a denial of service. To protect servers from overload, a simple bounding of the flow rate is inefficient. Various actions should be attempted including traffic shaping, load controlling, and policy management. Traffic should shape the flow that meets the server performance. This is done by delaying traffic excess using buffering, queuing mechanisms, and request rejection. Classifying traffic, however, is not enough to reduce or avoid overload on the supervised registry. For this, it should be complemented by a load controlling mechanism based on well-defined and reliable traffic metrics. Once a message is linked to its class, the set of load metrics associated to that class are assessed to decide which action should be taken in response to that message. Policy management is the administratively configurable part of the overload protection component. It defines the metrics values, describes the registry reactions, and specifies whether specific measures should be taken any time the load controller metrics reach their thresholds.

Denial of Service (DoS) attacks attempt to keep systems and applications from providing their services. Consequently, a server victim of such attack will appear unreachable to its users. Recently, DoS attacks have changed to more intense and damaging attacks including Distributed DoS attacks (DDoS) that use many compromised servers to launch coordinated DoS attacks against single or multiple targets. To protect servers against denial of service attacks, many proposals have been made without completely solving the issue. Protection mechanisms include packet filtering, automated attack detection, and security vulnerability fixing. With packet filtering, end-routers apply ingress packet filtering to allow only server supported protocols, deny security-critical services and suspicious identified source IP domains and services, and forged IP addresses. Routers should be monitored to update their filtering rules, as intruders' techniques and behaviors evolve.

Finally, security vulnerability fixing is achieved through the following actions: (a) the servers should be configured carefully and securely using trusted tools and personnel, (b) the unnecessary network services should be permanently deactivated, and (c) highly available services should be updated with all available security fixes as soon as the fixes are set up.

7.8   Conclusion

This chapter deals with e-service security. The notion of services has become essential to many fields of communications and information technology. The state-of-the-art in e-service provision through communication networks is moving from tightly joined paradigms towards loosely linked and dynamically related components. In this latter paradigm, project developers and service providers are pushing for the definition of techniques, methods, and tools as well as infrastructures to support the design, development, and operation of e-services. Moreover, standard organizations are investigating the protocol specifications and languages to help in the proper deployment of e-services.

The aim of this chapter is to examine the e-service paradigms, technical features, and security challenges. Some important schemes have been investigated, reviewed and discussed in order to show the effectiveness of such approaches as well as their limitations, if any exist.

References

Bellwood, T. *et al.* (2002). UDDI Version 3.0. Published specification (available at http://uddi.org/pubs/uddi-v 3.00-published-20020719.html).

Brooke, P. (2000). Risk assessment strategies. *Network Computing*, **11** (21).

Casati, F., S. Ilinicki, L. Jin, and M. Chan (2000). eFlow: An open, flexible, and configurable system for service composition. In Proceedings of WECWIS'00, Milpitas, CA, USA, June 2000.

Cerami, E. (2002). Web Services Essentials, 1st edn. O'Reilly & Associates.

ebXML (2001). ebXML Technical Architecture Specification v1.0.4 (available at http://www.ebxml.org).

ebXML (2001). ebXML Registry Services Specification v2.0 (available at http://www.ebxml.org/committees/regrep/documents/2.0/spects/ebRS.pdf.).

ebXML (2001). ebXML Collaboration-Protocol Profile and Agreement Specification v1.0.

Georgakopoulos, D., H. Shuster, A. Cichocki, and B. Baker (1999). Managing process and service fusion in virtual entreprises. Information Systems: **24** (6), 429–56.

Guemara-ElFatmi, S. and N. Boudriga (2003). A global Monitoring Process for securing e-government, 5th Jordanian Int. Electrical and Electronics Engineering Conf. (JIEEEC'03), Jordan, October 12–16, 2003.

Guemara-ElFatmi, S., N. Boudriga, and M. S. Obaidat (2004). Relational-based calculus for trust management in networked services, *Computer Communication Journal*, **27** (12), 1206–19.

Mehta, M., S. Singh, and Y. Lee (2000). Security in e-services and applications. In *Network Security: Current Status and Future Directions*, C. Douligeris and D. N. Serpanos (eds.), Wiley.

Nadalin et al. (2000). Web services security: SOAP message security, OASIS standard 200401 (available at http://docs.oasis-open/wss/2004/1oasis-soap-message-security-1.0).

Scannapieco, M., V. Mirabella, M. Mecella, and C. Batini (2002). Data Quality in e-Business Applications. In *Proceedings of CaiSE 2002*, Springer-Verlag, LNCS 2512, pp. 121–38.

Schuster, H., D. Georgapoulos, A. Cichocki, and B. Baker, Modeling and composing service-based and reference process-based multi-enterprise process. In *Proceedings of CaiSE 2000*, LNCS 1789, Springer-Verlag, pp. 247–63.

Snell, J., D. Tidwell, and P. Kulchenko (2002). Programming Web Services with SOAP 1st edn. O'Reilly & Associates.

SOAP (2000). http://www.w3/TR/SOAP

Tiwana, A. and B. Ramesh (2001). E-services: problems, oppportunities, and digital platforms, Proceedings of the IEEE 34th Hawaii International Conference on System Sciences, 2001.

Tsalgatidou, A. and T. Pilioura (2002). An overview of standards and related technology in Web services. *Distributed and Parallel Databases*, **12**, 135–62.

WSDL (2000). http://msdn.microsoft.com/xml/general/wsdl.asp

8 E-government security

Governments worldwide have dedicated the efforts to deliver their services in ways that meet the needs of businesses and citizens they serve in order to enable them to interact securely in places and at times that are convenient to them. It is commonly agreed that bringing trust and confidence is essential to increase the uptake of e-government services. This chapter provides key support to service providers wishing to provide e-government services in a trusted manner. It lays the foundations for enabling secure services that will really transform the way citizens and businesses interact with government.

8.1 Introduction

E-government is to provide a non-stop government information service. The goal is to create a friendly easy-to-use tool for the public and businesses to locate information and use services made available on the net by the government agencies. It aims to provide a large spectrum of public information, authorize a greater and better access to this information, and give more convenience to government services. E-government projects have been started in the mid 1990s in various countries. Each country has assigned its own project with varying focuses, applications, and security mechanisms. Nowadays, some of these projects are in their operational phase, while others are still in the design or prototyping phases. However, as computer power is growing to be cheaper and computer networks are becoming larger and more efficient, many threats against e-government have been observed lately. Threats unfortunately tend to reduce the efficiency of e-government and limit its promises. The threats are caused mainly by malicious acts that aim at disabling and hurting e-government services, which can result in major losses and damages. Therefore, the security of information is central to the building of confidence in e-government. Governments are urged to implement security in all e-government projects and consider the security issues in the early steps of design. In implementing e-government services, a growing debate is ongoing about the security issues in online services.

Security issues represent a major concern in order to: (a) comfort the citizens in using the e-government services, (b) enable administrations and agencies to access, share, and exchange information securely (Benabdallah *et al.*, 2002), and (c) reduce the cost and complexity of service offerings. The security issues to be addressed should consider essentially technical aspects including the authentication of individuals. They are about transforming the way a government interacts with citizens and businesses. They establish methods that are able to guarantee privacy and provide investigation. Finally, they provide a definition of the legal framework and the compliance of security services. Even though a large amount of work has been done to provide security for service publishing, user interaction, protection of transaction with e-government services, and defense against unauthorized access to government's resources, it appears that we need more methodologies to complete the framework for the efficiency of a large service provision.

To be successful, e-government services must build trust with businesses and citizens and common understanding with a variety of trusted third parties early in the process of building e-government projects, conducting project implementation, and operating and monitoring e-government services. The largest concern for all parties is to transact using a system that may negatively impact them. Trust also induces three issues of important interest to any online service that can be provided in e-government: (a) security protection of e-government resources, services, and transactions flowing to e-government sites, (b) guarantee of privacy whose aim is to protect personal information that the e-government collects about individuals, and (c) online authentication of users. When offering online services, a government must ensure that it is protecting all of the sensitive data, which may be involved, while meeting its citizens' needs for practicality, ease of use, and access availability. The technology of digital signatures, encryption, and PKI-related services coupled with the use of other methods of security including firewalls, intrusion detection systems, security monitoring systems, and network security protocols make the promises of e-government reasonable and achievable (Novak, 2002).

As e-government services grow in scale and scope, the number of sites, databases, and servers they use grows significantly. Therefore, protecting the privacy of citizens' personal information stored in these systems, while guaranteeing effective use of the information they contain, is a real challenge. Security is a costly task. It must be addressed during the specification of e-government projects, the design of solutions, and also during the operation of services (Boudriga, 2002; Wimmer, 2002; Fujioka, 1992; Rubin, 2001; Acquisti, 2004; Cranor, 1997; Sadeh, 2002).

8.2   E-government: concepts and practices

E-government represents the introduction of a large set of technological innovations as well as various government re-engineering procedures. It is defined as a way for governments to use the most innovative information and communication technologies, particularly web-based Internet applications, to provide citizens and businesses with more convenient access to government information and services in order to improve the quality of the services and provide greater opportunities to participate in the development of digital economy. Examples of these services include transactions between government and business, government and citizen, government and employees, and government and subunits/branches of government. E-government presents an essential motivation to move forward in the twenty-first century with higher quality, cost-effective government services and a better relationship between citizens and government. One of the most important aspects of e-government studies the ways of bringing citizens and businesses closer to their governments.

An e-government project is not really a simple information technology project. Although building infrastructure and putting services online will generally involve IT projects, e-government is about delivering results through better information management, protecting resources, guaranteeing privacy, and redesigning processes as organizational and cultural change enabled by Internet and business models are urging. Without this change, technology will just add to the cost of government without improving the results it delivers. Major e-government services include electronic tax payment, e-voting systems, online information (e.g., environmental data, health information), and electronic interaction systems (participation in government decision-making).

The readiness of citizens and businesses to use e-government services will depend on the trust that they have on the services and the benefit they can get from them. E-government services can be public or classified. There are four categories of e-government information and services: e-management, e-service, e-commerce, and e-decision making/e-democracy. An evaluation of the Australian local e-government indicated that there was progress in e-management, but little progress in the e-service, e-democracy, and e-commerce areas. These services have different levels of classifications: high, medium and low. These levels can in turn be broken into intermediary levels.

The challenges in e-government services' security include identifying users, authenticating users, storing public and classified information in the same Web sites, checking authorizations, auditing, signing transactions, resolving conflicts, keeping copies of information, and so on. Hence e-government security systems should be able to provide multiple authentication methods, authorization, credential issuance and revocation, audit, confidentiality, conflict resolution, accountability, availability, platform independency, privacy, data integrity, anonymity, scalability, single sign on, and so on. E-government scale leads to organizational transformation. Public agencies begin implementing e-government and governance initiatives. Organizational performance will be improved and services delivery will be better equipped to interact with citizens and provide services over the Internet. In addition, e-government is transforming organizations by breaking down organizational boundaries, providing greater access to information, and increasing the transparency of public agencies. This can enhance citizen participation and facilitate the democratic processes.

8.2.1 E-government assets

Below are the assets of any e-government system:

- *The personal data* Data related to a customer/entity used by any e-government service must be protected against loss, damage, and unnecessary disclosure on line. Measures of data protection should be relevant to privacy and legislation. It is important to note that, when the personal data are under the control of the customer, the customer is himself responsible for the protection of that data.
- *The corporate information* The information base of government in general and organizations offering e-government services in particular must be protected against loss, disclosure, and introduction of erroneous content.

- *The e-government service* By comprising the applications and delivery platforms, e-government services must be protected against threats targeting the availability and integrity of the other offered services.
- *Authentication credentials* Credentials used by the e-government platform must be protected against forgery or illegitimate use. Credentials should be frequently checked for multiple purposes including credential validity, trust levels (if any), and authorization.
- *Objects that represent monetary value* Objects with sensitive values must be protected against fraud. In particular, e-government transactions that are likely to result in cashable orders, or may relate to the delivery of goods that can be misappropriated, must be properly controlled and checked.

8.2.2   Challenges, limits, and obstacles to e-government

From a technical point of view, different risk types have been defined with relevance to data security in e-government. In a general classification, these risks can be grouped into the following three basic threats, which are not specific to e-government, but also apply to e-commerce:

- *Loss of confidence, trust, or credit* This is due to unauthorized information access, unavailability of services, uncompleted transactions, or weak protection of private data.
- *Loss of integrity* This loss is induced by unauthorized modification of data related to citizens or the availability of wrong information provided for e-government users.
- *Loss of availability of data and services* Such loss can occur due to damage in resource storage or functionality.

The aforementioned threats can be divided into inter-communication, intra-communication and system threats. Inter-communication threats may be passive or active. As described in Chapter 1, passive threats are manipulations, which cannot be proved, described as "leaking" out of information. Active threats are caused by manipulation of objects or relations. In comparison to passive threats, active threats can be recognized, but cannot be reinvented. Examples of active threats include message sequence manipulation (e.g., diverting or re-sending of messages) and modification (e.g., corruption of messages and routing information). Intra-communication threats are caused by the communication users themselves such as the repudiation of origin, repudiation of receipt of communication content, computer forgery, and damages of security marks (like access marks). System threats, on the other hand, are manipulations of resources like overloading, reconfiguring or reprogramming. Examples of system threats are the misuse of access rights, Trojan horses, viruses, spooling, and flooding.

From a non-technical viewpoint, additional security threats arise. In e-government systems, almost all data are security-sensitive. This is clear when considering that the consequences of misuses of stored citizens' data are more important than what they are in e-commerce. For example, the consequences of misusing stored citizens' data for committing crimes are more direct than the disadvantage of a credit card misuse when shopping on the Internet. Because they feel vulnerable when using e-government systems, citizens and enterprises want to have security solutions which provide subjective trust. Citizens have no opportunity to verify provided security and thus they may have to accept a certain security uncertainty. The difference with e-commerce is that citizens using e-government will not accept the same level of uncertainty as they might when using e-commerce transactions. Additionally, there are important differences between the scopes of e-commerce and e-government which influence the whole security context.

In e-government, public administration must consider the needs and the specific situation of all citizens. This causes a certain transparency of administration acts and citizen data for understanding and reproducing of individual cases. A proper e-government system has to prevent this transparency going to unauthorized parties. Additionally, making profit is the main stimulus for signing contracts in e-commerce. These contracts have allowed structures and are integrated in a well-structured legal framework. In the opposite, a large set of processes in e-government are not profit-oriented and are not necessarily well-structured like most processes in e-commerce.

Security threats and induced security requirements are the results of unstructured and semi-structured realities of administration processes. Analyzing security threats and requirements of an e-voting system, for instance, needs to consider the context of the e-voting components, the actors of the e-voting system, and several other special conditions. Specifying completely the general security threats and requirements in e-government may need an unacceptable load. However, security threats have to be investigated for the specific scope of each of the different e-government domains including: e-administration, e-democracy, e-voting, e-justice, e-assistance, etc.

Finally, notice that in each level of electronic processing of e-government (i.e., information management, communication, or transaction), appropriate security requirements need to be specified. A Web portal, which exclusively provides information, has fewer different security requirements than another e-government system, which has to support a complete set of transaction possibilities with highly sensitive data.

8.3   Authentication in e-government

Online authentication (or e-authentication) establishes the required level of confidence that the individual interacting with a government service is the individual he/she claims to be. The aim of online authentication is to provide citizens with the opportunity to access services on the Internet without having to appear in person every time they need it. Achieving this requires addressing the following issues: (a) what does an individual need to do to establish his/her identity, (b) once the identity is established, how can another party trust that identity without requiring full verification of the identity, (c) how can the fact be proved that an online transaction has really occurred between a citizen and the government, if the citizen denies it (EGOVNZ, 2004). One good example of objectives on deployment of e-governments projects is that made by New Zealand. These objectives are summarized below:

- Every citizen can have a single key for his/her online transactions across all e-government services.

- A citizen will only have to establish his/her identity once and this identity will be acceptable by all government agencies.
- Agencies collaborate to ensure that authentication processes are transparent to individuals. They can use the same authentication mechanism.
- Individuals will be provided with a trusted site that can be used to log-in to government services.

Three phases are required in order to achieve these objectives: (a) standardization, which assumes that all agencies should design, implement, and operate their authentication solutions in the same way and using the same standards, (b) collaboration, which assumes that agencies can share their authentication solutions and therefore reduce the design time and cost, and (c) uniformity, which assumes that a uniform approach is used for the online authentication of all services.

Typically, online authentication defines multiple levels of authentication in terms of the consequences of the authentication errors, damages, and misuse of credentials. As the consequences of an authentication error become more serious, the required level of assurance increases. Different rules and criteria should be made available for determining the level of assurance required for specific e-government services, based on the risks and their likelihood of occurrence of each service. After the assessment of risks and mapping the identified risks to the required level of assurance, the service designers can select then the proper technology to implement the solution that meets the technical requirements for each of the levels of assurance (Burr et al., 2004).

Among the levels of assurance, two levels have significant meaning: the lowest level and the highest level. The lowest level provides no identity proofing requirement, but it implements mechanisms that provide some assurance that the same claimant (i.e., the party to be authenticated) is accessing the protected service or resource. It allows a wide range of available authentication technologies to be used. Successful authentication for the lower level requires that the claimant proves through a secure authentication protocol that he/she controls its digital credentials. There are no provisions about the revocation or lifetime of credentials at the lowest level and relying parties may accept credentials that are digitally signed by a trusted entity or obtained directly from a trusted entity using a protocol where the trusted entity authenticates the relying party using a secure protocol.

The highest level of assurance is intended to provide the highest e-authentication assurance. For this level, authentication is based on the proof of possession of a key through a cryptographic protocol. A strong cryptographic authentication of all parties and all sensitive data transfers between parties is required. Only "hard" cryptographic tokens are allowed. This requires that a token should protect against threats to all the secret elements it contains. Credential service providers should provide a secure mechanism to allow relying parties to ensure that the credentials are valid. The mechanism may include revocation lists and online validation servers.

Intermediate levels of assurance may be based on the proof of possession of a cryptographic key. Authentication requires cryptographic mechanisms that protect this key against compromise. Some higher intermediate levels may require that, in addition to the key, the user must employ a password or biometric to activate the key. Different types of credentials can be used including a cryptographic key stored on a general-purpose computer, a symmetric key stored on a personal hardware device combining a nonce with a cryptographic key to produce an output that is sent to the verifier as password, or a digital certification.

With the registration process, an individual goes through identity proofing by a trusted registration authority (RA). If the RA is capable of verifying the individual's identity, the credential service provider (CSP) registers the applicant and generates the appropriate digital credential. The CSP is often an entity of the e-government while the RA may be an independent entity. Various RAs can coexist to provide registration based on the level of assurance that the digital credential can provide. However, various requirements need to be guaranteed. First, a trusted relationship should exist between all CSPs and ARs operating within the same e-government. Second, the records related to the registration must be maintained and kept protected against attacks either by the RA or the CSP. Third, the proofing process should be designed to a large degree depending on the assurance level of the authentication to be provided. Typically, the latter requirement include ways to ensure that the attributes occurring in a digital credential are sufficient to uniquely identify a single individual and that a subscriber (i.e., registered individual) cannot later repudiate the registration (in case for example of a dispute about a later authentication using the digital credential).

Threats can be divided into threats that involve attacks against the authentication protocol itself, threats involving attacks against the registration, and threats that may compromise the confidentiality of the attributes and information. Two types of threats need to be addressed to protect the registration within a secured e-government system: threats that compromise the registration process and threats that compromise the registration infrastructure. While the latter threat has been considered in Chapter 3, here we focus our attention on the former. Registration process threats can be of two types: (a) impersonation of a claimed identity, which is performed when an applicant claims incorrect identity while supporting his/her claim with a false set of attributes and (b) repudiation registration, which is realized when a subscriber denies registration, claiming that he did not register for the digital credential.

Authentication protocol threats include eavesdroppers observing authentication protocol execution for later analysis, fraud individuals presenting themselves as subscribers to verifiers to test credentials they presumed, illegal verifiers trying to handle legitimate subscriber claimants, impostor parties attempting to obtain sensitive user information, and attackers attempting to take control of an already authenticated session to collect sensitive information or insert invalid information (Burr *et al.*, 2004).

8.4 Privacy in e-government

Privacy is universally accepted as a fundamental right of individuals. Citizens and organizations should be assured that the information they submit to government agencies via all means will be handled truthfully. This includes personally identifiable data that government agencies may collect and store for the need of an e-government service. In providing available services to citizens and offering added-value functions, governments collect a wide range of personal information about their citizens (e.g., tax data, health records, law enforcement records, information occurring in driver licenses, etc.). With the move towards electronic data management and the development of the information society, governmental collection, storing, and processing of data have increased considerably. The development of e-government and electronic delivery of services have amplified government collection of personally identifiable sensitive data. Therefore, it becomes widely accepted that government's practices in collecting, retaining, and managing personal data about citizens pose a serious set of privacy concerns and challenges.

In the context of e-government, privacy is not just a question of what should be held secret, but the right to privacy is really the right to control the use of personal information that is released to other citizens, businesses, or agencies (Dempsey *et al.*, 2003). All over the world, the privacy of information about citizens and businesses is conducted by the principles of what is called "*fair information practices*." These principles, which were detailed and recommended by the *Organization for Economic Co-operation and Development* (OECD), represent the essential guidelines for responsible information practices that respect the privacy of individuals. The principles constitute the foundation of many national privacy laws, international agreements on data protection, and various industry codes of best practices. Depending on the country, such laws may apply to data about individuals collected by the government, to personal data gathered by private sector businesses, or both.

As it has been expressed by the OECD and some other international bodies, fair information practices should include, but not be limited to, the following:

- *Use limitation* Personal data should not be made known, made available or be used for purposes other than those specified in accordance with the "purpose specification." Personal data is used only with the clear permission of the data subject or by the authority of law.
- *Collection limitation* No more information should be collected than what is necessary to complete the transaction for which the data is collected. Any such data collected should be obtained by lawful means. Where appropriate, the collection should be obtained with the knowledge (and approval) of the data subject.
- *Data Security* Personal data should be protected by reasonable security mechanisms against security attacks, loss or unauthorized access, destruction, unauthorized utilization, modification, or disclosure.
- *Data quality* Personal data should be relevant to the purpose for which they are collected. They should be accurate and complete and be kept up-to-date.
- *Purpose specification* When personal data are collected, the purpose of the collection should be specified and the subsequent use limited to the fulfilment of that purpose. If the data are used in other services, the list of these services should be completely specified.
- *Openness* Means should be readily available to individuals to establish the existence and nature of databases, the main purposes of their use, and the identity of the entity responsible for the database.
- *Individual participation* An individual should have the right to obtain access to any data collected about him and held by an e-government entity. This right includes the

confirmation of whether an entity has data related to the individual and obtaining copies of the collected data.

There are obvious exceptions to some of the aforementioned principles in specific applications or countries. For example, in the case of law enforcement investigations or forensic investigations, it is not often possible to allow access to a suspect to the information that is collected for that purpose. Nevertheless, these principles provide a framework for dealing with the privacy issues produced by any government collection of personal information and handling the security requirements, risks, and assessment related to privacy. Moreover, the principles provide the support for privacy impact assessments (Free, 2001).

A privacy impact assessment (PIA) provides a framework for identifying and addressing privacy issues. A PIA is an assessment tool that describes, in detail, the personal information flows in a project, and analyzes the possible privacy impact of the project. The PIA constitutes an evaluation that is conducted to measure (*or assess*) how, for an e-government system, the implementation of new information policies, the acquisition of new computer systems, or the introduction of new data collection schemes will affect individual privacy. If the action is found to create a risk to privacy, the PIA recommends the modifications in the technology, policies, or schemes in order to mitigate the unfavorable effects on privacy. The PIA process attempts to ensure that privacy issues are identified and addressed by policy makers at the initial phases of the e-government project or policy (i.e., specification and design phases). It is important to address the privacy issues at the initial phases of the project as this will reduce the potential that the project will be found to have a negative impact on privacy after it has been implemented. Therefore, PIAs help avoid the high cost of redesign or rejection of projects. However, because the complete PIA is constructed progressively (phase after phase) during the project development, the PIA should be viewed as an evolutionary process that will become more refined as the project develops.

The formal result of the PIA is a privacy impact report. Typically, the contents of each report will vary depending on the nature of the e-service to be implemented, the status of privacy legal frame, and the level of security of the information to be collected. The contents include, but are not limited to, the following:

- A description of the subject organization's privacy policies and the assessment process that was used, the proposed project, the types of personal information that will be collected or used and how it will be disseminated or retained.
- An explanation of who will have access to particular categories of personal data. Employees should have access to the system only to the extent that is required for them to perform their duties. Procedures should be established to deter and detect browsing and unauthorized access.
- A privacy analysis that identifies how the new project or practice will impact individual privacy. This analysis should highlight areas that may violate privacy laws and stated policies.
- A risk assessment that lists the privacy risks that have been identified and an analysis of how these risks may affect individuals and the success of the project.
- A discussion of appropriate technical, procedural or other responses or safeguards that can be adopted to enhance privacy.

The analysis phase of the PIA investigates how information flows affect the choices individuals have regarding how information about them is handled, the potential degree of intrusiveness into the private lives of individuals, compliance with privacy law, and how the project fits into community expectations. Key issues to be addressed during the analysis phase of the PIA process include: (a) the compliance of the project with the privacy legislation, (b) the effect of the project on individuals changing their behaviour, and (c) the control of information about citizens.

Another task of a PIA is to determine which risks are critical to individual information. In assessing the seriousness of the privacy risks, the following should be addressed: (a) ensure that a minimum standard of privacy protection for persons affected by the project and analyze the situations where the project involves transfer of personal information across public or private sectors, (b) ensure that methods affecting privacy are transparent to individuals through adequate policies, and (c) check the flexibility in order to take into consideration the diversity of individuals affected by the project (Australia, 2004).

Countries seeking to promote e-government must protect the privacy of the information they collect, but do not need to build a complete privacy legal framework prior to conducting PIAs. The latter may be an important first step in dealing with privacy. The Canadian government was the first national government to make PIAs compulsory and required that all federal agencies perform PIAs for all services where privacy issues may be of concern. Canada has adopted a PIA policy that provides a consistent framework for identifying and resolving privacy issues during the design or re-design of government projects (Free, 2001). New Zealand is another early leader in the use of privacy impact assessments (Pacific, 2003).

In 2002, the USA adopted the E-Government Act, which requires federal government agencies to conduct privacy impact assessments before developing or procuring information technology or initiating any new collections of personally-identifiable information (E-GovAct, 2002). Under this new law, a PIA should address what information is to be collected, why it is being collected, the intended uses of the information, with whom the information will be shared, what notice would be provided to individuals, and how the information will be secured.

8.5   E-voting security

Electronic voting incorporates information and communication technologies in some or all of the election processes (Pràca, 2002). It is defined as any voting method where the voter's intention is expressed or collected by electronic means. E-voting systems may be classified in two main groups depending on where the casting of the ballot takes place: Kiosk voting and remote e-voting (Chaum, 2004a). The latter allows voting process to be performed from any place. It refers to an election process whereby people can cast their votes over the Internet from their homes, or possibly any other location where they can get Internet access. This type of voting system is called Internet voting system which is defined in (CIVT, 2000) as an election system that uses electronic ballots that would allow voters to transmit their voted ballot to election officials over the Internet.

Kiosk voting needs the voter to be present at specific locations such as polling stations to cast his/her vote since there is a need to use specific devices to record votes. In general devices used to record votes in kiosk voting are called *Direct Recording Electronic* (DRE) devices. It is assumed that the e-voting technique will replace traditional voting systems in the near future because several advantages can be achieved by such a technique. In fact, the e-voting solutions are technologically more sophisticated than the traditional pen and paper methods. The use of sophisticated techniques can lead to many advantages and challenges. The electronic tallying of ballots can be less labor intensive than the manual one. Therefore, it can be more accurate by avoiding many errors. The use of these techniques can also make the tallying process faster and results can be announced a very short time after the voting closes.

8.5.1   E-voting requirements

An e-voting system has its requirements and its limits that are largely detailed in the literature. Generally different assessments are performed to distinguish the feasibility issues from the security implementations and the limits of the proposed e-voting solutions. The main challenge for e-voting is to maintain secret suffrage (vote) and to prevent the casting of multiple votes by the same voter. Hence, different security and technical laws are intended to satisfy this aim. E-voting systems must satisfy different security requirements and non functional requirements. Among the most important security requirements, a voting system must satisfy the accuracy, democracy, verifiability, and privacy constraints. Non functional requirements are desired characteristics of a voting system. Among these requirements, we can enumerate convenience, flexibility, and mobility.

Remote e-voting avoids the need for the voter's presence in polling stations. This benefit is very important because it can increase the number of voters participating in the election.

8.5.2   E-voting limits

Because the consequences of an election are often important, the election results should not be threatened by malicious individuals whose aim is to corrupt these results. The fact is that when an e-voting is deployed, and the elector votes are transmitted across a network, the threats can be more dangerous, more numerous, and more frequent. The e-voting security constitutes a challenge that has to be taken up to guarantee the e-voting survival. Among the main limits of an e-voting system, we can mention the ones briefly described below.

Technical limits

Because the election system may be accessible worldwide, the potential number of attacks grows dramatically. The attackers are generally motivated by personal or financial gain. They can either be internal or external (Pràca, 2002). The internal attackers can be legitimate users, remote e-voting systems operators that exploit their privileged position, or others such as government employees that have access to the e-voting systems, but are not associated with the election process. External attackers can be hostile individuals, such as criminal or terrorist organisations, that aim to disrupt the government system or to access, corrupt, or steal data.

Several kinds of attacks can be employed within these two categories. One of the most famous types is the penetration of the e-voting system to modify the data stored in it in order to reveal certain confidential information or break the voter's anonymity. The introduction of malicious code in the e-voting system server constitutes another kind of well-known attack. This operation that can be made before or during the election process can undermine the votes' integrity and confidentiality. An exceptionally large number of votes may cause the e-voting system to become temporarily unavailable. This unavailability can also be caused by a particular kind of attack such as denial of service.

Social limits

The use of Information and Communication Technologies needs specific skills and know-how. Sometimes when trying to enhance the voting system security, one can increase the system complexity level, which may become inconvenient for voters. Moreover, the Internet technologies cannot be accessible for different persons. Then this point can undermine the equal access to the voting system and can prevent some persons from participation in the election processes. Moreover, the use of electronics in a process having important security and privacy issues such as the voting process should provide an important level of confidence in the system. Here, tests and security analysis of the voting system must be performed to enhance the confidence of voters in the voting system.

DRE limits

By using DRE machines, voters are not sure that their votes are correctly recorded by these machines. This is their important limitation. Moreover, different security studies have proved that the DRE software has its vulnerabilities that can affect the election results. In fact, in (Kohno et al., 2003), a security study of these machines demonstrates the existence of different vulnerabilities in the code of these machines, which allow different attacks and errors. In fact, malicious computer code, or malware (or malicious software), is often written in such a way that it is very difficult to detect. DRE software is complex, which makes it difficult to detect unauthorized modifications or the existence of such malicious code (Fischer, 2003).

Internet voting limits

With Internet voting, voters cast their ballots online, generally via a Web interface. With Web voting, the voter navigates to the proper election site using a Web browser on an ordinary PC and authenticates himself/herself to see the appropriate blank ballot form presented onscreen. The voter then fills out the ballot form and sends the completed ballot back to the election server. This type of voting is used by some corporations to elect their members. However, its use for public elections is not accepted due to the importance of the security requirements for this type of elections and due to different limits of Internet voting that we will enumerate below.

Internet voting is mainly based on the voter's computer and the Internet facilities and architecture. The problem here is that different vulnerabilities are known in these two architectures. Malicious code can be introduced through countless different channels to the voter's computer to interfere with the voting process in often undetectable ways. The voter may be prevented from voting, the privacy of the vote might be compromised, or the vote might be altered before transmission without the voter's knowledge. The massive vulnerabilities of standard personal-computer operating systems represent very serious concerns, in terms of hidden viruses, worms, Trojan horses, and further surprises unknowingly downloaded by the user and waiting to pounce on election day.

The weaknesses of the Internet as a platform for voting in elections include its vulnerability to many kinds of attacks such as the denial of service attacks, spoofing attacks, and man-in-the-middle attacks, which could lead to different problems that can affect the voting process and results. The most important problem here is that attacks on Internet voting systems can be launched remotely from any place, and might change the results of elections undetectably and, even if the attack is detected, it would be very difficult to recover the error.

Then the e-voting system cannot be an efficient and secure voting system. Different security considerations bring this type of voting into question. The primary ones are the danger that outside of a public polling place, a voter could be coerced into voting for a particular candidate and the existence of vote-selling problems since there is no physical control on the election process. Moreover, the remote voting is based on Internet technology, which is vulnerable to different attacks such as denial of service and the Domain Name Service attacks.

8.5.3 E-voting solution

Different solutions have been proposed to improve the electronic voting security including remote electronic voting systems. These solutions are, generally, based on the rapid progress of the computing, communication, and cryptographic techniques. All these solutions aim to satisfy most of the security and other voting system requirements.

Chaum scheme

To ensure the verifiability constraint while preserving the voter privacy and the uncoercibility constraint, David Chaum (Chaum, 2004b) has proposed a form of a receipt to be taken by the voter to check the validity of his vote recording and tallying. The proposed scheme provides an encrypted receipt that ensures the verifiability while reserving the voter privacy since the receipt does not reveal the vote. The vote is in an encrypted form. In this scheme, the proposed encryption is the visual cryptography whose definition is given in Ateniese et al. (1996). Other procedures are used to ensure different requirements, in particular, the voter anonymity. Among these procedures, one can consider the use of mix net, which is useful to assure a high level of anonymity. A mix net is an entity that, in addition to forwarding incoming messages, hides the relationship between incoming and outgoing messages. This is done by grouping together a number of incoming messages and shuffling them before forwarding them. In order to avoid content correlation between incoming and outgoing messages, all messages to a mix need to be encrypted in such a way that successive encryptions of the same message give different results. To assure more independence between messages, cascading a number of mixes to form a mix net was a solution that enhances the anonymity and the voters' privacy by hiding any relation between ballots and final results.

Another scheme was proposed recently that is based on Chaum's scheme, but it sidesteps the complexity of the visual cryptography by using a real and simple paper ballot. This scheme is the "prêt à voter" scheme, which is described in Chaum et al. (2004a). In the "prêt à voter" scheme, there is no use of visual cryptography, but a simple ballot form that preserves the encrypted receipt principle. For this scheme, there are different motivations. In fact, an important level of voters' anonymity is provided by the use of encrypted receipts and the mix nets as provided by a concept called encrypted onion. The verifiability requirement is then satisfied due to the use of the receipt that enables voters to check the validity of their vote recording and tallying. Also the universal verifiability is satisfied by using the Partial Random Check. The main limits of this scheme include: (a) it can allow double votes to occur and (b) it can allow adding invalid votes without being detected.

Blind signature schemes

The idea in such schemes (Chaum, 2004b) is that a voter prepares a ballot in clear text (i.e., a message stating for whom he votes). He then interacts with an authority that can verify that he is eligible to vote and has not already voted. If this is the case, the authority issues a blind signature on the ballot. This means that the voter obtains the authority's digital signature on the ballot, without the authority learning any information about the contents of the ballot. Finally, all voters send their ballots to the tallying authority that is responsible for counting votes. In order to preserve the privacy of voters, this must be done through an anonymous channel.

Blind signature schemes ensure a high level of voters' privacy because of the use of blind signature that preserves the voter anonymity by not enabling any authority to verify the voter identity and the content of the ballot at the same time. These schemes fail sometimes to assure the democracy constraint (Pràca, 2002). In fact, if the involved authorities collude, they can add invalid ballots for abstained voters. Hence we cannot guarantee that only eligible voters can vote; that is we can't guarantee the democracy constraint. Moreover, the use of anonymous channels remains an unclear issue because of the definition and the implementation aspects of these channels. In general, anonymous channels can be based on the use of mix net concepts, but here there is no necessity for blind signature.

Homomorphic encryption based schemes

To take benefit from the advances in encryption technology, some new e-voting schemes have proposed to use different cryptographic computations such as the Homomorphic encryption (*h-encryption*). Before the principle of these schemes is described, we should present some features used to accomplish these schemes. In fact, the *h-encryption* uses the threshold cryptography. Threshold cryptosystems distribute the functionality of cryptographic protocols to establish robustness. In the election paradigm, the tallying process can be shared among n voting authorities by using a threshold public-key encryption system. In this case there is only one public encryption key, while each of the n authorities has a share of the private decryption key. Each voter posts his/her vote encrypted with the public key of the authorities. The final tally is decrypted by the voting authorities jointly. Privacy of the votes and accuracy of the tally are assured provided at least a threshold of authorities is not faulty (or corrupted). In schemes based on *h*-encryption, and since the vote remains encrypted, it is useful to have a proof that the corresponding vote value is valid. The voter has to establish a Zero Knowledge proof allowing voting authorities to check the validity of his vote.

In *h*-encryption based schemes, voting options are coded as numbers 0, 1 or -1, and each voter encrypts his choice by a public key known by all voters. When submitting his encrypted vote, the voter must identify himself to prove that he is eligible to vote and has not voted before. Furthermore, he must prove the fact that his encryption contains a valid vote. Because all individual votes will remain encrypted and the proof is zero-knowledge, this does not violate the voter privacy.

To reveal the election result, different authorities multiply all encrypted votes to obtain an encryption of the votes' sum which will be decrypted to reveal the result. The private key needed for this has been secret-shared among a set of authorities. Each authority server holds a share of the private key. The shares have to be constructed through a threshold value "t" so that no information about the private key leaks as long as at most t servers are corrupt, or are broken by a hacker. On the other hand, if at least $t + 1$ servers behave correctly then a decryption operation can be executed.

8.6 Engineering secured e-government

8.6.1 E-government model

The e-government model that we propose has been discussed in Boudriga and Obaidat (2002) and Boudriga *et al.* (2002). It is based on thorough planning that includes laying out telecommunications infrastructure, developing information and communication technology pilot projects, preparing the required legislation, securing payments, creating a public key infrastructure for digital certification, and implementing on line government services. This model is represented by a workflow containing 10 tasks, a set of metrics that help control the main parameters in the e-government set-up, and a set of milestones used for analysis, verification, and triggering actions. The ten tasks are depicted by Figure 8.1 (where arrows describe temporal and input/output dependencies) and described as follows:

1. **Political decision to build e-government** (S_1) The development of e-government requires a political decision. Governments need to set up objectives in realizing their e-government system.

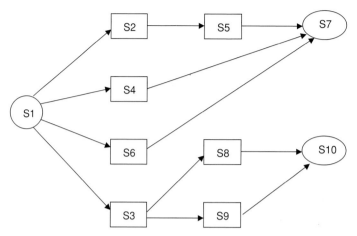

Figure 8.1   E-government model (Boudriga, 2002).

2. **Infrastructure evaluation** (S_2) Governments must determine if their telecommuni-cations network can handle the new traffic resulting from the use of these new services and whether it can provide a high level of service quality.

3. **Legal framework** (S_3) Governments need to formulate a legal framework for e-commerce, electronic document exchanges, and telecommunications regulations. Organizational framework should be established to define the legal relations between different ICT actors.

4. **Pilot IT projects** (S_4) To implement e-government, governments must start a number of pilot IT projects depending on their country's needs.

5. **Construction of an advanced technology network** (S_5) The network will allow access to multiple services including the next-generation Internet as a foundation to support the conversion to digital broadcast systems in order to create a global digital network.

6. **Reducing ICT illiteracy** (S_6) Decisions have to be taken to connect research centers, universities, and schools to the Internet in order to create community centers to promote the use of ICT by the population, and formulate incentives for companies to train their employees in ICT.

7. **Reducing digital divide** (S_7) Incentives should be formulated such as promoting low price, good quality computers, lowering Internet connection fees, and encouraging companies to train their employees in ICT.

8. **Developing digital payment** (S_8) The payment mechanisms should be adapted to the local practice. For example, if credit cards are not widely used by a local population, the government should come up with payment mechanisms that will promote on line transactions.

9. **Establishing a Certification Authority (CA)** (S_9) In developing on line government services, governments should formulate technical rules for handling secure transac-tions. Technical solutions should be based on building a local public key infrastructure

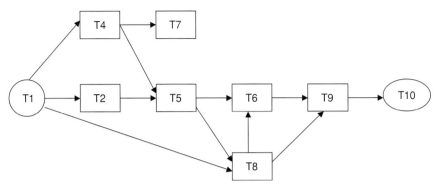

Figure 8.2   Security Model (Benabdallah *et al.*, 2002).

(PKI), developing local expertise in digital certification and signatures, and promoting, in particular, network security procedures and audits.

10. **Developing on line government services** (S_{10}) Incentive actions should be made to build a large spectrum of services starting with services associated with the major applications involved in conducting tax payment, trade, education, etc.

8.6.2   E-security model

In implementing the online services, government entities should define the types of security to make citizens and enterprises comfortable in using these services. Citizens and companies must have confidence that both the content and the infrastructure of the Internet are secure. These types of security should emphasize the protection of data privacy, access control, authentication, and network security. The security issues and mechanisms should be included in each task of the e-government model. These issues and mechanisms constitute the components of a workflow model performing a number of tasks described as follows and depicted by Figure 8.2:

Budget planning (T_1) A political decision is required to implement IT security and have a budget for it. This decision should emphasize the security of the communication networks, the info-structures, the e-services, and the customer protection.

Risk analysis (T_2) A security analysis is required to implement the above decision. In this task, the government needs to evaluate the threats to the existing info-structures, review the existing security policies and assess the qualified human resources in security.

Building security legal frame (T_3) The inclusion of a legal frame is necessary for ICT security. It can be achieved by the formulation of a set of laws aiming at: (a) protecting the user privacy rights, the intellectual property rights, online information, and the financial transactions, (b) enforcing the e-contracts and e-signature, and (c) defining clearly the rules, rights and duties of the ICT major actors. The role of security auditor may need to be legally defined.

Confidence assessment (T_4) An evaluation of the user's confidence in using the network to enforce ICT security is performed. The government must start with a number of pilot security projects. These projects should evaluate the confidence of the user in using the network, secure the virtual money platform, develop secure available e-services, and provide at least one digital signature and a PKI platform.

Planning (T_5) Security strategies can be built immediately after the pilot security projects are engaged. To this end, the government should: (a) identify the assets that have to be secured and define the level at which the assets will be protected, (b) set up a strategic security plan involving agencies and governmental departments, and (c) evaluate the cost of the security measures to implement the strategy.

Definition of security policies (T_6) In the construction of an advanced technology secure network, the government needs to define security policies in the transmission network, secure network access points to the network, and assure privacy, confidentiality, and use of VPN services.

Promotion (T_7) (Promoting of Security Good Habits) The government needs to make plans and incentive actions to train security specialists, promote security habits for the public at large, and organize workshops and conferences.

Credential practice (T_8) Practices should be clearly defined for the use of credentials. Technical solutions to set up a local public key infrastructure, as described in (S_9), will include the development of a secure Certificate Practice Statement (CPS), security auditing techniques, and secure PKI platforms.

Enhancing security (T_9) To enhance security for online services, the government should encourage companies to establish a security strategic plan, set security policies, and set up incident response teams with responding procedures.

Building a security observatory (T_{10}) The establishment of an IT security observatory needs: (a) means to collect and analyze information about vulnerabilities and threats, (b) skills to provide solutions to security problems and gaps, and to audit IT systems, and (c) resources to advise companies about the adequate security solutions.

The implementation of the IT security model requires a thorough planning to set the above tasks. These tasks are related to each other by objective achievement and time frame realization, as shown in Figure 8.2. The relationship between two different tasks is presented by an arrow in Figure 8.2 (Benabdallah, 2002). It shows the logical flow between their subtasks.

8.6.3 Implementing e-government

An e-government system can be considered as a monolithic multi-domain system. A simple attack such as a denial of service can have disastrous effect on the services provided by other interconnected systems making e-government services inoperative. Some authors predict that in a few years, the cyber-threats will be worse than the physical threats. To meet the objectives of e-government services, it is assumed that the security of governmental information and network is guaranteed. The security model to implement should address four major issues that are considered as an important foundation to the model. They correspond to tasks T_2, T_3, T_5, and T_6 of the aforementioned workflow and are described as below.

E-government risk analysis

Before setting security measures within e-government services, an analysis must be performed to determine:

(a) the assets that require protection, the corresponding types of threats and their probability of occurrence,
(b) the potential sources of threats and the cost from damaging the assets,
(c) the set of system vulnerabilities and the risk estimation of the threats they are related to.

Among the assets, one can find:

1. The information base of government departments and organizations offering e-government services. These assets can be threatened by loss, malicious disclosure, or introduction of erroneous content.
2. Any offered services that can be denied.
3. The authentication security policies, which can be threatened by forgery or malicious use.
4. Finally, all objects that represent monetary or other value that can be threatened by fraud. On the other hand, sources of threats can be internal (e.g., known users or employees) or external (from hostile outsiders). The hackers, who are generally motivated by financial gain, exploit e-government vulnerabilities such as weaknesses in the configuration or implementation of the operating system.

E-government legal framework

Governments have an important part in the implementation, promotion, and operation of e-government services. They should play an arbitrary and authoritative role. In addition, it should offer a trusted environment allowing different actors (i.e., users, service providers, etc.) to adopt these new types of services. All these e-government activities can be organized by specific laws. This strategy has been followed by different countries, where a set of laws to encourage e-services can be adopted including:

(a) the requirements needed to be satisfied by the technical specifications of electronic documents and electronic signature (contents, validities, use of electronic signature, etc);
(b) the requirements needed to be verified by the technical standards in order to guarantee a specific level of security for security elements (e.g., cryptographic algorithm, key size, etc.);
(c) the creation of a Digital Certification Root Authority and Certification Services Providers that are in charge of the emission, agreements, and verification of certificates or credentials;
(d) the protection of personal data (procedures, storage, certificate holder rights, etc.).

E-government security techniques

To construct its secure infrastructure, the government system should install selected techniques and mechanisms to ensure security and confidentiality. Security policies should be defined to allow several security levels and provide interoperability and coherence. It is suggested that the problem be considered at three hierarchical levels: the national level, intranet level and the server level. For each level, the threats, related risks and their corresponding security solutions are identified. A Virtual Private Network (VPN) seems to be a good solution to provide privacy and confidentiality. Governments have to promote the use of such a solution to secure the e-government services.

8.7   Monitoring e-government security

The e-government security monitoring process architecture that we present in this section is a waterfall-like model integrating four inter-operable sub processes. The sub processes are inter-dependent and can be viewed as a definition of the notion of life-cycle for the global security process.

8.7.1   Security monitoring life cycle

The four inter-operable sub processes that make up the life cycle of the security monitoring are: (a) security policy definition and security mechanism selection and implementation, (b) metrics definition and state characterization, (c) attacks detection and reaction to incidents, and (d) system restoration. Figure 8.3 depicts the sub processes composing the four processes and their inter-operation.

Security Policy Definition (SPD)

A process aims at defining a complete set of rules capable of keeping the system state at an acceptable level of security, if traffic engineering, access controllers, and service offering respect these rules. Rules are collected through questionnaires submitted to service and security system administrators. The SPD is first set up at the design of the e-government system to control access and traffic in the related trusted domains.

Design and Implementation (DI)

A sub process provides an efficient enforcement of the above security policy. Mechanisms can be characterized and acquired off-the-shelf, or designed as well. They can serve as event collectors and be organized within an ad hoc network designed on purpose.

Metrics Definition (MD)

This addresses the construction of a "complete" set of metrics in such a way that any evolution of the e-government security level toward an unacceptable situation is notified.

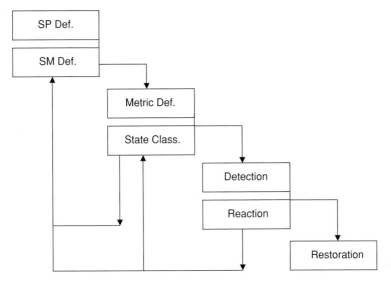

Figure 8.3   Monitoring life cycle.

Reducing the computational complexity of the management of the set of metrics may need to be addressed during this step.

Detection Process (DP)

This is a continuously running process that performs the evaluation of e-government states and the system damages. It can be one of the following processes: (a) a simple process that has to recognize certain types of metric values and directly deduce the occurrence of attacks, or (b) an intelligent process integrating the management of functions called multi-objective classifiers (MoCs) involving critical metrics, attacks' features, and damage assessment (Guemara-ElFatmi, 2003).

Detection and Reaction Process (DRP)

It helps to detect the occurring attacks and choose the appropriate reaction to the attacks discovered through detection based on a set of criteria including the service availability, solution cost, and response efficiency. A decision support system can be used to help achieve an acceptable and quick response to the detected attacks/anomaly. Reaction may include the definition of new metrics and MoCs and issue a set of information that the first process should use for security policy reshaping, system architecture re-definition, security mechanisms redesign, and metrics modification.

Restoration Process (RP)

This is a major task that first aims to restore the system state at an acceptable security level after an attack has occurred. Restoration may include the use of a multi-criteria decision support system because of the restoration cost, and downtime costs.

As presented in Figure 8.3, the processes involved in the monitoring life cycle may need to inter-operate. Some comebacks from one process can be redirected to one of its predecessors. For example, the definition of appropriate MoCs may need to get back to the design of mechanisms that implement the related metrics. The detection process may need to get back to the preceding process to check for metrics' information or MoCs re-computation. Comebacks are dictated by the need to refine analysis, change hypotheses, or generate new objects.

8.7.2   Monitoring tools

The efficiency of the monitoring process requires the existence of a number of tools. The following list enumerates different useful tools that should be appropriately located:

1. A sensors network helps to organize the set of event collectors, protecting their communication, and making easier any validation work on the metrics.
2. A set of graphic user interfaces providing activity reports, activity statistics, component configurations, sensor network administration, and security rule acquisition.
3. A set of routines that perform automated recovery actions (such as component isolation and interface reconfiguration) without the need for restarting the components composing the e-government system.
4. A rule-based decision support system that helps to achieve good definition of useful MoCs and appropriate responses to security incidents.
5. A set of security mechanisms including distributed firewalls, intrusion detection systems, access controllers, authorization systems, and cryptographic modules. These mechanisms should be part of the sensor network, or should be tightly connected to its components.
6. A set of databases logging all information required to detect attacks and perform decision support. Information includes intrusion signatures, attack features, and deduction rules.
7. A set of event journals reporting on and tracing all events detected by the security monitoring process. These journals should be appropriately protected.
8. A set of alarm generators that can be triggered by the occurrence of particular events or system behaviors. They are considered as prioritized notifications to the security administrators.

8.8   Advanced issues in e-government

Several issues need to be addressed to help to define and deploy secure e-government. We believe that a response support system is a challenging activity since it involves true monitoring and real-time reaction. M-government is becoming an interesting service since mobile communications are available from anywhere, to anyone, and at any time (Nicopolitidis *et al.*, 2003; Obaidat *et al.*, 2006).

8.8.1   Response support system

In addition to the management of metrics, the monitoring process performs the computation of MoCs, which combines metrics, risks, security rules, and events histories as explained in the previous sub section. Through MoCs management, the monitoring process also implements features that help to detect anomalies (unexpected behaviors of users or unexpected attributes variation). These features can use statistical information about the metrics, semantic relationship between resources accessed on the network, and probabilistic distribution characterizing the anomalies. In this sub section, we consider two issues that contribute to the efficiency of the monitoring process assessment: the response support system and the implementation/prototyping issues.

The response support system (RSS) is a semi-automatic process, in the sense that an incidence response team is involved with RSS in analyzing the critical security states and achieving responses. RSS can help to decide an appropriate response, if attacks are suspected through the occurrence of non secure states. This is achieved by the following steps:

1. Determining the attack types that have the highest probability to be at the origin of the anomalies among the candidate attacks that were revealed by the detection process.
2. Refining the determination of detected potential attacks through the analysis of metrics statistics and detection procedure.
3. Selecting appropriate actions to react. These actions could be done either in an automated manner or through human intervention. The selection may require the use of deduction rules including these that involve cost evaluation and consequence prediction.

This 3-step process assumes that various actions can be made available to react against detected attacks/misuses. A core set includes, in our opinion, the following set of actions:

1. A set of routines that performs automated recovery actions, propagation rules, packet filtering, or denying accesses;
2. A set of actions that can ensure stopping a service offering (temporarily or definitely, if needed) or changing service configurations;
3. A set of actions invoking modifications of sensors configuration, security policies, and security mechanisms;
4. A set of government oriented actions that can strengthen administration processes by delocalizing resources or redefining authorities over resources.

Therefore, a response may appear as a solution for one or more optimization problems involving limits to be observed (on cost, delay, infrastructure, etc.). Guidelines are useful to set up and assignment of ad-hoc teams is beneficial.

8.8.2   From e-government to m-government

Mobile devices are now taking significant roles in our business life. The number of these devices has increased tremendously during the recent decade. New generations of mobiles can provide many types of services. They can be used for transferring data and video and exchanging large scale business transactions. M-government is in its early stage of development. It may be defined as the activity of involving the utilization of all types of wireless technology, mechanisms, applications, and protections for improving benefits to citizens and businesses involved in e-government services. Various issues are driving the move from e-government to m-government activities including recent changes in the technological infrastructure and the improvement in mobile telecommunication service offerings.

The degree of penetration of mobile devices has generated important demands on m-government services and mobile users have expressed the need to have government services delivered and accessible anywhere and anytime on their mobile devices. Important third-generation applications have already been deployed in different areas (e.g., Japan, Asia/Pacific, the USA, and Western Europe) and have shown that 3G applications are expected to use large bandwidth. Implementing m-government will therefore be the cause of a series of new challenges in addition to those naturally shared by the e-government deployment efforts. Lanwin establishes in (Lanwin, 2002) a list containing the main challenges. In the following, we will discuss those which are most relevant to m-government.

Infrastructure development

The information technology infrastructure for m-government is both physical and logical. The physical infrastructure refers to the technology, equipment, and network resources required when deploying m-government. Logical infrastructures such as institutional arrangements and software systems make m-government transactions possible. Even though m-government is in its initial stage, various software packages are available for m-government services. *Packet-Writer, Pocket-Blue*, and *Pocket-Rescue* represent a few examples of m-government software developed. The infrastructure should be able to allow integration and interoperation of security mechanisms.

Payment infrastructures

They are essential to the deployment success of m-governments. A very typical obstacle for consumers to transact online is a feeling of mistrust in sending their credit card information over the mobile device, since it can be copied (mainly on the radio links) and replayed. Moreover, certification infrastructures and payment infrastructures are still inappropriate for use through mobile devices. For example, certificate management, certificate validation, and signature verification are still difficult to perform. Sophisticated payment operations such as delegation of payment and anonymous payment are also difficult to guarantee security.

Privacy and security general issues

The most significant concern that citizens have about m-government is that their mobile phone numbers, transactions, messages, and activities will be traced. Wireless networks are vulnerable because they use public airwaves to send signals. The interception of all traffic on the mobile Internet is relatively easy to achieve by knowledgeable adversaries. Efficient measures must be set up to overcome the mistrust, and assure mobile users that privacy is protected on all communication segments and information processing systems. Therefore, the specification, design, and planning phases of the m-government project, privacy and security issues should be considered so that developers will be able to select appropriate mobile devices.

Accessibility

The success of m-government depends mainly on the quality of access mechanisms. Such mechanisms should be able to provide the identification and authentication of citizens as mobile users. They should provide the authentication of the citizens when they access any m-government service based on different credentials. The mechanisms should be transparent to the technologies used by the mobile devices and should comply with open standards in order to increase citizen participation, provide citizen-oriented services, and allow easy access to m-government information in alternative forms, possibly using video and voice communications.

8.9 Conclusion

This chapter deals with e-government security. Governments all over the world have moved to offer services to their citizens electronically as such paradigm saves money, time and provides convenience and better accuracy. However, in order to do so, each government has to put in place, in addition to the proper communication and information infrastructure, robust security schemes in order to protect privacy and secrets of citizens and governments. It is commonly agreed that building confidence is essential to encourage citizens to use e-government services. This chapter covers the main concepts and practices of e-government systems including assets and challenges. Authentication and privacy as related to e-government have been dealt with in detail. Moreover, we addressed e-voting security, its requirements, limits, solutions and techniques. We dedicated a section on how to engineer a secured e-government model. We also reviewed the major aspects of monitoring an e-government's security system. Finally, we reviewed and investigated the chief advances in e-government systems.

References

Acquisti, A. (2004). *Receipt-Free Homomorphic Elections and Write-in Voter Verified Ballots*, CMU-ISRI-04-116 (http://www.heinz.cmu.edu/~acquisti/papers/).

Ateniese, G., C. Blundo, A. De Santis, and D. R. Stinson (1996). Visual Cryptography for General Access Structures (http://citeseer.ist.psu.edu/ateniese96visual.html).

Australian Government (2004). *Managing Privacy Risk: An introduction guide to privacy impact assessment*, Nov. 2004.

Benabdallah, S., S. Guemara-ElFatmi, and N. Boudriga (2002). Security issues in e-government models: what governments should do, IEEE Int. Conf on Syst. Man. and Cybernetics (SMC 2002), Hammamet, Tunisia.

Boudriga, N. and M. S. Obaidat (2002). Driving citizens to information technology. MDF-4, world bank: (http://www.worldbank.org/mdf/mdf4/papers/boudrig-obaidat.pdf).

Boudriga, N. and S. Benabdallah (2002). Laying out the foundation for a digital government model, case study: Tunisia. In *Advances in Digital Government: Technology, Human Factors, and Policy*. W. J. McIver, Jr. and A. K. Elmagarmid (eds.), Kluwer.

Burr, W. E., D. F. Dodson, and W. T. Polk (2004). *Electronic Authentication Guideline*, NIST special publication 800–63 version 1.0.1.

California Internet Voting Task Force (2000). *California Internet Voting Task Force. A Report on the Feasibility of Internet Voting* (http://www.ss.ca.gov/executive/ivote/final_report.pdf).

Chaum, D., P. Y. A. Ryan, and S. A. Schneider (2004a). *A Practical, Voter-verifiable Election Scheme*, Technical Report Series CS-TR-880, Dec. 2004, http://www.cs.ncl.ac.uk/

Chaum, D. (2004b). Secret-ballot receipts: true voter-verifiable elections. *IEEE Security and Privacy*, **2**(1): 38–47 (http://www.voterverifiable.com/article.pdf).

Cranor, L. F. and R. K. Cytron (1997). Sensus: a security-conscious electronic polling system for the Internet. Proceedings of the Hawaii International Conference on System Sciences, Jan. 7–10, 1997, Wailea, Hawaii, USA.

Dempsey, J., P. Anderson, and A. Schartw (2003). *Privacy and e-government*. A Report to the United Nations Department of Economic and Social Affairs as background for the *World Public Sector Report: E-Government*, http://www.internetpolicy.net/privacy/20030523cdt.pdf.

E-Government unit, New Zealand (2004). *Authentication in e-government: best practice framework for authentication* (available at http://www.e-government.govt.nz/authentication).

Fischer, E. A. (2003). *Election Reform and Electronic Voting Systems (DREs): Analysis of Security Issues*. CRS, http://www.epic.org/privacy/voting/crsreport.pdf

Freedom of Information and Privacy Office (2001). *Privacy Impact Assessment Guidelines*. Freedom of Information and Privacy Office, Management Board Secretariat, Ontario, Canada. http://www.gov.on.ca/MBS/english/fip/pia/index.html

Fujioka, A., T. Okamoto, and K. Ohta (1992). A practical secret voting scheme for large scale elections, *Advances in Cryptography – AUSCRYPT '92, Lecture Notes in Computer Science 718*, Springer-Verlag, pp. 244–251.

Guemara-ElFatmi, S. and Noureddine Boudriga (2003). A global monitoring process for securing e-government, 5th Jordanian Int. Electrical and Electronics Engineering Conf. (JIEEEC'03), Jordan, October 12–16, 2003.

Kohno, T., A. Stubblefield, A. D. Rubin, and D. S. Wallach (2003). *Analysis of an Electronic Voting System*. Johns Hopkins Information Security Institute Technical Report TR-2003-19 (http://avirubin.com/vote).

Lanwin, B. (2002). A Project of Info Dev and The Center for Democracy and Technology: The E-Government Handbook for Developing Countries. http://www.cdf.org/egov/handbook/2002-11-14egovhandbook.pdf.

Nicopolitidis, P., M. S. Obaidat, G. P. Papadimetriou, and A. S. Pomportsis (2003). *Wireless Networks*. Wiley.

Novak, K. (2002). Digital Signature and Encryption Technology: What it Means to Local Governments and Citizens, the Public Entity Risk Institute's Internet Symposium: Safe and Secure – Cyber security and Local Government, http://www.riskinstitute.org.

Obaidat, M. S., G. I. Papadimitviou, and S. Obeidat (2006). Wireless local area networks. In *Handbook of Information Security*. H. Bidgoli (ed.), Vol. 1, pp. 617–36, Wiley.

Pràca, D. (2002). Electronic Voting Schemes. Dept. of Computer Science, Faculty of Mathematics, Physics and Informatics, Comenius University, Bratislava, Zuzana Rjašková, (http://www.brics.dk/~jg/).

Pacific Privacy Consulting and Xamax Consultancy (2003). *Authentication for e-government: privacy impact assessment report* (available at http://www.e.govt.nz/services/authentication/authent-pia-200312/authent-pia-200312.pdf).

Rubin, A. (2001). Security Considerations for Remote Electronic Voting over the Internet, http://avirubin.com/.

Sadeh, N. (2002). *M-Commerce: Technologies, Services and Business Models*. Wiley.

US government, (2002). *E-government Act 2002*, Public Law 107–347-Dec 17, 2002 (available at http://www.cdt.porg/legislation/107th/e-gov/).

Wimmer, M. and B. von Bredow (2002). A holistic approach for providing security solutions in e-government, Proceedings of the 35th Hawaii International Conference on System Sciences, 2002.

9 E-commerce security

The Internet is dramatically changing the way that goods (tangible and intangible) and services are produced, delivered, sold, and purchased. Due to this development, trade on the Web becomes an essential requirement for enterprises. From e-commerce to m-commerce, which has become a major service nowadays, every enterprise works hard to find out a way to sell and buy that can satisfy its requirements. Several payment protocols have been developed. The security of servers, transactions, and payment operations has become a major issue for the success of business on the Internet.

9.1 Introduction

E-commerce security has become a serious concern for enterprises and citizens who rely on distributed digital processing in their daily operations. From a customer's perspective, the purpose of an e-commerce system is to enable the customer to locate and purchase a desired good (tangible or intangible) or service over the Internet when he/she is interested in getting it. Its function is to provide a virtual store. From a merchant's perspective, the key function of an e-commerce system is to generate higher revenues than the merchant would achieve without the system. To this end, the e-commerce system must recreate or utilize existing data and business processes and provide other processes to facilitate electronic purchase and provide product information, inventory systems, customer service, and transaction capabilities including credit authorization, tax computation, financial settlement, as well as delivery. Additional functions of an e-commerce system are to help redefine and enhance an enterprise's capability, customer-service capability, and delivery effectiveness. An e-commerce system is one of the areas of an enterprise's infrastructure that is open to customers via the Internet and is linked with other information technology systems that affect customer service.

Creating a complete on-line selling environment can require considerable time, money, and technical expertise. It also requires a large effort of risk analysis and securing processes as demonstrated in the preceding chapters. Many businesses are slowed down at the first or second step of the typical 3-step process to build an effective e-commerce Web presence. The 3-step process is syntactically described as follows.

Step One Develop a content site and handle purchasing transactions off-line. During this step, the enterprise can easily develop a simple version of its Web site in a cost effective manner. However, this limits the Internet function to promotion, while no revenue opportunity is involved. The security issues involved during this step are easy to handle.

Step Two Develop an on line catalog and handle transactions off-line. During this step, there is no need for sophisticated technology. The catalog can manage a large variety of products and provide an acceptable level of security. However, catalog building adds cost, without the possibility of reducing this expense through use of on line transactions.

Step Three Develop an on line catalog and manage transactions on line. This can manage large product assortment and complete sales at lower cost. However, the catalog building is expensive and on line transaction management requires sophisticated technology and extensive security efforts.

The transaction management (TM) handles credit and debit-card transactions on behalf of the merchant and the end customer. It may contain a common payment application programming interface that is used for all payment types and functions such as receive, approve, deposit, and refund. TM handles the necessary authorization requests and logging of the transaction. It also handles the establishment and completion of transaction information with the merchant, credit-card company (or issuer), and customer. The transaction server manages the payment process, from communicating with the consumer to connecting the merchant's financial institution (or acquirer). Records of transactions must be maintained to facilitate conflict resolution and later reporting. The TM should also contain a component to process digital certificates from an organization using certificate-authority software or follow-on security technologies.

Transaction protocol security (such as SSL and SET) are in charge of securing electronic transactions and provide open multiparty standard protocols for conducting secure banking card and debit-card payments over the Internet. Some of these protocols provide message integrity, authentication of all financial data, and encryption of sensitive information. Registration systems reduce the risk in e-commerce by establishing trust through authentication and non-repudiation using SET standards, which, in turn, control cost efficiencies and open new opportunities for commerce. Payment protocols require components placed at the end of the customer's location, the merchant's transaction-system place (whether on merchant premises or service provider environment), and the financial institution's site.

Concern about privacy, integrity, and security of online transactions has slowed down adoption of e-commerce technologies as a normal way of doing business. To gain acceptance and confidence of their participants, all organizations must achieve a high level of control. Establishing control in e-commerce requires developers and managers to expand the traditional view of internal control to encompass the activities of customers, suppliers, and other users of their electronic platforms. Privacy, authentication, and denial-of-service attacks are three classes of risk, especially prevalent in e-commerce. Using the appropriate control practices, one can suggest possible ways of controlling these risks, which demand new types of assurance of services in e-commerce.

Security breaches and frauds may cost business and consumers significant losses. Although technical defects contribute to security breaches, fraud is often made possible since many trading processes applicable in traditional commerce are easily damaged when conducted over the Internet. Although computer scientists and practitioners have provided technical tools that enhance computing and networking security, issues related to Internet-based e-business processes remain incomplete. There is a growing need for powerful tools and rigorous methods in the design and verification of correct e-process systems that operate over the Internet.

9.2 E-commerce security requirements

Recently, the use of e-commerce systems has grown at a phenomenal rate. A large spectrum of products (tangibles and intangibles) is sold on the Internet, with payments made essentially by debit or credit cards. In addition, there is an increasing concern related to the security of the payment systems used to process online transactions. Confidentiality of payment card information due to disclosure of this information to malicious adversaries could enable them to perform fraudulent transactions at the customer's expense.

9.2.1 General form of the e-commerce process

Typically, there are two types of e-commerce paradigms; business-to-business and business-to-consumer. The business-to-business e-commerce refers to a company selling or buying from other companies. This is actually not new, as many businesses have already been doing it since the 80s by means of Electronic Data Interchange (EDI). In fact, since the 80s, organizations have been using EDI to conduct business transactions electronically. Some of these transactions include sending/receiving of orders, invoices and shipping notices. EDI is a method of extending the organization's computing power beyond its boundaries. But the high cost and maintenance needs of the networks required by EDI are out-of-reach for small and medium sized businesses. With the introduction of the Internet, companies, regardless of size, can communicate with each other in a cost-effective manner. Companies that do so use it in several ways, depending on whether they are a manufacturer or supplier.

The business-to-customer e-commerce refers to a company selling its products or services to the customers using the Internet as the communication medium. Using e-commerce to market and sell can complement the traditional shop-front method. Although there are some businesses that rely solely on the virtual shop front as they do not have a physical store for walk-in customers (e.g., virtual library, hotel reservation application). With the business-to-consumer e-commerce, the company first establishes a Web site on the Internet, where it can put up information about its products and services, allow customers to order these products from the Web site and provide customer support services and protection. In order to get customers to the Web site, the company must inform the public about its existence using e-service means of advertising. To keep customers visiting the Web site more than once, the company must also update it regularly with news about products or promotions.

The main phases of the trading process as discussed by different studies available in the literature are described as follows (Guttman et al., 1998):

- *Information acquisition* During this phase, the customer collects information about merchants and the goods they offer on their Web sites and evaluates them. Intermediate agents can be involved during this phase to help in formation description, acquisition, and product comparison.
- *Negotiation* During this step, the customer and the merchant negotiate the terms of the contract regarding the methods of payment and delivery. Negotiation may have different forms depending on the nature of the e-commerce application and may include activities such as bargaining.
- *Execution phase* During this phase, the customer fulfills his payment obligations by providing the needed information and the merchant has to deliver the purchased product as defined in the contract.
- *After-sales phase* During this step, problems such as returning delivered goods, in case of non satisfaction, replacing delivered with defects, or fixing and repairing good problems can be addressed.

Security of online transactions includes also authentication, identification, non repudiation, and transaction integrity. A payment system involves four roles at least: a *payer*, a *payee*, an *issuer* and an *acquirer* (Hassler, 2001). The underlying payment transaction model is depicted in Figure 9.1, which shows the interactions of four roles:

Payer – The payer is an authorizer of a payment means supported by an issuer. Ordering a payment may be done using a card, a token, or a certificate. The payer is the customer or buyer in an electronic commerce scenario.

Payee – The payee is a merchant providing goods, services, and/or information and receiving electronically the payment for something purchased by the payer. Usually, the payee is simply referred to as the vendor, merchant, or seller in an electronic commerce scenario.

Issuer – The financial instrument that supports issuing payment cards (or means) by using cryptographic technologies which guarantees the association with "real money." Its role is to provide the payer and the payee with instances of monetary value which are used in payment protocols to transfer "real money" from the payer to the payee.

Acquirer – This is a financial institution (a bank, for example) which transforms the cryptographic objects involved in the payment into "real money" on behalf of the payee.

The transaction model involves also what is called *payment gateway*, which is a system that provides communication services and interfaces to the acquirer to support the authorization and capture the transactions, and certificate authority that provides the creation and distribution of digital certificates for cardholders, merchants, and payment gateways, in the case where the certificate paradigm is included in the authentication process involved in the payment. (Setbk2, 1999).

Traditional transactions are used to encapsulate database operations so as to provide atomicity, consistency, isolation and durability. They provide clean semantics to concurrent execution of operations. What differentiates transactions in e-commerce from the earlier environment is a number of issues that we summarize as follows. The first issue is the possible hostility of the open environment. From the transactional point of view, this means

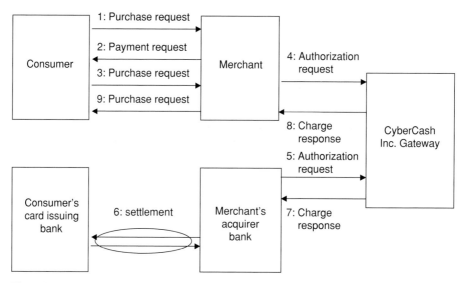

Figure 9.1   Payment model.

that there is a risk that the parties involved in an e-commerce transaction might be forged. Another issue is the vulnerability of the terminal and mechanisms used by the attacked. They can be misused or stolen. Therefore, the transactional mechanisms used to conduct e-commerce transactions should not rely on the device identity such as phone number or IP address and it should not allow usage of customer's identity based on the device identity.

9.2.2   Security requirements

The security requirements for each role described in Figure 9.1 vary from one role to another. However, it appears that acquirer and issuer have very close requirements. In the following we examine individually the requirements of each role.

Client

Transaction confidentiality, especially the information occurring in the payment card, is a major security need for a client. The nature of the transaction may require confidentiality. Transaction integrity is another security requirement since clients would not want an attacker to change the price, the delivery address, or the description of the merchandise really purchased. A client needs to be sure that he/she is transacting with a trustworthy merchant since it is relatively easy to be lured by sites that can appear to sell goods, but are actually collecting confidential information. Also, a client needs a mechanism to ensure that a malicious merchant or an adversary will not be able to reutilize previously accepted payments (i.e., reply attack). Finally, clients need to establish a non-repudiation mechanism.

Merchant

Requirements desired by merchants include, but are not limited to: (a) the evidence that a client has accepted to pay the price associated with the product on sale and that the client's card is valid, (b) the authentication of the client to make sure that he/she is the legitimate holder of the card presented for payment and that he/she has really agreed on the payment, (c) the impossibility to modify the details of a transaction to purchase a product that the merchant has authorized, and (d) a malicious adversary should not be able to present a copy of an old payment to request another delivery of the purchased product. Moreover, the acquirer should be able to claim that the merchant has obtained a second payment using an old transaction.

Issuers and acquirers

Requirements needed by the issuers and acquirers aim to ensure that:

1. Neither clients nor merchants can deny their participation in a purchase transaction.
2. An issuer (or acquirer) should be able to prove that it is the client who authorized the payment and that he is the owner of the card supporting the transaction. This is to prevent a situation where a client denies making a transaction and the issuer may be forced to refund the same amount to the client.
3. No one can modify the details of a transaction, once its details have been confirmed. This means that it should not be possible for a merchant to change the information occurring in a transaction after it is agreed on.
4. A malicious user should not be able to utilize a transaction to replay a payment.

9.2.3   Available security protocols

Various security protocols have been developed for e-commerce. The major protocols include:

1. The *Secure Socket Layer* (SSL) protocol SSL was developed in 1994 by Netscape to provide secure communication between Web browsers and Web servers. SSL provides server authentication, data integrity, and client authentication (Thomas, 2000).
2. The *Transport Layer Security* (TLS) protocol This was introduced by the Internet Engineering Task Force in 1995 (Dierks and Allen, 1999).
3. The *Secure Electronic Transaction* (SET) protocol SET was developed by Visa, MasterCard, and other companies to facilitate secure electronic commerce transactions and provide confidentiality of payment card information, data integrity, authentication of both merchant and cardholder, and authorization of transactions (Setco, 1997).
4. The *3-D Secure Protocol* This has been developed by Visa recently (Visa, 2002). It provides cardholder authentication for merchants using access control servers and the Visa Directory Server.

SSL and TLS are by far the most used protocols to provide security for transactions made over the Internet. SET is a complex protocol involving a large set of operations for each transaction processing, requires every participating entity to cryptographically sign transmitted messages and totally encrypt sensitive information, and allows merchants to verify the integrity of the order information using the notion of dual signature. It is clear that 3-D secure protocol does not meet all the security requirements. In fact, it aims only at providing cardholder authentication and associated non-repudiation to reduce or eliminate the risk of card-not-present, where the merchant is required to take liability for a doubtful transaction.

9.3 Transaction security with SSL/TSL

The aim of this section is to analyze the SSL/TLS and SET protocols in terms of how well they satisfy the security requirements outlined in the previous section. This is needed in order to facilitate the design of new protocols to enhance the security of transactions.

9.3.1 SSL/TLS features

SSL and TLS are currently the most widely used protocols for providing security for the client/merchant Internet link. To provide its services, SSL is divided into two layers: the handshake protocol and the record layer. The handshake protocol is responsible for initializing and synchronizing cryptographic state between the communicating entities, while the record layer provides confidentiality and authentication as well as the protection against replay attacks. Typically, in order to establish a connection, SSL requires five steps (Freier *et al.*, 1996):

1. The customer (or its browser) first sends a *ClientHello* to the e-commerce site. *Client-Hello* includes information such as SSL version, data compression method to use, session Id, and a random number.
2. The server sends a *ServerHello* message. Then it sends a *ServerKeyExchange* message containing the server's public key. Finally, it sends a *ServerHelloDone* message to indicate that it has finished its initial negotiation.
3. The customer sends its certificate, if needed by the server. Then he/she sends a *Client-KeyExchange* message containing the key information that will be used to generate the master secret key and keys that will be subsequently used for encryption. The client also sends a *CertificateVerify* message to prove that he/she has the corresponding private key in the certificate.
4. The client sends a *ChangeCipherSpec* message to indicate the starting point of a protected channel. Then he/she sends a *ClientFinish* message containing a hash of the handshake messages exchanged. The message is encrypted and authenticated.
5. The server sends back a *ChangeCipherSpec* message and a *ServerFinish* message with the same semantics as in step 4.

It appears clearly from the description of SSL function that:

1. SSL protects transaction confidentiality by using symmetric encryption. It protects the confidentiality of transmitted data against interception attacks and provides integrity protection for transferred data.
2. SSL uses the server certificate as the basis for server authentication. To this end, the client can check the server authentication by verifying its ability to decrypt information encrypted using the server's public key. In addition, SSL can provide client authentication if the client has a public key signed using a certificate issued by a CA trusted by the server. SSL provides protection against third party replay attacks by using a random number during handshake.
3. SSL provides no non-repudiation services; that is, neither client nor merchant have any cryptographic evidence that a transaction has taken place.

The transport layer security (TLS) protocol (as introduced in 1995 by the IETF) works in a similar way to SSL, but it presents some differences (Dierks and Allen, 1999; Rescola, 2000) that we describe as follows:

1. For message authentication, TLS relies on the computation of message authentication codes.
2. For certificate verification, TLS assumes that the signed party information includes only the exchanged handshake messages. Conversely, the information in SSL consists of a two round hash of the handshake messages (the master secret and the padding).
3. For key material generation, TLS employs a pseudorandom function to generate the materials using a master secret, a label in which the name of the key is specified, and a seed as initial inputs. On the contrary, SSL uses a complex scheme to generate the material.

9.3.2   Security limitations of SSL/TLS

Despite their wide use, SSL and TLS have some drawbacks that can be summarized as follows:

1. Transaction information is protected against interception attacks only while it is being transmitted. Therefore, sensitive information such as client's account information is available to the merchant. Hence, the clients need to trust the merchant and have to rely on the security of the merchant's Web server. If the merchant server is penetrated, a large number of user account details could be compromised.
2. SSL/TLS provides integrity protection for transferred data, but it offers no protection against modification of transaction information by corrupted merchants or clients.
3. SSL/TLS protocol uses the server certificate as the basis of server authentication. Nevertheless, there remain some risks of server masquerading. Man in the middle attacks can be introduced easily by using a sniffing application to intercept the communications between two entities during the initialization step. If an SSL/TLS connection is in use, the attacker can simply establish two secure connections, one with the client and the other with the server. Thereby, the attacker can read and modify the information sent

between the two parties and can convince client and server that they are communicating through a secure channel.

4. The client authentication scheme gives rise to a serious threat, since anyone who has access to the client's PC and knows the corresponding PIN or password to decrypt the private/secret keys may be able to perform a transaction on behalf of the client, especially when the merchant uses the client identity to access records containing client personal information (e.g., account details and address).

5. SSL simply provides a secure means of communication between clients and servers, but does not provide long-term evidence regarding transactions.

9.4 Transaction security with SET

Secure transaction systems are critical to the success and deployment of e-commerce. The design of SET has involved a large number of companies including IBM, Microsoft, Netscape, RSA, and Verisign. SET is not itself a payment system, instead, it is a set of security protocols and data structures (or formats) allowing users to employ the existing credit card payment infrastructures on the Internet, in a secure manner. The purpose of the SET protocol is to establish payment transactions that provide confidentiality of information, guarantee the integrity of data related to the payment instructions for goods and services, and authenticate both the cardholder and the merchant. SET provides three services: a secure communications channel among all parties involved in a transaction, trust by the use of X.509v3 digital certificates, and privacy since the sensitive information in a transaction is only available to parties when and where necessary. For this to be achieved, both cardholders and merchants should register with a certificate authority first, before they can be involved in an e-commerce transaction. Once registration is done, cardholder and merchant can start performing their transactions, which involve five basic steps in this protocol:

1. The customer browses the website and selects the goods to purchase. Then the customer sends the order and payment information, which includes two parts in one message: the purchase order (say part a) and the card information (say part b). While the former information part is for the merchant, the latter is for the merchant's bank only.

2. The merchant forwards part b to its bank to check with the issuer for payment authorization.

3. On receipt of the authorization from the issuer, the merchant's bank sends it to the merchant.

4. The merchant completes the order, sends confirmation to the customer and captures the transaction from his/her bank.

5. The issuer finally prints a credit card bill (or an invoice) to the customer.

9.4.1 Protocol overview

SET relies on cryptography and digital certificate to ensure message confidentiality and security. Message data is encrypted using a randomly generated key that is further encrypted using the recipient's public key. This is referred to as the "*digital envelope*" of the message and is sent to the recipient with the encrypted message. The recipient decrypts the digital envelope using a private key and then uses the symmetric key to unlock the original message. Digital certificates (or *credentials*) are digital documents attesting to the binding of a public key to an individual or entity (as described in Chapter 4). Both cardholders and merchants must register with a *certificate authority* before they can engage in transactions. The cardholder thereby obtains electronic credentials to prove that he is trustworthy. The merchant similarly registers and obtains digital credentials. These credentials do not contain sensitive details such as credit card numbers. Later, when the customer wants to make purchases, he/she and the merchant exchange their credentials. If both parties are satisfied then they can proceed with the transaction.

For a merchant to conduct transactions securely with millions of subscribers, each consumer would need a distinct key assigned by that merchant and transmitted over a separate secure channel. However, by using public key cryptography, that same merchant could create a public/private key pair and publish the public key; allowing any consumer to send a secure message to that merchant. This is why SET uses both methods in its encryption process. The secret key cryptography used in SET is the well-known Data Encryption Standard (DES), which is used by financial institutions to encrypt PINs.

9.4.2   SET process and security

The SET protocol utilizes cryptography to provide confidentiality of information, ensure payment integrity, and enable identity authentication. For authentication purposes, cardholders, merchants, and acquirers will be issued digital certificates by their supporting authorities. It also uses *dual signature*, which hides the customer's credit card information from merchants and also hides the order information from banks. The steps performed to protect the privacy process are listed below, see Figure 9.2:

1. **The customer opens an account** The customer obtains a credit card account, such as MasterCard or Visa, with a bank that supports electronic payment and SET.
2. **The customer receives a certificate** After a suitable verification of identity, the customer receives an X.509v3 digital certificate, which is signed by the bank. The certificate verifies the customer's RSA public key and its expiration date. It also establishes a relationship, guaranteed by the bank, between the customer's key pair and his/her credit card. A merchant who accepts a certain variety of cards must be in possession of two certificates for two public keys: one for signing messages and one for key exchange. The merchant also needs a copy of the payment gateway's public-key certificate.
3. **The customer places an order** This is a process that may involve the customer first browsing through the merchant's Web site to select items and determine their prices. The customer then sends the list of the items to be purchased from the merchant, who returns an order form containing the list of items, their individual prices, a total price, and an order number.
4. **The merchant is verified** In addition to the order form, the merchant sends a copy of his certificate, so that the customer can verify that he/she is dealing with a valid store.

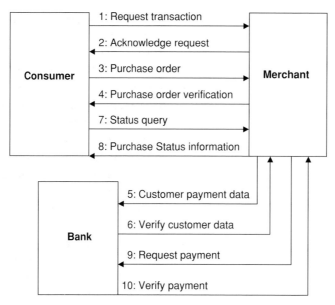

Figure 9.2   SET transaction model.

5. **The order and payment are sent** The customer sends both an order and payment
    information to the merchant, along with the customer's certificate. The order confirms
    the purchase of the items in the order form. The payment contains credit card details.
    The payment information is encrypted in such a way that it cannot be read by the
    merchant. The customer's certificate enables the merchant to verify the customer.
6. **The merchant requests payment authorization** The merchant sends the payment
    information to the payment gateway, requesting authorization that the customer's avail-
    able credit is sufficient for this purchase.
7. **The merchant confirms the order** The merchant sends confirmation of the order to
    the customer.
8. **The merchant provides the goods or service** The merchant ships the goods or pro-
    vides the service to the customer.
9. **The merchant requests payment** This request is sent to the payment gateway, which
    handles all of the payment processing.

9.4.3   Certificate operation

Two operations can be considered in this context: certificate issuance and registration of
actors.

(a) **Certificate Issuance** The cardholder certificates function as an electronic representation of the payment card. Because they are digitally signed by a financial institution,
they cannot be altered by a third party and can only be generated by a financial institution. A cardholder certificate does not contain the account number and expiration date.
Instead the account information and a secret value known only to the cardholder's software

are encoded using a hashing function. If the account number, expiration date, and secret value are known, the link to the certificate can be demonstrated, but the information cannot be derived by looking at the certificate. Within the SET protocol, the cardholder supplies the account information and the secret value to the payment gateway where the link is verified.

A certificate is only issued to the cardholder when its issuing financial institution approves it. By requesting a certificate, a cardholder declares the intent to perform commerce via electronic means. This certificate is transmitted to merchants with purchase requests and encrypted payment instructions. Upon receipt of the cardholder's certificate, a merchant can be assured that the account number has been validated by the card-issuing financial institution. Because they are digitally signed by the merchant's financial institution, merchant certificates cannot be altered by a third party and can only be generated by a financial institution. These certificates are approved by the acquiring financial institution and provide assurance that the merchant holds a valid agreement with an acquirer. A merchant must have at least one pair of certificates to participate in the SET environment.

An acquirer must have certificates in order to operate a certificate authority that can accept and process certificate requests directly from merchants over public and private networks. An issuer also must have certificates in order to operate a certificate authority that can accept and process certificate requests directly from cardholders over public and private networks. Issuers receive their certificates from the payment card division.

(b) Registration   As described in the previous part of this subsection, the cardholder and the merchant have to register with a CA before they can transact. And, the registration processes have to be secure enough, since they involve sensitive details.

Cardholder Registration As defined by SET, this process includes six messages to be exchanged between the cardholder and the issuer (or CA):

1. The cardholder initiates a request to the CA.
2. After the CA receives the initiating message from the cardholder, it replies by a message including the CA's public key-exchange key certification signed by root CA, CA's signature certificate, and the initial request encrypted using CA's private key.
3. The cardholder randomly generates a symmetric key $K1$, which is used to encrypt the request, and sends this along with a digital envelop including key $K1$ and its credit card number.
4. The CA determines the cardholder's issuing bank by the credit card number and returns the appropriate form, which is signed by the CA, along with CA's signature certificate.
5. The cardholder generates a public/private signature key pair, two symmetric keys $K2$, $K3$ and a random number $S1$. It creates a message with his filled registration form, public key, and $K2$, and its digital signature. This message is encrypted using $K3$ and sent with a digital envelop including $K3$ and card number.
6. The CA verifies the information and issues a secret value using a random number $S2$ it generates and $S1$. This secret value added to the account number and the expiration date are used in the computation of a hash function to generate a secret number. The CA signs the certificate, includes the secret number and the cardholder's public signature key and sends this certificate encrypted using $K2$ along with its signature certificate.

Merchant Registration The registration of the merchant is simpler than the cardholder. Four messages are needed. The first two messages are almost the same as those occurring with the cardholder registration, except that in the second message the registration form is sent. The merchant has to generate two public/private key pairs – one for signature and the other for key-exchange, instead of one pair as made with the cardholder registration.

Two problems can be observed with registration. First, the cardholder is not required to generate a fresh signature key pair, but may register an old one. This may represent a risk that the old one could be compromised. Second, an adversary working for a CA can give every cardholder the same secret value. This introduces some risk that a criminal can impersonate the cardholder. The first insecurity can be repaired in the cardholder's implementation. The second one can be fixed by the use of strong hash functions.

SET is safe since it addresses all the parties involved in typical credit card transactions: consumers, merchants, and the banks. It is difficult to deploy widely since it needs all the participants to have some part of the software, even very expensive hardware. It may be in the interests of the credit card companies and banks, but it looks quite different from the perspective of merchants and consumers.

In order to process SET transactions, the merchants have to spend a large amount of money in equipment and services when they already have what can be considered as sufficient security provisions (i.e., in SSL). Despite the fact that SET is a very comprehensive and complicated security protocol, it has to be simplified to be adopted by every party involved, otherwise, it might be abandoned.

SET protocol introduces an important innovation: the dual signature. The purpose of the dual signature is to link two messages that are intended for two different recipients. In this case, the customer wants to send the order information (say OI) to the merchant and the payment information (say PI) to the bank. The merchant doesn't need to know the customer's credit card number, and the bank doesn't need to know the details of the customer's order. The customer is afforded extra protection in terms of privacy by keeping these two items separate. However, the two items must be linked in a way that can be used to resolve disputes if necessary. The link is needed so that the customer can prove that a payment is intended for the correct order and not for some other goods or services.

The need for the link can be highlighted in the following way. Assume that the customer sends the merchant two messages: a signed OI and a signed PI and that the merchant passes the PI to the bank. If the merchant can capture another OI from this customer, the merchant could claim that the later OI goes with the PI. The linkage prevents this. The customer takes the hash (using SHA-1) of the PI and the hash of the OI. These two hashes are then concatenated and the hash of the result is taken. Finally, the customer encrypts the final hash with his or her private signature key, creating the dual signature.

Now suppose that the merchant is in possession of the dual signature (DS), the OI, and the message digest for the PI. The merchant also has the public key of the customer, taken from the customer's certificate. Then the merchant can compute the two quantities. If these two quantities are equal, the merchant has verified the signature. Similarly, if the bank is in possession of DS, PI, the message digest for OI, and the customer's public key, the bank can verify the signature. Summarizing this, we can list the steps as below:

1. The merchant has received OI and verified the signature.
2. The bank has received PI and verified the signature.
3. The customer has linked the OI and PI and can prove the linkage.

Acquirers can authenticate the merchant certificate request and issue a certificate that provides assurance that the merchant holds a valid agreement with the acquirer. Therefore, cardholders and payment gateways can authenticate merchants by verifying the signatures on the merchant's certificate and by validating the certificate's chain, if any. On the other hand, payment certificates are issued to provide assurance that a payment gateway has been authorized. Merchants can authenticate payment gateways by verifying the signatures on the payment gateway's certificate and validating the certificate's chain. Moreover, since SET allows the cardholder to use the payment gateway's public key for encrypting the symmetric key used to encrypt the payment instruction, the cardholder system needs the ability to authenticate the payment gateway. The merchant provides the cardholder with the payment gateway's encryption certificate. The cardholder can validate this certificate and then be assured that the payment's gateway is legitimate and that the payment information is kept confidential.

9.5 Securing electronic payment

The objective of an electronic payment system is to transfer a monetary value from the payer to the payee by using a payment protocol and a financial institution or network which links the exchanged data to some economic real world value. The financial network may be built of individual financial institutions (i.e., banks or authorized service providers).

9.5.1 Payment classification

Payment systems are classified based on the presence or absence of a direct communication between the payer and the payee. In the case of indirect communication the payment itself involves only the initiator of the payment and the issuer and acquirer. Indirect payment systems are usually considered as part of home banking in the context of digital payments on the Internet. Most current digital payment systems implement direct payments. Another classification is dependent on the time when the monetary value is actually taken from the payer. Two classes can be distinguished, pre-paid and post-paid systems. Pre-paid systems are commonly referred to as cash-like, or token-based. Post-paid systems are usually called account-based.

Direct cash-like payment systems use the token money model. In direct cash-like systems, the payer withdraws money from the issuer under the form of tokens and gives the payment tokens to the payee who in turn deposits them with its acquirer. Direct account-based systems are similar to the conventional cheque systems. They allow the payer to hand over a payment authorization to the payee, who presents the payment authorization to its acquirer. Then, the acquirer cashes it in from the issuer and the latter indicates to the payer the completion of the payment. In the indirect push payment model the payer instructs the issuer to transfer funds to the payee's account at the acquirer. This model operates like a traditional bank transfer model. The payee is just notified of the incoming payment.

Existing electronic payment systems can be broadly classified into two categories: macropayment systems and micropayment systems. Macropayment systems, the most widely used form of payment, are represented by credit cards and electronic money transfers. They are generally designed to support transactions involving medium–high valued payments. In contrast, micropayment systems are often characterized by low-value transactions. Examples of e-commerce micropayment systems include Millicent and PayWord (Fisher, 2002). Millicent is a secure lightweight protocol for electronic commerce. It uses a form of electronic commerce called *scrip*. PayWord is another micropayment scheme for low-value transactions. It works as follows. First the user sets up an account with a broker. Then he receives a certificate which must be renewed periodically. The certificate authorizes the user to generate a chain of hashes using a hash function. Each hash represents a payword and has the same value. The final chain is the concatenation of the repeated hashes. The certificate also insures that the broker will cash in the paywords.

Typical examples of macropayment systems include credit-card based systems, which provide payment through the existing credit card payment infrastructure. In credit-card transactions sensitive information is exchanged. The information being transmitted online is encrypted by using SSL or TLS. Credit-card payment over the Internet is designed for high cost purchases and includes high cost transaction processing. Because the value of macropayment transactions is high, these systems are characterized by a requirement for strong security measures. The theft of a user's credit card number could, for example, result in the loss of a significant amount of money by users and merchants. Additionally, macropayment systems typically induce high overheads in terms of computational resources, set up costs, and processing fees.

As opposed to macropayments, micropayments are lightweight payment schemes designed to support low valued transactions. With micropayments, users obtain digital cash from an authorized broker using conventional payment methods such as credit cards. The digital cash can be used to pay for services directly. Thus, micropayment schemes can operate as a hierarchy of issuers and provide for lightweight validation of cash without the need to have cash validated always by a central authority. As a result of their focus on low-value transactions, micropayment systems have lower security requirements than their macropayment counterparts and thus are characterized by lightweight encryption techniques. Informally speaking, encryption techniques need to be only strong enough to make the costs of breaking them greater than the potential monetary benefits. Lightweight encryption techniques are sufficient for micropayment-based payment systems. Such encryption schemes decrease the latency of a payment transaction and have significantly lower computational costs.

Five key phases can be identified in a commercial transaction:

1. *Getting means of payment* This phase entails using the appropriate means of paying for objects and obtaining digital cash in a given currency.
2. *Service discovery* During this step, the client discovers the available services and selects one or some of them based on a set of factors including price.

3. *Payment negotiation* When an e-service has been selected by a customer, the client can negotiate payment based on specific parameters such as payment means and authentication mechanism.
4. *Service utilization* During this phase, the customer utilizes the selected service, while making on-going payments.
5. *Termination* This phase includes the action performed after the utilization of service has ended. Actions involve reclaiming any unspent money or obtaining a proof of payment and service use.

It is important to note that for any transaction, not all of these stages can be applicable and the order may differ from one transaction to another. For example, client applications may do service discovery and price negotiation prior to obtaining the appropriate currency and use the service subsequently. Hence, these stages should simply be viewed as a set of tasks that must be supported in order to provide comprehensive support for a large spectrum of commercial transactions.

9.5.2   Anonymity

In order to make electronic payment systems largely accepted, a good level of anonymity has to be provided. Similar to physical cash, the privacy of the customers should be protected. Unfortunately, the anonymity of electronic money can be misused by criminals for illegal purchases. In order to make anonymous electronic cash systems acceptable to governments and banks, specific mechanisms for revoking anonymity under well-defined conditions have been introduced. Such revocation must be possible only for a trusted third party, named trustee. In fact, users, banks, and government may agree that some kind of anonymity control has to be provided. For example, it should be possible to trace the owner of a coin or trace the coins originating from a specific withdrawal. This will only be possible in cooperation with specific trustees, who will only reveal information if, for example, the bank obtained an approval by a judge. According to Davida *et al.* (1997), generally the following requirements are necessary for anonymity control of electronic money:

1. Electronic money should be anonymous and unrelated to legitimate users.
2. Anonymity should be revocable, but only by an authorized trustee and only when it is necessary.
3. The trustee should not have any power other than tracing. He can order the revocation of anonymity for certain transactions.
4. The bank should not be able to impersonate users (even in collaboration with third parties).
5. A trustee must be involved only when revocation is required. He should remain unconnected, otherwise.

Two types of anonymity control models have been discussed in the literature (Davida *et al.*, 1997): the owner tracing protocol and the coin tracing protocol. The owner tracing protocol reveals the identity of the owner of a specific coin. In this protocol, the bank gives to the trustee the information it received during the deposit protocol. The trustee then returns information which can be used to identify the owners. Owner tracing prevents illegal purchases and money laundering, as it allows the authorities to identify the involved users. A coin tracing protocol traces the coin that originated from a specific withdrawal. In this kind of protocol the bank gives to the trustee the information about a specific withdrawal. The trustee returns information that will appear when the coin is spent. Coin tracing allows the authorities to find the destination of suspicious withdrawals. Therefore, it can be used to identify the seller of illegal goods.

9.6   M-commerce and security

Recent advances in wireless technology have increased the number of mobile devices and have given speed to the rapid development of e-commerce activities using these devices. These advances have also contributed to the creation of new transaction processes that are made through, wireless communications and wired e-commerce technologies, which are called m-commerce systems. M-commerce enables a new mode of information exchange and purchases. From the customer's point of view, m-commerce represents convenience while it is associated with a potential increase of opportunities for the merchant.

Because of the characteristics of mobile devices and the nature of the wireless link, the emerging m-commerce operates in an environment that is different from e-commerce systems conducted over the wired environment. While e-commerce involves commercial transactions between two or more entities by moving at least some portion of the transaction to electronic setting, m-commerce goes further by moving some of the computational setting to mobile platforms. M-commerce can be defined as any type of economic activity that is considered (by law or by the business community) as e-commerce and that is performed by a mobile wireless terminal using at least one party (Veijalainen and Tsalgatidou, 2000).

An increased number of new m-commerce applications and services have started to be offered (e.g., location-based services). At the same time applications and services already offered by e-commerce are becoming available for mobile devices. Mobile applications, however, differ from standard e-commerce applications on the following major features:

1. *Ubiquity* This is an essential advantage of m-commerce. Customers can acquire any information whenever and wherever they want, regardless of their location.
2. *Reachability* Business entities are able to arrive at customers anywhere anytime. With a mobile terminal, a customer can however limit his/her reachability to particular individuals or at particular periods of time.
3. *Localization* The knowledge of the customer's location at a particular moment can add significant value to m-commerce since many location-based applications or services can be provided.
4. *Personalization* M-commerce applications that are accessible may need to be personalized in ways appropriate to users. This can reduce the amount of information stored in the mobile device.

5. *Dissemination* Information can be distributed to a large set of customers since the wireless infrastructure supports simultaneous delivery of data within a given geographic area.

As m-commerce extends the current e-commerce into a more direct and personalized environment, it renovates the business activity by providing opportunities of additional value added services. Several classes of mobile applications have been identified to represent a good opportunity, including mobile finance applications, mobile advertising, mobile inventory management, and location based applications. As wireless technology evolves, it is expected that more innovative m-commerce applications will occur (Varshney, 2001).

Nowadays, the wireless applications protocol (WAP) is playing an important role in the development of mobile applications since it bridges the gap between the mobile environment and the Internet and requires complex security mechanisms for m-commerce applications to be developed. WAP is an open and global standard for mobile solutions. It integrates Internet and GSM technologies so that contents of web servers can automatically be reformatted and moved to mobile devices. A component of WAP is the wireless markup language (WML), which represents the WAP's equivalent to HTML. WAP defines a micro browser that displays content pages in WML-format that can be transmitted to the mobile device using the WAP communication protocol. WAP addresses various issues including low bandwidth and the high latency of the wireless networks, as well as the resource constraints of the mobile devices.

9.6.1   M-commerce features

M-commerce is a complicated process that requires a series of operations. Transactions within m-commerce typically involve mobile customers, merchants, banks, mobile network operators, and other entities concerned with mobility and localization. Several issues differentiate m-commerce from e-commerce environments including the concepts of communication autonomy of the mobile terminals. This means that the mobile devices are not always reachable through the network and that it is natural that they can become disconnected. From the transactional point of view, this means that transactional mechanisms should not assume continuous capability of the terminals to communicate with, but should expect that there would be periods during which a mobile terminal cannot be located. This concern suggests that transactional mechanisms must be tightly linked to the security mechanisms and location capabilities.

Technical restrictions of mobile devices and wireless communication complicate the practical use of m-commerce. Mobile terminals demonstrate a greater extent of mobility and flexibility, but present limitations such as reduced display, small multifunction keypad. This slows down the development of user friendly interfaces and graphical applications for mobile devices. Mobile devices are also limited in computational power and storage capacity. These drawbacks do not allow a real support of complex applications and transactions.

There are still more challenges that face the rapid development of mobile commerce. These challenges include:

- *User trust* From the transactional point of view, each party participating in a transaction has to be able to authenticate the other entities involved in the transaction in order to be sure that the received messages are not modified or replayed by an intruder. Because the wireless links present additional vulnerabilities, users of m-commerce applications are more concerned with security. Mobile customers need to be sure that their financial information is secure and that the wireless transaction they are involved in is safe.

- *C-autonomy* Communication autonomy of the mobile devices means that the devices are not always reachable through the network and that they may remain disconnected. This can happen due to many motivations (e.g., user decision, coverage shortage, and power limitation). This means that transactional mechanisms in m-commerce should not assume continuous capabilities of the terminal to communicate. This also may require the need to design the transactional mechanisms closely with the security measures.

- *Hidden computing* To overcome the computational capacity and storage limitations, mobile devices may need to use pervasive computing and transparently interact with the infrastructure. This interaction can include downloading information and thus requires security and performance.

9.6.2   M-commerce transactions

Transactions in m-commerce environments are used to transmit messages and provide atomicity, consistency, isolation, and durability (ACID). Ensuring ACID or similar properties is hard to achieve in m-commerce due to the aforementioned limitations of the mobile terminals and frequent disconnections. A mobile transaction is atomic if either all the operations specified in the transaction are successfully executed or the executed operations are compensated (i.e., their effects deleted). Three types of atomicity can be defined for e-commerce: money atomicity, goods atomicity, and certified delivery (Tygar, 1996). They can be applied to mobile environments. Money atomicity states that the m-commerce protocols must guarantee that funds are transferred from one party to another without the possibility of the creation or destruction of money. Goods atomic protocols are money atomic and should be able to provide an exact transfer of goods for the money paid. Finally, a certified delivery protocol should be goods atomic and allow both the merchant and the customer to prove exactly which merchandise is received if the money is transferred. Ensuring atomicity in m-commerce is a complicated process. Mobile devices may be unreachable or may be unable to participate completely in the mobile transaction. In addition, atomicity needs to be jointly handled with security measures.

Consistency requires that all information extracted from an electronic transaction and stored should be consistent with historical data and the transaction content. A customer, for example, should not be able to withdraw money from an account characterized by a negative balance. Isolation of a mobile transaction ensures that the execution of all the actions specified in the transaction do not interfere with the operations occurring in other transactions being executed at the mobile devices or at the merchant server. To provide consistency and isolation, various measures need to be taken including protection against anti-replay attacks, enforcement of distributed constraints, identification of transactions and sub-transactions, adoption of a strong serialization process, installation of server-side security mechanisms, etc.

The durability property of a mobile transaction requires that once a mobile transaction has completed its execution, its results become permanent even on the occurrence of failures at the server level. This property can be mainly guaranteed by protecting the security of the database components where the transaction results are stored (Tsalgatidou *et al.*, 2000).

WAP transactions in cellular networks (e.g., GSM) are insecure unless additional security mechanisms are added. This is because both IP-based and GSM systems have some vulnerabilities. There are many ways of increasing security of WAP. One way to provide *end-to-end encryption* from the WAP client to the content server is to have two separate secure channels. The first channel is a wireless TLS channel from the WAP client to the WAP gateway and the second channel is a TLS channel from the WAP gateway to the content server. It is important to notice that this method does not imply end-to-end security. One problem raised by this approach is that the communicating parties, the WAP client and the content server, can not actually know whether there is end-to-end encryption. Moreover, the traffic flow between the WAP client and the content server is decrypted and encrypted at the WAP gateway of the operator allowing a malicious operator to listen secretly to and modify the data. Thus, data integrity and confidentiality properties are not fulfilled by the two secure channels approach.

Another method for the content provider to supply end-to-end security is to have specific WAP gateways and network access points, to which the mobile users can connect. Wireless TLS can be utilized between the WAP clients and the WAP gateway and TLS can be used between the WAP gateway and the content server. This approach reduces the problems related to trusting the operator. However, the main disadvantage of this method is that the users have to maintain a list of phone numbers for each content provider that they desire to access. The content providers must in turn accept the extra costs involved in maintaining a WAP gateway. Hence, the method is too complex to be used with a large set of applications. Although the approach satisfies the integrity and confidentiality properties, it does not ensure non-repudiation.

Current commercial products have developed different mechanisms to provide partly or totally ACID property. Typical solutions use cryptographic algorithms and address a variety of scenarios including the following (Veijalainen and Tsalgatidou, 2000):

1. The security mechanisms are integrated into the software and hardware of the terminal in such a way that the terminal has a credit card capability. This scenario is adopted by the solution developed by Nokia, Merita, and Visa.
2. A credit card reader is included with the handheld terminal. This scenario can be found included in the solution provided by Motorola and MasterCard.
3. A Smart trust type solution card, where the PKI private key is stored on the SIM card and is utilized for authentication and non-repudiation. This approach is adopted in the Sonera smart card solution.

Available commercial products present various weak points. In smart trust solutions, the usage of the private key is protected by the utilization of a PIN with a few digits (4 digits in general); thus, the real level of authentication and non repudiation depends on how the PINs are kept secret by the customer. For the first type of solution (as presented by Nokia), the access to the device is protected by some identification mechanism. The length and number of the PINs in this solution can be large and hardly memorized. The devices however are vulnerable to theft. The *Sonera smart trust solution* provided in the third case appears to be the safest scenario. The main advantage of this approach is that it can scale with the largest mobile phone market base. Disadvantages with the Sonera smart trust approach include the following: (a) it is a proprietary solution, (b) it does not interoperate with other wireless PKI solutions and (c) from a user point of view, the mobile phone can only be used within the application provider's space.

9.7   Conclusion

This chapter has presented the main concepts and foundations of e-commerce security systems. E-commerce security has become a great concern for enterprises and citizens who rely on the Internet, local networks and information systems in their daily operations. From the customer's point of view, the major goal of an e-commerce system is to enable him to trace and purchase a desired product whether it is tangible or intangible or a service over the Internet when he is interested to get it. Basically, its function is to provide a virtual store. From a merchant's point of view, the chief function of an e-commerce system is to produce higher revenues than using traditional methods. Clearly, it is important that any e-commerce system must recreate or utilize existing data and business processes and provide other processes to facilitate electronic purchase and provide product information, inventory systems, customer service, and transaction capabilities including credit authorization, tax computation, financial settlement, as well as delivery in a secure manner that protects the privacy and money of people involved in the process. This chapter sheds light on e-commerce security requirements and protocols used for e-commerce systems with special emphasis on SSL, TLS, and SET protocols. Moreover, the chapter reviewed security of e-payment systems and its various aspects. A detailed section was dedicated to m-commerce foundations, transactions, features, and its security.

References

Davida, G., Y. Frankel, Y. Tsiounis, and M. Yung (1997). Anonymity control in e-cash systems. In *Financial Cryptography '97*, Springer-Verlag, LNCS 1318, pp. 1–16.

Dierks, T. and C. Allen (1999). The TLS protocol version 1.0 – RFC 2246. IETF, Jan. 1999.

Fisher, M. (2003). Towards a generalized payment model for Internet service. Master thesis, Technical University of Vienna.

Freier, A. O., P. Karlton, and P. C. Kocher (1996). The SSL protocol version 3.0, Netscape 1996.

Guttman, R. H., A. G. Moukas, and P. Maes (1998). Agent-mediated electronic commerce: a survey. MIT Media Lab, June 1998.

Hassler, V. (2001). *Security Fundamentals for E-commerce*. Artech House Publishers.

Rescola, E. *SSL and TLS – Building and Designing Secure Systems*. Addison-Wesley.

Thomas, S. (2000). *SSL and TLS Essentials – Securing the Web*. Wiley.

Tygar, J. D. (1996). Atomicity in electronic commerce. In *Proceedings of the 15th PODC Conference*; 1996, ACM/IEEE, pp. 8–26.

Tsalgatidou, A., J. Verijaleinen, and E. Pitoura (2000). Challenges in mobile electronic commerce. In Proceedings of 3rd International Conference on Innovation through Electronic Commerce (IeC2000), Manchester, UK, Nov. 14–16, 2000.

Varshney, U. and R. Vetter (2001). A framework for the emerging mobile commerce applications. In *Proceedings of the 34th Hawaii International Conference on System Sciences*, 2001, IEEE Computer Society.

Veijalainen, J. and A. Tsalgatidou (2000). Electronic commerce transactions in a mobile computing environment, Proc. Int. Conf. on information society in the 21st century emerging technologies and new challenges, Nov 5–8, 2000, Japan.

Visa/Mastercard (1997). Secure Electronic Transactions Standard. Book 1: Business Description, 1997. Available at http://www.setco.org

Visa/Mastercard (1997). SET Secure Electronic Transaction Specification. Book 2: Programmer's Guide, May 1997.

Visa International Service Association (2002). 3-D Secure Protocol Specification: System overview version 1.0, July 2002.

10 Wireless LANs security

Security of wireless networks has become an important issue recently due to the increased dependence of individuals and organizations on these systems in their daily life. The goal of this chapter is to present the major trends and techniques in the security of wireless local area networks as well as to review the needs for securing access to such systems as any breach to such systems may entail loss of money, risk to the secrets of companies and organizations, as well as national security information. We will review the types of attacks on wireless networks. One section is dedicated to the review of services of any reliable security system that include confidentiality, non repudiation, authentication, access control, integrity, and availability. We will also shed some light on the chief aspects of the Wired Equivalent Privacy (WEP) Protocol and security aspects of mobile IP. The major weakness of the WEP protocol will be investigated. Then, we will review the features of the newly devised WPA protocols that proved to have superior security characteristics. Finally, we shed some light on Virtual Private Networks as related to wireless LAN security.

10.1 Introduction and rationale

The growth of Wireless Local Area Networks (WLANs) since the mid 1980s was triggered by the US Federal Communications Commission (FCC) determination to authorize the public use of the Industrial, Scientific and Medical (ISM) bands. This decision abolished the need for companies and end users to obtain FCC licenses in order to operate their wireless products. From that time on, there has been a substantial expansion and achievements in the area of WLANs. The rapid growth in the area of WLANs can be partly credited to the need to support mobile networked services and applications. There are many occupations these days that require employees to physically move while using a device, such as a hand-held PDA or laptop computer, which exchanges information with other user devices or a central node. Among such professions are rental car workers, healthcare workers, police officers, nurses, physicians, and military people. Fixed (wired) networks require a physical connection between the communicating nodes and parties, a fact that creates great complexity in the implementation of practical equipment. Hence, WLANs are considered the best choice for such applications (Nicopolitidis *et al.*, 2003; Papadimitriou *et al.*, 2004; Obaidat and Sadoun, 1999).

Moreover, WLANs do not need a complex infrastructure and expensive running costs. A wireless LAN does not need cabling infrastructure, significantly lowering its overall cost. In situations where cabling installation is expensive or impractical such as in historic buildings, or the battlefield, WLANs seems to be the only realistic method to employ for networking. In addition, the lack of cabling means a lot of saving on the cost of cabling and reduced installation time. Faulty cables in fixed networks are a major concern. Moisture which is responsible for erosion of the metallic conductors and accidental cable breaks can put down a wired network. Hence, using WLANs can reduce both the downtime of the network and eliminate the costs associated with cable replacement.

WLANs have many applications including LAN extension, cross-building interconnection, nomadic access and ad hoc networking. WLANs reduce installation costs since they use fewer cables than fixed (wired) LANs. A WLAN can be used to link devices that operate in the non-cabled area to the organization's wired network. Such an application is referred to as a LAN extension. Another area of WLAN application is nomadic access. It provides wireless connectivity between a portable terminal and a LAN hub. One example of such a connection is the case of an employee transferring data from his laptop computer to the server of his office upon returning from a trip or meeting. Another example of nomadic access is the case of a university campus, where students and working personnel access applications and information offered by the campus through their portable computers (Obaidat and Sadoun, 1999; Stallings, 2000; Stallings, 1999; Denning, 1983; Obaidat and Sadoun, 1997; Obaidat, 1997; Obaidat, 1993; Bleha and Obaidat, 1991; Bleha and Obaidat, 1993; Stallings, 2002).

WLANs are also used for ad-hoc networking applications such as a conference room or business meeting where the attendants use their laptop computers or PDAs in order to structure a temporary network so as to communicate information during the meeting. Other examples include battlefield and disaster site applications where the soldiers and the rescue workers can communicate information with each other. WLAN technology can also be used to interconnect wired LANs in nearby buildings.

High bit error rate is considered one major drawback of wireless medium transmission when compared to wired medium. High bit error rate is due to the nature of the wireless medium, which can be up to ten times that of wired LANs. The chief causes of the high BER are: (a) atmospheric noise, (b) physical barriers found in the path of the transmitted and received signal, (c) multipath propagation that is caused by propagation mechanisms such as reflection, diffraction, and scattering, which make signals travel over many different paths, (d) interference caused by nearby networks and systems, and (e) Doppler shift due to station mobility, which is caused when a signal transmitter and receiver are moving relative to each other. In such a case, the frequency of the received signal will not be the same as that of the source. When they are moving closer to each other, the frequency of the received signal will be higher than that of the source, and when they are moving away from each other the frequency will decrease. Inward interference is due to devices transmitting in the frequency spectrum used by the WLAN. Nevertheless, most WLANs nowadays employ spread spectrum modulation, which works over a wide bandwidth. Narrowband interference only influences part of the signal, hence causing just a few errors, or even no errors, to the spread spectrum signal. Conversely, wideband interference, such as that caused by microwave ovens operating in the 2.4 GHz band, can have devastating effects on radio transmission. The outward interference arises when WLAN signals disturb the operation of neighboring WLANs or radio devices, such as navigational systems of airplanes or

Figure 10.1 Hidden and exposed terminal scenarios: (a) "hidden" and (b) "exposed."

ships. Nonetheless, due to the fact that most WLANs employ spread spectrum technology, outward interference is usually considered unimportant (Nicopolitidis *et al.*, 2003; Obaidat and Sadoun, 1999).

Two related problems that are caused by the lack of fully connected topology between the WLAN nodes are the "hidden" and "exposed" terminal problems, depicted in Figure 10.1. The "hidden" terminal problem is due to the situation where a station A, not in the transmitting range of another station C, senses no carrier and starts a transmission. If C was in the middle of a transmission, the two stations' packets would smash together (collide) in all other stations (B) that can hear both A and C. The opposite of this problem is the "exposed" terminal problem. In this case, B reschedules transmission since it listens to the carrier of A. Nevertheless, the aim of B, C, is out of A's range. In this case B's transmission could be successfully received by C, however this does not occur since B reschedules due to A's transmission.

One major difference between WLANs and wired LANs is the fact that collision detection is difficult to implement in WLANs. The reason is that a WLAN node cannot listen to the wireless channel while sending, as its own transmission would flood out all other incoming signals. Clearly, using the protocols that employ collision detection is impractical in WLANs (Nicopolitidis *et al.*, 2003; Obaidat, 2006).

Power management is of great concern in WLANs as the laptop or PDA is usually powered by a battery that has a limited time of operation. Clearly, various mechanisms and protocols should be taken in order to save on the energy consumption of the mobile nodes without sacrificing the quality of service of the received signal.

It is worth mentioning here that when preparing for a WLAN's installation one must consider the factors that affect signal propagation. In a regular building or even a small office, this task is challenging. The signal propagates in all directions when an omni-directional antenna is used. However, walls, fixtures, trees, and even people can considerably change the propagation patterns of WLAN signals. Such problems are usually investigated prior to the installation of related WLAN equipment.

Security is considered a major challenge and concern in WLAN systems as radio signals may propagate outside the geographical area of the organization that owns the WLAN and all that an intruder needs to do is to approach the WLAN operating area to snoop on the information being exchanged between communicating parties. Yet, for this situation to occur, the impostor needs to possess the network's access code. One way to enhance security of a WLAN is to encrypt traffic; however, it increases the overhead and cost. WLANs can also be a victim of electronic sabotage. Most WLAN systems utilize CSMA-like protocols where all nodes have to stay quiet as long as they sense a transmission in progress. If a hacker establishes a node within the WLAN domain to continually send out packets, all other nodes are disallowed from transmitting, which brings down the network.

It is important to mention here that WLANs appear to be safer to human health than cellular phone systems as radio-based WLAN devices and components function at a power level that ranges between 50 to 100 milliwatts, which is significantly lower than the 600 milliwatt to 3 watt range of a traditional cellular phone. Newer versions of cell phones operate at 125 mw. In the case of infrared-based WLANs, danger to human safety is even lower and diffused infrared (IR) WLANs present no risk under any situation. WLANs have become widely used and very popular due to their many advantages and applications. However, security of such systems is of great concern as data sent over them can be easily broken and compromised. The security concerns in wireless networks are much more serious than in fixed networks as data sent on a wireless network is basically of a broadcast nature and everyone can hear it. Hence, serious countermeasures should be taken in order to protect the communicated data. In general wireless and wireless networks are subject to substantial security risks and issues. Among these are (Obaidat and Sadoun, 1999; Stallings, 2000): (a) unauthorized access by unwanted party, (b) threats to the physical security of the network, and (c) privacy.

10.2   Attacks on WLANs

The reliance of organizations and individuals on computer communication networks and information systems including wireless networks has increased enormously in recent years. The success of many organizations and firms depends heavily on the effective, good and secure operation of these systems. The total number of IT systems and networks installed in most companies has increased at an extraordinary rate. Organizations, companies and governments store important and confidential data on products, bank accounts, marketing, credit records, property and income tax, trade secrets, national security documents, and confidential military data, among others. Accessing such data and information by intruders or unauthorized users may cause loss of money or discharging secrets of trades or governments to opponents or adversaries (Obaidat, 1999).

Attacks on information and networks can be classified into two categories: (a) passive attacks and (b) active attacks (Obaidat and Sadoun, 1999; Stallings, 2000; Stallings, 1999; Denning, 1983; Obaidat, 1997a; Obaidat, 1997b; Obaidat, 1993). Passive attacks are naturally snooping on the transmitted signal. The intruder attempts to access the data being sent. Here, we can identify two subcategories: (a) traffic analysis and (b) release of message contents. In the latter class, the intruder gets to a file being transferred or an e-mail message. In a traffic analysis type of attack, the intruder is able to find out the place and ID of communicating hosts and watch the frequency and length of encrypted information being communicated. The collected information can be very useful to the intruder as when analyzed it can expose important information in deducing the kind of information being communicated (Obaidat, 1999; Stallings, 2000; Stallings, 1999; Denning, 1983; Obaidat, 1997a; Obaidat, 1997b; Obaidat, 1993).

In active attacks, the intruder alters data or creates fraudulent streams. Such kinds of attack can be classified into the following subcategories: (a) reply, (b) masquerade, (c) denial of service, and (d) modification of messages. A "reply" active attack entails the passive capture of a data unit and its succeeding retransmission in order to create an undesired access. A masquerade attack occurs when an intruder pretends to be someone else. For instance, authentication can be composed and replayed after a legitimate verification sequence has occurred. Denial of service happens when an authorized user is blocked from accessing the system by modifying the system's database by the intruder. Finally, modification of messages happens when some portions of a genuine message are altered or messages are deferred or recorded to produce an illegal result.

Most of the time, passive attacks are not easy to identify, yet, there are procedures that can be employed to prevent them. Alternatively, it is difficult to prevent active attacks. The chief categories of attacks on wireless networks can be summarized as below (Obaidat and Sadoun, 1999; Stallings, 2000; Stallings, 1999; Denning, 1983):

- Jamming In this case, the authorized traffic cannot reach customers or access points because unauthorized traffic overpowers the frequencies. The intruder can use particular devices to overflow the 2.4 GHz frequency. This type of denial of service can start from outside the service area of the access point, or from other wireless devices mounted in nearby networks.
- Interruption of service In this case, the resources of the system are damaged or become inaccessible.
- Fabrication Here, the intruder affects the authenticity of the network. The intruder puts in fake objects such as a record to a file.
- Modification This is basically an assault on the integrity of the network. The attacker not only gets access to the network, but also changes data by modifying database values, statements and parameters of a computer program.
- Interception This type of attack includes wiretapping or eavesdropping so as to collect data/information from the network's traffic. Eavesdropping is not difficult in wireless networks as the signal travels without cables and anyone equipped with the proper RF transceiver in the range of transmission can snoop on the data. Such devices are inexpensive. The sender or intended receiver may not notice that their messages have been snooped at. Furthermore, if there is no proper electromagnetic shielding, the traffic of a wireless system can be eavesdropped from outside the building where the network is operating. Clearly, there is a need to encrypt the traffic using efficient encryption schemes that can not be broken. Until recently wireless LANs have not had good security schemes as the WEP (Wired Equivalent Privacy) was broken many times. However, these days we have efficient security schemes such as the WPA (Wi-Fi Protected Access) and IEEE 802.11i protocols.
- Client-to-Client Attacks Wireless clients that run TCP/IP protocols such as file sharing are at risk of the same misconfiguration problem as fixed ones. Moreover, replication of IP or MAC addresses whether intentional or accidental may cause interruption of service.
- Attacks against Encryption Until recently, the IEEE 802.11b standard, which is often called the Wi-Fi Wireless LANs standard, has used an encryption technique that is called Wired Equivalent Privacy (WEP). The latter scheme has been broken by many individuals, which put the future of wireless LANs' widespread use at risk. However,

this security problem was resolved by introducing a new security scheme called Wi-Fi Protected Access (WPA) technique that has proved to be excellent. WPA is pretty much an earlier version of IEEE 802.11i that has excellent security characteristics.

- Misconfiguration In a wireless LAN environment, unless the network administrator configures each access point properly, these access points remain at the risk of being accessed by intruders or unauthorized individuals.

- Brute Force Attacks This is against the passwords of Access Points (APs). The overwhelming majority of access points use a single password or key that is shared by all connecting wireless clients. Clearly, intruders may attempt to compromise this password by guessing all possibilities. When the intruder guesses the password he can gain access to the access point and compromise the security of the entire system. Furthermore, if the password or the key is changed on a regular basis, the security of the system will be at great risk, especially when employees leave the company. Also, it is worth mentioning that managing a large number of access points and clients makes matters worse.

- Insertion Attacks In this case, a new wireless network is deployed without following security procedures. Moreover, it may be caused by the installation of an unauthorized device without appropriate security review. To give an example of this, a corporation may not be familiar that some of its employees have set up wireless facilities on the network. Such a crook access point will compromise the database of the organization. Therefore, it is necessary to implement a policy to secure the set up of all access points. Moreover, the network should be checked on a regular basis to make sure that there are no unauthorized devices connected to it.

The chief characteristics that should be maintained in any network security system are (Papadimitriou and Obaidat, 2006; Meddeb *et al.*, 2006; Obaidat and Sadoun, 1999; Stallings, 2000; Stallings, 1999; Denning, 1983; Obaidat, 1997a; Obaidat, 1997b; Obaidat, 1993; Bleha and Obaidat, 1991; Bleha and Obaidat, 1993; Stallings, 2002; Obaidat, 2006):

- Integrity Integrity means that procedures such as changeover, placing, or erasure of data can only be performed by authorized users. In this context, we identify three aspects of integrity: authoritative actions, defense of resources, and error control.

- Confidentiality This refers to the need to limit access to the network to authorized users. Other kinds of access can be a privileged one where viewing, printing, or even knowing the presence of an object is allowed.

- Denial of service In this case, a legitimate user should not be denied access to objects to which he has legitimate access. Such access concerns both service and data. Denning mentioned that the success of an access control is based on two concepts: (a) user identification and (b) protection of the access right of users (Denning, 1983).

In general, computer networks including wireless LANs have security problems due to the following reasons:

- Complexity Since these systems are complex, reliable and protected operation is a challenge. Additionally, computer networks may have different stations with different operating systems, which complicate matters further.

- Sharing Because network resources are common, more users have the possibility of accessing networked systems than a single station.
- Multiple point of attack While a file stays physically on a remote host, it may pass via many nodes in the system before arriving at the user.
- Anonymity An intruder is able to attack a LAN from hundreds of miles away and hence never needs to touch the network or even come into close contact with any of its users.
- Unknown path Generally, in computer networks paths taken in order to route a packet are rarely known ahead of time by the network user. Moreover, these users have no control on the routes taken by their own packets. Routes taken rely on several factors including load condition, traffic patterns, and cost.

10.3   Security services

The main categories of security services are described briefly below (Obaidat and Sadoun, 1999; Obaidat and Sadoun, 1997; Obaidat, 1997; Obaidat, 1993; Stallings, 2002; Obaidat, 2006):

- Authentication The service is meant to make certain that the message is from a genuine source. It guarantees that each corresponding individual is the entity that it claims to be. Moreover, this service should make sure that the link is not meddled with in a way that a third individual mimics one of the authorized individuals.
- Access control Access control should be correct and clever enough so that only authoritative parties can use the network. Moreover, this precision must not deny allowed parties from using the network.
- Confidentiality This service guards the data carried by the network from passive attacks. It also includes the guarding of a single message or a specific field of a message. Moreover, it includes protection of the traffic from intruders who try to analyze it. Clearly, there is a need to put in place some measures that avoid intruders from observing the frequency and duration of use, as well as other patterns of use and traffic characteristics.
- Integrity We should distinguish between connectionless and connection-oriented integrity services. The latter deals with a flow of messages and guarantees that the messages are sent correctly without replication, alteration, rearranging, or reply. The connectionless integrity service deals only with the safeguard against message modification. Any efficient security scheme should be able to identify any integrity difficulty and if a breach of integrity is detected, then the service must report this matter immediately. A software scheme or a human intervention method should resolve this issue. The former approach is supposed to fix the problem without human intervention.
- Availability Certain types of attacks may well result in failure or decline in the availability of the system. Automatic schemes can solve some of these problems while others need some kind of physical method.

- Non repudiation Here, this service averts the sending or receiving side from contesting the sent or received data/message. In other words, when a message is delivered, the sender can prove that the message was in fact received by the assumed recipient.

10.4   Wired equivalent privacy (WEP) protocol

The goal of WEP is to offer the level of confidentiality that is comparable to that of a wired LAN. In other words, Wired Equivalent Privacy (WEP) protocol was intended to give a similar level of privacy over wireless networks that one may get from a wired network. WEP scheme can be used to guard wireless networks from eavesdropping. Moreover, it is meant to prevent unlawful access to wireless networks. The protocol relies on a secret key that is shared between an access point and a wireless node. The key is used basically to encrypt data packets before sending them. Most 802.11 implementations use a single key that is shared among all mobile nodes and the access points. The IEEE 802.11 standard does not specify how the standard key is created. WEP depends on a default set of keys that are shared between access points and wireless LAN adapters (Papadimitriou and Obaidat, 2006 IEEE 802.11b).

Wireless LANs users want to have the system secure enough so that an intruder should not be able to: (a) reach the network using analogous wireless LAN equipment, and (b) gather wireless LAN traffic by eavesdropping or other schemes for additional analysis (Stallings, 2002). In IEEE 802.11 wireless LANs, access to a network's resources is not permitted for any user who does not know the present key. Eavesdropping is prohibited using the WEP mechanism. A pseudo random number generator is initialized by a shared secret key. The WEP scheme has the following chief characteristics: (a) it is reasonably robust as a brute-force attack on this scheme is not easy since every individual frame is sent with an initialization vector that reinitializes the Pseudo Random Number Generator for each frame, (b) it is *self-synchronizing* – since just like in any LAN, wireless LAN stations operate in connectionless conditions where packets may get lost. The WEP scheme re-synchronizes at each message (Checkpoint, Cisco, Rsasecurity, Sans, Netmotionwireless, (IEEE 802.11b)). Figure 10.2 shows an authenticated frame.

The WEP employs the RC4 encryption mechanism, which is often called the stream cipher. The RC4 is basically a stream cipher similar to the encryption scheme implemented in the Secure Socket Layer (SSL) to protect access to web sites. WEP scheme is part of the IEEE 802.11 standard and it describes how encryption should support the verification (authentication), integrity, and confidentiality of packets sent using wireless systems. The 802.11 committee has chosen then the RC4, a proven encryption scheme.

It is known that the open system authentication is the default authentication for the 802.11 standard. This approach relies on the default set of keys that are shared between the wireless access points and wireless devices. A client with the correct key only can correspond with any access point on the network. Data are encrypted before sending them, and an integrity check is performed to guarantee that the packets are not altered on their way to the destination. Only the client with the right key can decrypt the sent data avoiding unlawful users from accessing the information.

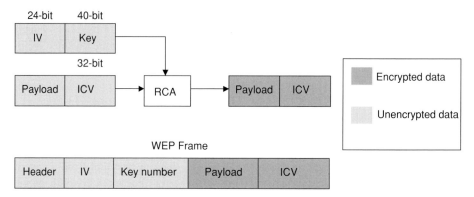

Figure 10.2   An authenticated frame.

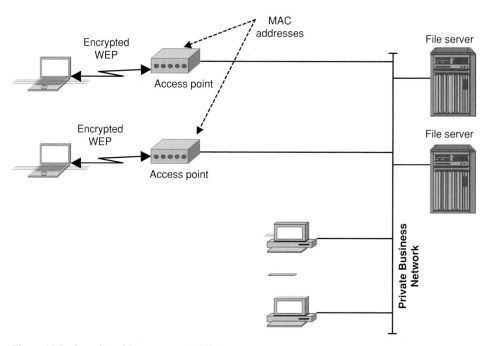

Figure 10.3   Security with access control list.

The access control list can offer a nominal level of security. Manufacturers of Wireless LANs provide security to these systems by using the access control list, which is based on the Ethernet MAC addresses of the clients. The list is made up of the MAC addresses of all of its clients. Only clients whose MAC addresses are found in the list can access the network. Figure 10.3 shows the WEP based security with the access control list (Walker, 2002; Checkpoint, Cisco, Rsasecurity, Sans, Netmotionwireless, IEEE 802.11b, ISAAC, Walker, 2001; Ukela, 1997).

The 802.11 standard specified two schemes for using the WEP. The first scheme offers a window of four keys and a station or an access point can decrypt packets enciphered with

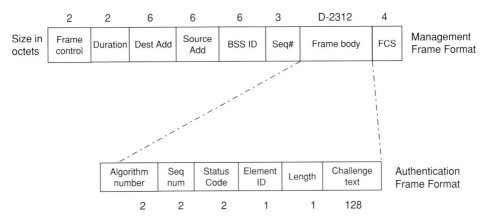

Figure 10.4 Authentication management frame.

any of the four keys. Transmission is restricted to any one of the four manually entered keys that are often called the default keys. The other method is named the key-mapping table where each unique MAC address can have its own distinct keys. The use of a separate key for each client alleviates the cryptographic attacks found by others. The drawback lies in the requirement to manually configure all of these keys on each device or access point.

The station that needs to authenticate in the shared key authentication scheme sends an authentication request management frame signifying that it desires to use the shared key authentication. The responder replies with the test text, which is the validation management frame to the originator. The Pseudo Random Number Generator PRNG with the common secret and the random initialization vector produces this test text. Once the originator receives the challenge management frame from the responder, it reproduces the contents of the challenge text into the new management frame body. The fresh management frame body is next encrypted by the common shared secret together with the new Initiating Vector (IV) chosen by the originator. Next, the frame is mailed to the responder, who decrypts the arrived frame and verifies that the Cyclic Redundancy Code (CRC) Integrity Check Value (ICV) is legitimate, and that the challenge text corresponds to the one that was sent in the first message. If there is matching, then the verification is successful and the originator and the responder switch roles and replicate the process to guarantee mutual authentication. Figure 10.4 illustrates the authentication management frame process. The element identifier recognizes if the challenge text is incorporated. The length field recognizes the length of the challenge text, which comprises a random challenge string (IEEE 802.11b, ISAAC, Walker).

10.5 Problems with the WEP protocol

Shortly after its deployment, WEP's cryptographic weaknesses started to be uncovered. Many autonomous studies conducted by various academic institutions and corporations found that even with WEP enabled, third parties can break WLAN system security. An intruder with the proper simple equipment can gather and analyze enough data to get back the common encryption key. Even though such security violations might take days on a small Wireless LAN where traffic is not heavy, it can be performed in hours on a busy commercial network. Regardless of its weaknesses, WEP offers some level of security when compared with no security at all and can be used by the casual home user for the sake of preventing eavesdroppers. For large and sophisticated users, WEP can be reinforced by setting it up in combination with other security schemes such as Virtual Private Networks (VPNs) or 802.1x authentication with dynamic WEP keys.

The Wired Equivalent Privacy (WEP) protocol has several problems that until recently affected the spread of wireless local area networks. The main problems are described briefly below.

10.5.1   Keystream reuse

WEP guarantees data confidentiality through RC4 (a stream cipher) scheme. The RC4 operates by turning the public IV and user secret key into a keystream, a long string of pseudorandom bits. Encryption happens by XORing the keystream with the plaintext. In order for decryption to occur, the same keystream must be generated (though the IV and secret key). The keystream is then XORed against the ciphertext to produce the original message (Nicopolitidis *et al.*, 2003; IEEE 802.11b). The basic problem with this process is that encrypting two messages using the same public IV and the same key (in other words creating the same keystream) is that information is exposed about both messages that make them easy targets for malicious attacks. Unfortunately, WEP's design fosters the occurrence of keystream reuse, specifically with the key and the IV (Nicopolitidis *et al.*, 2003; Stallings, 2000).

*   The Key

One of the most underlying problems of the WEP protocol directly results from key usage within a network. The key used in a network usually stays unchanged. Clearly, this is a problem because intruders have more opportunity to find out the secret key of a user over the time span in which the key is being used. Given that the key is seldom changed, the IV is typically the cause of keystream reuse. Therefore, the key offers little help to the IV in averting the creation of keystream reuse (Nicopolitidis *et al.*, 2003; Obaidat, 2006).

In WEP, the key is shared among all network users. WEP obliges that users of a specific network must have the same key. Security is very much weakened due to the fact that exposing the key increases with the increase of the number of users; the more people who know the "secret" increase the chances that the secret will be revealed. Instead of monitoring the activity of one user to uncover the key of a network, an intruder can monitor several users to uncover the same key (Netmotionwireless).

Public places such as airport waiting rooms, libraries, hotels, and cafés are especially vulnerable. Users may be deceived into believing that physical separation from their work environment thwarts any corporate network attacks through their mobile station. Unfortunately, public portals provide hackers an ideal setting from which to intrude to the network. A user connecting to his corporate network from a public network enables intruders to "sniff" the network and access all packets sent in plaintext on the wireless LAN. Attackers have the ability to gain access to the corporate network through this mechanism without having to deal with firewalls or other security measures normally installed in a physical network environment (Obaidat, 2006).

Some network administrators, in an effort to minimize the risk of shared knowledge of the key, decide to keep the secret from the end users and configure the user's machine themselves instead. However, the secret key is still configured within the user's computer. Administrators can also configure their network systems to ban usage in public wireless LAN locations; nevertheless, this is done at the expense of mobility and flexibility on the user's part. While changing the key will certainly enhance security, this should be done manually. Multiple users share a common key so replacing "compromised key material" can be a laborious and difficult task. Changing a key forces all users to rearrange their wireless network drivers. Clearly, this is not viable and infeasible for large corporations. Such changes are unusual and are predicted to continue to be so in the future. Unfortunately, these infrequent changes (if any) give intruders added time needed to analyze wireless traffic (Obaidat, 2006).

- • The Initializing Vector (IV)

The architectural defects of the Initializing Vector (IV) increase the risk of attacks. The WEP uses a 24-bit IV in sending the plaintext portion of a message, roughly ensuring the reuse of an IV for transmitting data. A busy Access Point (AP) will use up its IV within a quarter of a day (Obaidat, 2006; Nicopolitidis *et al.*, 2003).

As stated earlier, duplicates of IVs can result from its architectural flaws. Moreover, because the secret key generally remains unchanged, the reuse of IVs will definitively occur. This is considered a problem since duplicates of IVs are detectable by intruders since IVs are public. Although the WEP standard recommends the changing of IVs after every packet, there is no given information on how to select IVs. Numerous schemes of attack are possible once an IV is discovered (Obaidat, 2006; Nicopolitidis *et al.*, 2003; Ukela, 1997).

Reuse of IV gives an attacker the ability to decrypt the encrypted data without knowing the encryption key. Two packets encrypted with the same IV can be easily decrypted. The following algebraic manipulation show how this attack may happen:

$$C_1 = P_1 \otimes RC4(v, k)$$
$$C_2 = P_2 \otimes RC4(v, k)$$
$$C_1 \otimes C_2 = P_1 \otimes RC4(v, k) \otimes P_2 \otimes RC4(v, k)$$
$$C_1 \otimes C_2 = P_1 \otimes P_2$$

In the first two equations, two ciphertexts, C1 and C2 are calculated by bitwise XORing the plaintexts P1 and P2 and the keystream RC4 (v, k) where v is the IV and k is the secret key. As the above manipulation illustrates, two ciphertexts that use the same IV annul the keysteam, resulting in the XOR of the plaintexts (Obaidat, 2006; Nicopolitidis *et al.*, 2003). The addition of two plaintexts is equal to the addition of two ciphertexts generated by the same key.

10.5.2 *Message authentication*

In order to ensure that data does not get modified during transit, the WEP protocol makes an integrity checksum using the CRC-32 algorithm. However, the checksum is insufficient to prevent an intruder from changing a message. As apparent by the several attacks that can occur by the CRC-32 checksum failings, access control of the wireless LAN networks is insecure.

A number of flaws have been found in WEP that seriously weaken the security of wireless networks. WEP is vulnerable against the following main attacks: (a) active attacks that insert new traffic from unlawful mobile nodes, (b) active attacks that decrypt traffic based on tricking the access point, (c) passive attacks that decrypt traffic based on statistical analysis, and (d) dictionary-building attacks that permit real-time automated decryption of traffic after some analysis (Obaidat, 2006; Nicopolitidis *et al.*, 2003).

The active attack that injects traffic occurs due to the situation where an intruder knows the correct plaintext for one encrypted message. Through this info, the intruder can form correct encrypted packets. This entails forming a new message, computing the CRC-32, and conducting bit flips on the real encrypted message. Then the packet can be transmitted to the access point or to a mobile node and accepted as a legitimate packet.

One active attack is based on traffic decryption by fooling access points. In this case, the intruder makes a speculation about the header of the packet; not the packet's substance. Fundamentally, all that is required to guess is the target IP address. The intruder then can change the values of specific bits in a way to alter the target IP address to send the packet to a node in his control, and to send it using a crook mobile station. The packet can be decrypted using the access point and sent unencrypted via routers to the attacker's node, exposing the plaintext. It is worth mentioning here that it is likely to alter the target port on the packet to be port 80, which permits the packet to be forwarded through most firewalls.

Passive intrusion is based on traffic decryption where an eavesdropper can capture all wireless traffic pending an IV collision. The intruder can get the XOR of two plaintext messages by XORing two packets that employ the same Initialization Vector (IV). Keep in mind that IP traffic is frequently expected and has redundancy, which can be utilized to remove many possibilities for the contents of messages. Statistical analysis can be employed to obtain good guesses about the contents of one or both of the messages. If the intruder is able to perceive the entire plaintext for one message, then it is feasible to find out the plaintexts of all other messages that have the same IV. One possible example of this intrusion happens when the intruder employs a host on the Internet to transmit traffic from outside to a host on the wireless network facilities. The intruder can know the contents of such traffic; therefore, the plaintext will be recognized. If the intruder captures the encrypted version of the message sent over an IEEE 802.11 WLAN, then he will be able to decrypt packets that employ the same IV.

One type of intrusion is called "the table-based attack," where the small space of possible initialization vectors permits an intruder to construct an encryption table. In such a case, when the plaintext for the packet is recognized, the intruder can associate the RC4 key stream generated by the IV, which can be employed to decrypt all other packets that use the same Initialization Vector (IV). Obviously, the intruder over time can construct a table of inclusion vectors and the matching key streams. As soon as such a table is constructed, the intruder will be able to decrypt all packets transmitted over that wireless link. These attacks can be realized by low-cost equipment. Thus, it is suggested not to depend totally on WEP and consider using additional security techniques (Nicopolitidis, 2003; ISAAC).

Despite the fact that it is not easy to decode a 2.4 GHz digital signal, off-the-shelf hardware devices that can observe wireless network signals can easily be obtained by intruders. Off the shelf IEEE 802.11 wireless devices are now available with programmable firmware, which can be reverse-engineered so as to insert traffic. Intruders can deliver this firmware and put it up for sale for interested parties including competitors and enemies at high cost.

10.6   Wi-Fi protected access (WPA)

Wi-Fi Protected Access (WPA) has been developed by the Wi-Fi Alliance in conformity with the Institute of Electrical and Electronics Engineers (IEEE) and released in May, 2003. Basically, WPA improves significantly the security robustness versus the previous WEP standard. By significantly increasing the protection available for wireless LANs, WPA permits enterprises to move forward immediately and begin benefiting from the enhanced productivity, convenience, and cost savings of wireless networking and mobile computing. Due to the fact that WPA is a subset of the next-generation 802.11i specification, it presents a comfortable, immediate, and standards-based answer to wireless local area networking security. Moreover, WPA is both forward- and backward-compatible and is intended to be installed as a software upgrade on both wireless access points (APs) and clients.

Wi-Fi Protected Access (WPA) had several design goals: (a) to be robust, and an interoperable security substitute for WEP, (b) to be software upgradeable to already existing Wi-Fi certified products, (c) to be appropriate for both home and business settings, and (d) to be obtainable right away.

In order to reach these objectives, two main security improvements needed to be completed. WPA was created to offer an enhanced data encryption, which was vulnerable in WEP and user authentication, which was basically missing in WEP.

In order to enhance data encryption, WPA exploits its Temporal Key Integrity Protocol (TKIP), which offers important data encryption enhancements such as the per-packet key mixing function, message integrity check (MIC), an extended initialization vector (IV) with sequencing rules, and a re-keying scheme. By using these improvements, TKIP answers all WEP's known weaknesses.

WEP has more or less no user authentication scheme. In order to reinforce user authentication, WPA implements 802.1x and the Extensible Authentication Protocol (EAP). Collectively, these enhancements offer a framework for better user authentication. Moreover, this framework employs a central authentication server, such as RADIUS (Remote Authentication Dial In User Service), to verify each user on the network before he uses it, and also utilizes "mutual authentication" in order not to have the system's user inadvertently join a rogue network that may take his network identification info.

WPA uses Temporal Key Encryption Protocol (TKIP) for encryption, creating a key in a different way from WEP. WPA will use the IEEE 802.1X protocol for access control.

TKIP is designed to minimize the problem of the encryption key reuse. It makes encryption more complex as well as improves intrusion detection. The TKIP uses a 128-bit temporal key that is shared among clients and Access Points. TKIP combines the temporal key with the user's MAC address. A large 16-octet initialization vector (IV) is then added to yield a key that will be used to encrypt the data. Through this method, temporal keys are changed about every 10 000 packets, adding huge security to the network. An integrity check is also done (Stallings, 2002).

In order to initiate encryption, WPA uses a key, which is a sequence of numbers entered by the user. In WEP, wireless clients and access points (APs) are manually configured with the same 40-bit key. Keys are static as well as quite short, and distributed manually. This lack of automated key management contributes to the problem of having keys with never-ending life duration. Furthermore, the encryption scheme exposes part of the key stream in the initialization vector (IV), which is intrinsically flawed.

Such limitations made wireless networks prone to passive confidentiality breaches, where an intruder could read data on the network. Moreover, active integrity breach attacks, where intruders insert traffic in the system, affect negatively the data integrity. Also, these days inexpensive tools, which enable intruders to sniff the network's traffic and then recover the WEP encryption keys, are easily available. WEP does not have the means to identify the source of each packet.

Such inherent weaknesses of the original 802.11 specification make it easy to intruders to deploy "rogue" wireless LAN access points (APs) that can expose corporate resources to attacks. It is possible to solve such problems using more than one method including deployment of the newly introduced Wi-Fi Protected Access for within the enterprise and using VPN for public network (remote) access. Other possibilities are using a proven Virtual Private Network (VPN) solution with WEP or combining standard WEP functionality with IEEE 802.1X with or without a VPN. However, we believe that the first scheme, using WPA, is more effective and convenient.

WPA integrates several technologies to tackle all known 802.11 security flaws. It offers enterprises an inexpensive and scalable resolution to protect corporate WLANs without the need to have the expensive VPN/firewall technology. WPA gives an efficient user-based authentication using the 802.1X standard and the Extensible Authentication Protocol (EAP). Moreover, it offers a robust encryption scheme by using 128-bit encryption keys and the Temporal Key Integrity Protocol (TKIP). The message integrity check (MIC) prevents intruders from gathering and modifying or falsifying data packets.

Such a combination defends the secrecy and integrity of WLAN transmissions as well as ensuring that only legitimate users gain access to the network. Furthermore, it significantly improves the security and manageability through the automatic key distribution, exclusive master keys for each user and each session, and exclusive, per-packet encryption keys.

TKIP was designed as a software improvement to existing WEP-based 802.11 devices, and it tackles all the weaknesses of the static keys used in WEP. Here, encryption keys are 128 bits, which make them more robust. Moreover, TKIP employs per user, per session per packet keys by cryptographically hashing the preceding packet initialization vectors of data packets with the session key. This key can then be employed by existing RC4 WEP hardware engines in order to obtain robust security characteristics without the need to add new hardware. The blend of 802.1X authentication, authentication protocols, dynamic keys, and TKIP improvements allows enterprises to realize wireless LANs enhanced data security and integrity.

In order to deploy an enterprise WPA, we will need:

- An authentication server such as a RADIUS server exploring an efficient EAP type like the TLS, TTLS, or PEAP.
- A pre-shared key (PSK) in order to offer an inexpensive option for small network environments such as home wireless LANs.
- WPA- enabled access points.
- WPA-enabled laptop, PDA, or any other wireless device.

It is worth mentioning that some access points (APs) may contain a software-based mechanism such as a switch that permits mixed WEP or WPA-only stations. This allows network managers to mix WPA-enabled devices with non-upgraded 802.11 devices as they move to an all-WPA system. Clearly, devices that are not upgraded stay as an easy target to attacks.

WPA offers an instant and inexpensive WLAN security answer with robust authentication and encryption. Moreover, it offers an easy replacement path from 802.11 with WEP to 802.1X to Wi-Fi Protected Access with TKIP, and eventually to 802.11i security answer.

10.7   Mobile IP

Current internet protocol versions do not support host mobility. These were designed such that moving hosts were not considered: a node's point of attachment to the network remains unchanged at all times, and an IP address identifies a particular network. To support a mobile host with current methods, reconfiguration is necessary any time a mobile host moves. This is an unacceptable solution as it is time consuming and error prone, thus, the rise of Mobile IP.

Mobile IP is a standard suggested by a working group within the Internet Engineering Task Force (IETF). It was meant to allow a mobile node to use two IP addresses: a fixed home address and a care-of address that is changed at each new point of access. In Mobile IP, the *home address* is fixed (static) and is used, for example, to recognize TCP links. The *care-of address* is changed at each new point of connection. It signifies the network number and hence recognizes the mobile node's point of attachment with respect to the network configuration. The home address makes it seem that the mobile node is constantly able to receive data on its *home network*, where Mobile IP needs the existence of a network node called the *home agent*. Every time the mobile node is not connected to its home network the home agent receives all the packets sent to the mobile node and makes the necessary procedures to hand them over to the mobile node's present point of attachment (Nicopolitidis *et al.*, 2003).

Home agent in Mobile IP environment forwards packets from the home network to the care-of address by creating a new IP header, which holds the mobile node's care-of address as the destination IP address. Then the newly established header encapsulates the original packet, making the mobile node's home address to have no effect on the encapsulated packet's routing until it reaches the care-of address. This kind of encapsulation is also called tunneling. It is clear to note that the key to Mobile IP is the cooperation of three separable procedures: (a) discovering the care-of address, (b) registering the care-of address, and (c) tunneling to the care-of address.

The home and foreign agents usually broadcast agent announcements at ordered periods. If a mobile node needs to obtain a care-of address and does not desire to wait for the periodic announcements, then the mobile node can broadcast or multicast a solicitation that can be responded to by any foreign or home agent that gets it. It is worth mentioning that home agents employ agent announcements to let them be known, even if they do not provide any care-of addresses. Nevertheless, it is not easy to relate preferences to the different care-of addresses in the router announcement. An agent advertisement makes the following functions: (a) permits the detection of mobility agents, (b) records one or more available care-of addresses, (c) notifies the mobile node about specific features provided by foreign agents, such as substitute encapsulation techniques, (d) allows mobile nodes to find out the network number and mode of their link to the Internet, and (e) permits the mobile node to be acquainted.

Mobile nodes employ router solicitations to find out any change in the position of mobility agents found at the present point of connection. If announcements are not any more evident from a foreign agent that beforehand had given a care-of address to the mobile node, then the mobile node must assume that foreign agent is not any more within range of the mobile node's network interface. In such a case, the mobile node must start hunting for a new care-of address, or may use a care-of address recognized from announcements that it is still in receipt of. The node may decide to wait for another announcement if it does not get any newly advertised care-of addresses, or it may transmit an agent solicitation.

Mobile IP is an Internet protocol designed with the main objective of supporting host mobility. It is meant to enable a host to keep on connected to the Internet regardless of its geographical location. Mobile IP can track a mobile node without the need to modify its permanent IP address. Moreover, Mobile IP enables a node to maintain Internet connection during its movement from one Internet access point to another. Network access is guaranteed always and from all locations. Home and local resources would be accessible all the time and e-mail would by no means be missed. There would not be any excuse for being short of productivity because of the absence connectivity.

It is important to state here that the term mobile means that the user is connected to one or more applications across the Internet and the access point changes continuously. Obviously, this is dissimilar from when a traveler employs his Internet Service Provider (ISP) account to be hooked to the Internet from different sites during his movement or trip (Papadimitriou et al., 2004). Mobile IP is the amendment to the normal IP so that it permits the client to transmit and receive the datagrams regardless of where it is connected to the network. The prime security concern in using this scheme is the redirection attacks. The latter happens when a cruel client gives bogus information to the home agent in the mobile IP network. In such a case, the home agent is advised that the client has a new care-of address. Therefore, all IP datagrams sent to the genuine client are forwarded to the malicious client.

The main features associated with Mobile IP (Obaidat, 2006; Nicopolitidis et al., 2003) are listed below:

1. No geographical restrictions This implies that a user can take a laptop computer or a PDA anywhere without missing the connection to the home network.
2. No modification on the present IP address and IP address format The current IP address and address format stays unchanged.
3. No need for a physical connection This means that Mobile IP finds local IP routers and links automatically.
4. No need for modifications to other routers and hosts Except for the mobile nodes/ routers, the rest of the routers and hosts will use present IP. It leaves transport and higher protocols unchanged.
5. Supports security Verification is done to make sure that the rights of authorized users are not affected.

Mobile IP was intended to withstand two types of attacks: (a) a nasty agent that may answer old registration messages and disconnect the node from its network, and (b) a node that may act as if it is a foreign agent and transmit a registration demand to a home agent to redirect traffic that is meant for a mobile node to itself. Message verification and correct use of the ID field of the registration request and reply messages are usually employed to shield mobile IP from such attacks (Arbaugh, 2000; Obaidat, 2003). Every registration request and reply has an authentication extension that includes the following fields:

- Type This is an 8-bit field, which specifies the type of verification extension.
- Length An 8-bit field, which specifies the number of bytes in the authenticator.
- Security Parameter Index A 4-byte field used to name security context among a pair of nodes. The design of the security framework is made in a way so that the two nodes share the same secret key and parameters pertinent to the authentication mechanism.
- Authenticator A code that is added to the message by the sender employing a shared secret key. The recipient uses the matching code to ensure that the message has not been altered. The default authentication technique is the keyed-Message Digest-5 (MD-5), which generates a 128-bit message digest. It is worth mentioning here that MD5 was designed in 1994 as a one-way hash algorithm that uses any length of data and generates a 128-bit "fingerprint" called often message digest.

The chief services that are supported in Mobile IP are explained briefly below:

- Encapsulation This is the procedure of including an IP datagram within another IP header that holds the care-of address of the mobile node. The IP datagram is left unchanged during the enclosing process.
- Decapsulation The mechanism of removing the outmost IP header of the received packets in order to have proper delivery of the enclosed datagram. Clearly, this is the reverse of the encapsulation operation.
- Agent Discovery Home and foreign agents send their accessibility on each link to where they can provide service. A recently arrived mobile node can transmit an invitation on the link to find out if any potential agents are in attendance.
- Registration When the mobile node is on the road, it enrolls its care-of address with its home agent so that the home agent recognizes where to relay its packets. Based on

the network design, the mobile node could either enroll directly with its home agent, or indirectly by the aid of the foreign agent.

It is important to know what needs to be done in Mobile IP in systems that have IPv6. IPv6 incorporates several characteristics for streamlining mobility support that are not found in IPv4, such as Stateless Address Auto configuration and Neighbor Discovery. Moreover, IPv6 tries to considerably simplify the procedure of renumbering that could be significant to the future routing of the Internet. Due to the fact that the number of mobile nodes hooked to the Internet will likely increase significantly, effective support for mobility will be a vital step in the future of the Internet. Clearly, the need to pay attention to supporting mobility has become a crucial issue.

Mobility Support in IPv6, as devised by the Mobile IP working group, goes along the design for Mobile IPv4. It holds the views of a home network, home agent, and the utilization of encapsulation to carry packets from the home network to the mobile node's present point of connection. Whilst detection of a care-of address remains a requirement, a mobile node can arrange its care-of address by means of Stateless Address Autoconfiguration and Neighbor Discovery. Hence, foreign agents do not need to support mobility in IPv6.

IPv6 mobility uses a lot the idea of route optimization specified for IPv4 especially that of delivering binding updates exactly to correspondent nodes. When it becomes familiar with the mobile node's current care-of address, a correspondent node can hand over packets directly to the mobile node's home address without the need of assistance from the home agent. Optimization of routes is expected to considerably enhance the performance of IPv6 mobile nodes. It is reasonable to necessitate this additional functionality of all IPv6 nodes since processing binding updates can be realized as a minor change to IPv6's use of the destination cache.

On the contrary to the manner in which route optimization is specified in IPv4, in IPv6 correspondent nodes do not tunnel packets to mobile nodes; they use IPv6 routing headers that realize a variant of IPv4's *source routing* alternative. Various early proposals to support mobility in IPv4 devised a similar use of source routing alternatives; however, there are two major obstacles prohibiting their use. These are: (1) IPv4 source routing alternatives need the destination of source-routed packets to follow the reversed path to the source back along the designated midway nodes. In other words, malicious nodes using source routes from distant sites inside the Internet could mimic other nodes, an issue worsened by the absence of verification schemes; (2) Current routers show poor performance when treating source routes. The outcomes of installing other protocols that employ source routes have not been constructive.

Nevertheless, the objections to employ source routes do not relate to IPv6 because of its thorough specification that removes the need for source-route reversal and allows routers to disregard alternatives that do not require their awareness. Therefore, correspondent nodes can utilize routing headers at no risk. Clearly, this permits the mobile node to simply find out when a correspondent node does not have the right care-of address. These packets that are handed over by encapsulation as an alternative to source routes in a routing header should have been transmitted by correspondent nodes, which need to get required information from the mobile node. Among the other major options supported by IPv6 mobility are: (a) automatic home agent detection, (b) soft handoffs, which in Mobile IPv4 is described for foreign agents as part of route optimization, (c) presence with Internet way in filtering, and (d) renumbering of home networks (Obaidat, 2006; Nicopolitidis *et al.*, 2003).

10.8 Virtual private network (VPN)

A VPN is defined as a communication environment in which access is managed so as to permit peer connections only within a defined community and is designed using some form of partitioning of a shared basic communications medium. The latter communication environment offers services to the network on a non-discriminatory basis. Ferguson and Huston defined a VPN as a private network constructed within a public network infrastructure, such as the global Internet (Ferguson, 2000; Obaidat, 2006; Papadimitriou *et al.*, 2004; Rosenbaum *et al.*, 2003). Others define a virtual private network (VPN) as a network that permits two or more private networks to be linked over a publicly accessed network. It is comparable to a wide area network (WAN) or a strongly encrypted tunnel. The main characteristic of VPNs is that they use public network infrastructures such as the global Internet instead of expensive, private leased lines while having similar security and encryption characteristics as a private network.

VPNs accomplish their task by permitting the user to tunnel all the way through the wireless network or other public network in a way that the tunnel participants get advantage from at least the same level of confidentiality and features as when they are connected to a privileged wired network.

VPN resolution for wireless access is currently an inexpensive alternative to WEP. VPNs are already being widely deployed to offer remote employees with secure access to the organization's network via Internet. For the remote user application, VPNs offer secure, devoted links often called "tunnels" over an insecure network.

Virtual Private Networks (VPNs) have grown as a tradeoff for enterprises wishing the ease and cost-effectiveness provided by shared networks, but necessitating the good security provided by private networks. While closed WANs utilize separation to guarantee that data is secure, VPNs use an arrangement of encryption, authentication, access management, and tunneling in order to offer access only to allowed groups and individuals and to shield data while in transit (Papadimitriou and Obaidat, 2006). In order to imitate a point-to-point link, data is wrapped with a header that offers the needed routing info permitting it to travel the public internetworks to arrive at the destination. In order to emulate a private link, data being transmitted are encrypted for secrecy. Packets that are interrupted on the public network are not decipherable with no encryption keys (Nicopolitidis *et al.*, 2003; Hunt and Rodgers, 2003; Strayer, Brahim, Arora *et al.*, 2001).

Typical private networks implement connectivity between network entities using a set of links composed of devoted links. These are typically leased from telephone service providers as well as specially installed wiring systems. The capability of these links is obtainable continuously despite it being predetermined and nonflexible. Traffic on these networks is owned only by the organization deploying the network. This means that there is a guaranteed degree of performance related to the network. However, this comes with a cost. The main negative aspects of this scheme (Arora *et al.*, 2001) have to do with money and the time needed to install and maintain the dedicated links. Furthermore, the management of private networks lies totally on the organization, thus the investment is not easy to justify for small or middle sized corporations.

Clearly, a VPN can decrease expenses by substituting multiple communication links and legacy equipment by a single link and one device for each location (Nicopolitidis, 2003; Papadimitriou, 2004). To expand the reach of a company's Intranet(s), a VPN over the Internet guarantees two benefits: cost effectiveness and global access. Nevertheless, there are three concerns related to VPNs. These are: manageability, security, and performance. A brief description of each benefit is given below (Nicopolitidis *et al.*, 2003; Obaidat, 2006; Papadimitriou and Obaidat, 2006).

Manageability

The administration of a VPN should tackle the fast changes in the organization's telecommunication necessities without incurring high costs.

Security

To secure access to the transmitted data, the latter should be encrypted as the Internet is not a secure network.

Performance

Due to the fact that Internet Service Providers (ISPs) hand over IP packets to destination based on a "best effort" attempt, the overall transport performance of a VPN over the Internet cannot be guaranteed and it is variable. Additionally, encryption and authentication may reduce performance considerably.

The main benefits of Virtual Private Networks (VPNs) are (Obaidat, 2006; Papadimitriou and Obaidat, 2006; Arora *et al.*, 2001):

- Security By using sophisticated encryption and authentication techniques, VPNs can save data from being accessed by intruders.
- Scalability These networks enable organizations to employ the Internet infrastructure in ISPs simply and at affordable cost. Clearly this permits corporations to considerably improve capacity without the need to add much additional new infrastructure.
- Compatibility with Broadband Technology VPNs permit telecommuters to reach the resources of their organization networks using high speed Internet techniques such as DSL and Cable Modem technology, which offer users considerable flexibility and effectiveness. Furthermore, such broadband links offer a cost-effective solution to connect remote offices and sites of the company. VPNs are currently set up on many enterprise networks.
- VPNs require few administration procedures.

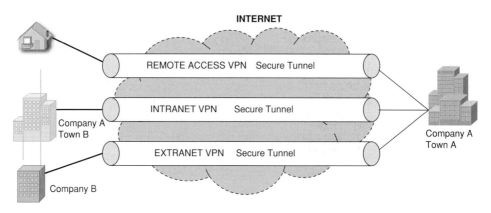

Figure 10.5 Types of VPN services.

- Traffic to the internal network is out-of-the-way until VPN authentication is carried out.
- WEP key and MAC address list management as well as other security schemes turn out to be optional because the security measures are established by the VPN channel itself.

Despite the above many advantages, VPNs suffer from the following disadvantages when applied to wireless LANs (Harding, 2003; Papadimitriou and Obaidat, 2006; Hunt and Rodgers, 2003; Strayer, Brahim, Arora *et al.*, 2001):

- They do not support multicasting and roaming among wireless networks.
- They are not totally transparent.

10.8.1   Forms of VPN services

There are several types of VPN implementations that serve different needs. Corporations may need their VPN to provide dial-up service or to allow third parties such as customers or suppliers to reach specific components of their VPN (Nicopolitidis *et al.*, 2003; Papadimitriou and Obaidat, 2006; Hunt, 2003). In general, VPNs can be categorized into three main categories: Intranet, Extranet and Remote Access VPNs; see Figure 10.5. A brief description of each class is given next.

Intranet VPNs This type of VPN hooks up a number of LANs (intranets) situated in different geographic locations over the common network infrastructure. Normally, such service is employed to connect several geographic locations of a single company (Nicopolitidis *et al.*, 2003; Arora *et al.*, 2001; Obaidat, 2006; Braun, 2001). It allows the sharing of information and resources between isolated employees. For instance, a local office can access the network at the head office, usually including major resources like product and client databases. Because an Intranet VPN is created by connecting two or more trusted corporate LANs, which are usually protected by firewalls, many of the security fears are eased.

Extranet VPNs This kind of VPN widens access to business computing resources to partners, such as customers and suppliers, allowing access to shared information (Wright, 2000). These users are constrained to limited areas of the Intranet, and usually signified as the De-Militarized Zone (DMZ). The firewall and authentication scheme are supposed to distinguish among the organization's workers and other users as well as recognize their access rights. Therefore, the organization's employee links must be bound to the organization's Intranet, while known third party connections must be relayed to the DMZ (Hunt and Rodgers, 2003). There exists a variety of Extranet configurations that vary in terms of their rank of security and access. The major traditional Extranet types are (Hunt and Rodgers, 2003; Strayer, Brahim, Arora *et al.*, 2001; Ribeiro, 2004):

- Private Extranet Here access to a private Extranet is strictly for members only, with no use made of shared networks. In a way, such a configuration cannot be considered a VPN, as it is actually private.
- Hybrid Extranet This type corresponds in operation to a private Extranet, with the exception that it uses a shared network to offer connection. Here, admission to private resources is restricted to related resources only.
- Extranet Service Provider This scheme is provided by an ISP that constructs Extranet services based on its backbone network. Clearly, this family is a kind of provider-provisioned VPN.
- Public Extranet This class offers data, which is accessible worldwide such as an organization that offers a public FTP or Web site that can be accessed freely by any user. In such circumstances, public servers cannot be employed to compromise private components of the Extranet.

Remote Access VPNs

This type of VPN links telecommuters and mobile users to business networks. Typically, a VPN allows remote users to work as if they were working on their office computers. Installing a remote access VPN may end up in substantial savings, avoiding the need to handle a large number of modems, and substituting the need for toll-calls to these modems with calls to local ISP accounts. By using DSL, Cable Modems or other high speed Internet access means, some of the restrictions on performance that are associated with remote access can be removed (Hunt and Rodgers, 2003). Furthermore, WLANs enable computers to achieve network connection with much bandwidth without a physical connection (Ribeiro, 2004; Ferguson and Huston, 2000; Rosenbaum, 2003).

When installing remote access VPNs, several security concerns should be considered and special precautions should be taken to guarantee that the corporate network is not vulnerable due to an insecure remote user.

It is important to mention that the increased flexibility of wireless networks brings up several security issues. Due to the omnipresent nature of the communication channels, clients of the infrastructure should be verified so as to prevent unauthorized users getting access to network resources. In contrast, because any wireless apparatus may snoop on the communication channel, encryption of data should be compulsory if a certain level of data defense and secrecy is desired.

Tunneling is defined as the encapsulation of a specific data packet into another data packet so that the inner packet is opaque to the network over which the outer packet is routed (Strayer). The need for tunneling comes up when it is not suitable for the inner packet to move directly across the network for different purposes. For instance, tunneling can be employed to move multiple protocols over a network based on some other protocol, or it can be used to conceal source and destination addresses of the original packet. When tunneling is used for security purposes, an unsecured packet is put into a secure and normally encrypted packet (Nicopolitidis, 2003; Strayer).

Tunneling allows network traffic from many sources to travel via separate channels across the same infrastructure, and it enables network protocols to pass through heterogeneous infrastructures. Moreover, tunneling enables traffic from many sources to be distinguished in order to be routed to specific destinations and get particular levels of service. The two components that are able to distinctively establish a network tunnel are: (a) the endpoints of the tunnel and (b) the encapsulation protocol employed to transport the original data packet in the tunnel (Obaidat, 2006; Papadimitriou and Obaidat, 2006; Arora *et al.*, 2001; Rosenbaum *et al.*, 2003).

Tunneling is considered the major important scheme utilized by VPNs. The chief idea after this concept is that a portion of the route between the source and the destination of the packet is decided regardless of the destination IP address. In the context of Broadband access networks, the importance of tunneling is due to the following facts:

First, the destination address field of a packet sent in an access VPN may point to a non-globally unique IP address of an organization's internal server. Such an address should not be uncovered to the Internet routers as these routers do not know how to route these packets.

Second, most often a packet transmitted by a user of an access VPN should be routed first to the ISP of this user, and only then from the ISP to the company's network. The first link of the routing between the host and the ISP cannot be done based on the target IP address of the packet, even if this address is globally unique (Obaidat, 2006; Papadimitriou and Obaidat, 2006; Arora *et al.*, 2001; Rosenbaum *et al.*, 2003).

There are two chief kinds of tunneling schemes that are employed by VPNs (Papadimitriou and Obaidat, 2006; Hunt and Rodgers, 2003): End-to-End Tunneling and Node-to-Node Tunneling. A brief description of both techniques is given next.

- End-to-End Tunneling This scheme is also often called "transport model" tunneling. Here, the VPN devices at each end of the connection are in charge of tunnel establishment and encryption of the data transported between the two sites. Thus, the tunnel may expand through edge devices like firewalls to computers transmitting and receiving traffic. An example of a protocol that uses end-to-end tunneling is the Secure Sockets Layer/Transport Layer Security (SSL/TLS). This mechanism is considered very secure as data never emerges on the network in clear-text shape. Nevertheless, implementing encryption at the end-hosts adds to the complexity of the procedure of imposing security strategies. The gateways that are supposed to be in charge of enforcing security schemes are used only for routing packets to their target in this situation, and hence they

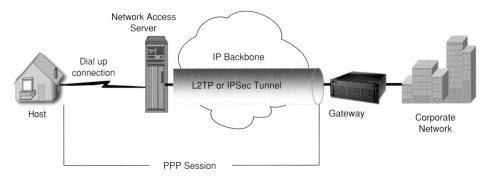

Figure 10.6 Compulsory tunneling scenario.

hold no knowledge of the content or objective of the traffic. Clearly, this is considered tricky for the filtering programs installed at the gateway.

• Node-to-Node Tunneling In this tunneling scheme, transport in the LANs remains unaffected as it is supposed that internal traffic is unreachable from outside. When the signal arrives at the gateway, it is encrypted and delivered via a dynamically created tunnel to the corresponding device on the destination LAN, where data is then decrypted to retrieve its original format, and sent over the LAN to the intended destination. Obviously this has an extra security benefit in that an intruder using an analyzer at some point on the network between the two tunnel servers can find out the IP packets with the sender and recipient addresses equivalent to those two servers. Keep in mind that the real true source and destination addresses are concealed in the encrypted payload of these packets. Due to the fact that this information is hidden, the would-be intruder does not get any hint as to which traffic is going to or from a specific node, and thus will not identify which traffic deserves trying to decrypt. Clearly, this also avoids the need for Network Address Translation (NAT) in order to translate between public and private address spaces, and moves the accountability for doing encryption to a central server. Thus, rigorous encryption work does not have to be done by nodes.

In terms of the disadvantages of the node-to-node tunneling, we have the following two disadvantages (Nicopolitidis *et al.*, 2003; Obaidat, 2006; Papadimitriou and Obaidat, 2006):

1. Poor scalability. This means that the number of tunnels needed for a VPN rises geometrically as the number of VPN nodes increases. This problem has serious performance consequences for large VPNs.
2. Sub-optimal routing. Because tunnels act for only the end-points and not the path occupied to arrive at the other end of the tunnel, the lanes taken across the common network may not be the best possible. This may introduce performance issues.

One can also differentiate between two different tunneling modes: voluntary and compulsory tunneling. In the first scheme, the tunnel is created upon the demand of the user for a determined goal. In the latter scheme, the tunnel is established without any action from the user, and without giving the user any option in the matter (Nicopolitidis *et al.*, 2003; Papadimitriou and Obaidat, 2006). Figures 10.6 and 10.7 illustrate these concepts, which

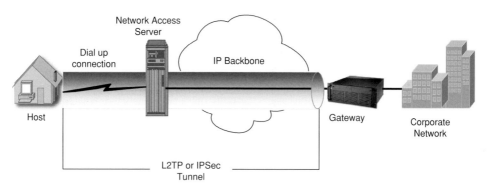

Figure 10.7 Voluntary tunneling scenario.

show that in both figures a host is trying to connect to a company's network using a dial-up connection to a network access server.

Chapter 12 of this book gives a detailed study on all aspects of virtual private networks.

10.9 Summary

To conclude, wireless networks hold great potential, but also pose great technical challenges. As more and more individuals and organizations employ wireless networks in their daily work, protecting privacy and confidentiality will be a vital goal.

Wireless LANs provide new services and applications that wired LANs cannot offer; however, they also bring in new security worry due mainly to the fact that the signal propagates without physical transmission medium. Nearly all wireless network systems are at risk of being attacked and compromised. Regrettably, fixing the problem is not an easy task. The introduction of Wi-Fi Protected Access (WPA) scheme is considered a milestone in the progress of wireless LANs as before its release many organizations were very concerned about employing WLANs in their operations due to the vulnerability of these systems. Although WPA does not solve all security concerns of WLANs, there are ways by which we can mitigate these concerns by proper integration of standards, technologies, management, policies, and operating environments. Given a few RC4 packet keys in the WPA scheme, it is likely to uncover the Temporal Key (TK) and the Message Integrity Check (MIC) key. Although this is an impractical strike on WLANs, it demonstrates that some parts of WPA are weak. Clearly, the entire security in WPA depends on the secrecy of all packet keys. Therefore, it is vital to keep each and every packet key secret. It is worth stating here that biometric-based security techniques have the potential to improve the security levels of all networks including wireless LANs. Recently, we have started to observe an ever-growing interest in this technology due to its vast potentials and robustness.

References

W. A. Arbaugh, N. Shankar, and Y. C. Justin Wan (2002). Your Wireless Network has No Clothes, available at:http://www.cs.umd.edu/~waa/wireless.pdf

Bleha, S. and M. S. Obaidat (1991). Dimensionality reduction and feature extraction applications in identifying computer users. *IEEE Transactions on Systems, Man and Cybernetics*, Vol. **21**, No. 2, 452–6.

Bleha, S. and M. S. Obaidat (1993). Computer user verification using the perceptron. *IEEE Transactions on Systems, Man and Cybernetics*, Vol. **23**, No. 3, 900–2.

Brahim, H., G. Wright, B. Gleeson, R. Bach, T. Sloane, A. Young, R. Bubenik, L. Fang, C. Sargor, C. Weber, I. Negusse, and J. Yu, Network based IP VPN Architecture using Virtual Routers, Internet draft: draft-ietf-l3vpn-vpn-vr-00.txt.

Braun, T., M. Guenter, and I. Khalil (2001). Management of quality of service enabled VPNs. *IEEE Communications Magazine*, 90–98, May 2001.

Checkpoint http://www.checkpoint.com/products/vpn1/vpnwp.html

Cisco http://www.cisco.com/warp/public/779/largeent/learn/technologies/VPNs.html

Denning, D. (1983). *Cryptography and Data Security*. Addison-Wesley.

Ferguson, P. and G. Huston (2000). "What is a VPN?," A White paper, available online at http://www.employees.org/~ferguson.

Harding, A. (2003). SSL virtual private networks. *Computers & Security*, Vol. **22**, No. 5, 416–420.

Hunt, R. and C. Rodgers (2003). Virtual Private Networks: Strong Security at What Cost? Available at http://citeseer.nj.nec.com/555428.html.

IEEE 802.11 Working Group, available at: http//grouper.ieee.org/groups/802/11/index.html.

IEEE 802.11b Wired Equivalent Privacy (WEP) Security, at: http://www.wi-fi.com/pdf/Wi-FiWEPSecurity.pdf

ISAAC. Security of WEP Algorithm, at: http://www.isaac.cs.berkeley.edu/isaac/wep-faq.html

Meddeb, A., N. Boudriga, and M. S. Obaidat (2006). IPsec: AH (Authentication Header) and ESP (Encapsulating Security Payload). In *Handbook on Information Security*, Vol. **1**, Wiley, pp. 932–43.

Netmotionwireless http://www.netmotionwireless.com/resource/ whitepapers/security.asp

Nicopolitidis, P., M. S. Obaidat, G. I. Papadimitriou, and A. S. Pomportsis (2003). *Wireless Networks*. Wiley.

Obaidat, M. S. (1993). A methodology for improving computer access security. *Computers & Security*, Vol. **12**, 657–62.

Obaidat, M. S. and D. T. Macchairolo (1993). An on-line neural network system for computer access security. *IEEE Transactions on Industrial Electronics*, Vol. **40**, No. 2, 235–41.

Obaidat, M. S. (1997). An evaluation simulation study of neural network paradigm for computer users identification. *Information Sciences Journal-Applications*, Vol. **102**, No. 1–4, 239–58.

Obaidat, M. S. and B. Sadoun (1997). Verification of computer users using keystroke dynamics. *IEEE Transactions on Systems, Man and Cybernetics*, Part B, Vol. **27**, No. 2, 261–9.

Obaidat, M. S. and B. Sadoun (1999). Keystroke Dynamics based authentication. In *Biometrics: Personal Identification in Networked Society*, A. Jain, R. Bolle, and S. Pankanti (eds.), Kluwer, pp. 213–230.

Obaidat, M. S. and G. I. Papadimitriou (2006). Fundamentals of Wireless LANs. *Handbook on Information Security*, Vol. **1**, Wiley, pp. 617–636.

Papadimitriou, G. I., M. S. Obaidat, C. Papazoglou, and A. S. Pomportsis (2004). Design alternatives for virtual private networks. Proceedings of the 2004 Electronic Government and Commerce: Design, Modeling, Analysis and Security, EGCDMAS 2005 (M. S. Obaidat and N. Boudriga, eds.), pp. 95–105, Setubal, Portugal, August 2004.

Papadimitriou, G. I. and M. S. Obaidat (2006). Virtual Private Networks (VPNs) Basics. In *Handbook on Information Security*, Vol. **3**, Wiley, pp. 596–611.

Ribeiro, S., F. Silva, and A. Zuquete (2004). "A Roaming Authentication Solution for Wifi using IPSec VPNs with client certificates," TERENA Networking Conference, June 2004.

Rosenbaum, G., W. Lau, and S. Jha (2003). Recent directions in virtual private network solutions. IEEE International Conference on Networks (ICON 2003), September 2003.

Rsasecurity http://www.rsasecurity.com/rsalabs/3-6-3.html

Sans http://rr.sans.org/wireless/wireless_list.php

Stallings, W. (1999). *Cryptography and Network Security: Principles and Practice*, 2nd edn., Prentice Hall.

Stallings, W. (2000). *Network Security Essentials: Applications and Standards*. Prentice Hall.

Stallings, W. (2002). *Wireless Communications and Networks*. Prentice Hall.

Strayer, W. and R. Yuan, Introduction to virtual private networks, Available online at http://www.awprofessional.com/articles/

Arora, P., P. Vemuganti, and P. Allani (2001). Comparison of VPN Protocols – IPSec, PPTP, and L2TP. Project Report ECE 646, Fall 2001, available at: http://ece.gmu.edu/courses/ECE543/reportsF01/arveal.pdf.

Ukela, S. (1997). Security in Wireless Local Area Networks, available at: http://www.tml.hut.fi/Opinnot/Til-110-501/1997/wireless_lan.html

Walker, J. (2001). Overview of 802.11 Security. Available at: http://grouper.ieee.org/groups/802/15/pub/2001/Mar01/ 01154r0P802-15_TG3%

Walker, J. (2002). *Unsafe at any Key Size: An Analysis of the WEB Encapsulation*, Tech. Report 03628E, IEEE 802.11 Committee, March 2002. Available at: http//grouper.ieee.org/groups/802/11/Documents/DocumentHolder/0-362.zip

Wright, M. A. (2000). Virtual private network security. *Network Security*, July 2000, 11–14.

Part IV

Protecting enterprises

Introduction to Part IV

The use of communication technologies has become a crucial factor that is able to considerably improve and affect the productivity of an organization. The need to secure information systems and networked infrastructures is now commonplace in most enterprises. This is essentially due to the importance of the information transmitted across communication networks and stored in networked servers. As a consequence, strong links are being built between security and the enterprise business activity and various tools have been made available for enterprises. These tools include, but are not limited to, filters and firewalls, intrusion detection systems, anti-malicious software systems, virtual private networks and risk management systems.

Intrusion detection systems analyze system and user operations in computer and network platforms in search for an activity that can be considered undesirable from a security point of view. Because of the complicated structures of attacks, data sources for intrusion detection include audit information, network traffic, application logs, and data collected from monitors controlling system behavior. Generated alerts are correlated in order to reduce the number of false alarms, detect efficiently multi-action attacks, and propose responses to the detected intrusions. On the other hand, risk management, which is the discipline that deals with the determination of vulnerabilities and threats, is an important aspect in securing enterprises. It integrates a list of architectures, techniques, and models to evaluate properly whether a current state of an enterprise is encountering threats. Risk management also analyzes the threats that are able to reduce the enterprise's system production and estimates the risk and impact of such threats.

There has been a significant interest in computer malicious programs during the last decades. As the number of these programs keeps on increasing, an efficient software system is needed to fully protect the enterprise's applications and software without an excessive requirement of system manager intervention. Virtual private networks represent another set of securing measures that protect the communications of the enterprise while they traverse untrusted networks such as the public telecommunication networks.

In Chapter 11, a global view is proposed to the reader through a presentation of the intrusion classification. Several approaches for the detection of malicious traffic and abnormal activities are addressed including pattern matching, signature-based, traffic anomaly based, heuristic-based, and protocol anomaly based analyses. A model is proposed to describe events, alerts, and correlation. It defines the fundamentals of most intrusion detection methodologies currently used by enterprises. A survey of the main concepts involved in the model is presented. The chapter also discusses the definition and role of the correlation function.

Chapter 12 presents the basics and techniques of virtual private networks (VPNs). It also reviews VPNs services that include Intranet, Extranet and Remote Access VPNs tunneling. Security concerns that arise when transmitting data over shared networks using VPN technology are also addressed in detail. The protocols used in VPNs such as PPTP and L2TP as well as security aspects are also discussed.

Chapter 13 discusses malware definition and classification. It describes the ways that major classes of malware, such as viruses, worms, and Trojans, are built and propagated. It also describes the major protection measures that an enterprise needs to develop and presents a non-exhaustive set of guidelines to be followed to make the protection better.

Chapter 14 investigates the characteristics that a risk management framework should possess. It discusses the typical risk management approaches that have been proposed. The chapter highlights some of the structured methodologies that are developed based on a set of essential concepts including vulnerability analysis, threat analysis, risk analysis, and control implementation. The chapter also stresses the limits and use of these approaches as well as the role of risk analysis and risk assessment techniques.

11 Intrusion detection systems

Intrusion detection systems analyze an e-based computer and network system and user operations in search of activity considered undesirable from a security point of view. Because of the complicated structures of attacks, data sources for intrusion detection may include audit information, network traffic, application logs or data collected from file system alteration monitors. Generated alerts are correlated in order to reduce the number of false alarms, detect efficiently multi-action attacks, and propose responses to intrusions.

11.1 Introduction

Intrusion detection is the process designed to monitor, analyze, and correlate the information events that occur in a network or a computer system, in order to detect malicious computer and network activities, find signs of intrusions and trigger (or propose) immediate responses to protect the system under monitoring. An intrusion is defined as an attempt to compromise the confidentiality, integrity, availability of a system or to go around the security mechanisms of the system. An intrusion is performed by an adversary accessing the system remotely, an authorized user trying to gain additional privileges that they are not allowed to have, or an authorized user misusing the privilege he is granted. Intrusion detection allows enterprises to defend their information systems against threats.

Although the current intrusion detection technologies cannot provide a complete protection against attacks, it enhances protection capabilities of enterprises and completes the myriad of security solutions. Intrusion detection products, however, are different from other security products. For example, firewalls offer active protection against attacks while an intrusion detection system (IDS) can generate alerts when suspicious activity is observed. It also can detect signs of attacks, identify attacks, and implement passive mechanisms. An IDS inspects all incoming, internal, and outgoing network traffics and identifies suspicious patterns that may indicate an attack or an attempt to attack an information system by adversaries. Intrusion detection systems involve a large spectrum of detectors to collect all related events, methods to correlate events and identify attacks, and interfaces to report and react on detected intrusions. An IDS can also take some active responses such as modifying an access list or shutting down a communication connection.

The role of an intrusion detection system is threefold. First, it eliminates the unneeded information (for a complexity reduction of its function) from the audit information it collects. Second, it either presents a synthetic view of the security-related actions taken during normal usage of the system, or a synthetic view of the current security state of the system. Third, it takes a decision to evaluate the probability that these actions or this state can be considered as signs of intrusions or vulnerabilities. A countermeasure component of the IDS can then take the corrective actions to either prevent the actions from being executed or changing the state of the system back to a secure state.

Several attractive features can characterize the use of IDSs within an enterprise. These include: (a) an IDS can act as a quality control for the security planning, design, assessment, operation, and administration, (b) an IDS provides all information about intrusions that have been performed to help diagnosis improvement and mitigation of intrusion causes, (c) an IDS can help implementing the process of forensic investigation using correlation activities and log files, and (d) an IDS can detect signs of attack and other violations that cannot be prevented by traditional security solutions. In addition, intrusion detection systems can be classified typically into two categories: network-based and host-based systems. A network-based IDS monitors the network packets in transit and analyzes them, while a host-based IDS monitors operating system logs and data files. Various types of commercial products were made available in the market recently. They implement different monitoring, detection, and response approaches.

Nonetheless, four typical drawbacks can be noticed for intrusion detection systems. They are:

1. An IDS is just an active system that is always monitoring the security state of a system; it is not proactive and cannot prevent attacks, but it can detect a composite attack before the attack terminates.
2. An IDS cannot provide a full protection of the system it is monitoring. It is just an additional, but essential, security solution.
3. An IDS cannot cope with scalability problems. It cannot handle a huge set of alerts and provide real-time analysis efficiently.
4. An IDS cannot protect against all types of attacks mainly because of two reasons. First, the evolving nature of attacks (like viruses) makes it difficult for an IDS to detect newly created attacks. Second, the complication of the attack structure authorizes actions that are not detectable using events.

Three criteria have been proposed in the literature (Porras, 1998) to evaluate the efficiency of an intrusion detection system. They are: accuracy, performance, and completeness. Accuracy is concerned with the proper detection of intrusions and the reduction of false alerts. Inaccuracy occurs when an IDS reports as intrusive or anomalous a legitimate action. Performance is a criterion that characterizes the rate at which audit events are processed. Poor performance induces the non efficiency of real time detection. Finally, the concept of completeness defines the property of an IDS which is able to detect all intrusions. Incompleteness occurs when the IDS fails to detect an intrusion when it occurs. Several additional criteria can be added to provide a better evaluation of IDSs. These criteria define the desirable characteristics to deploy and use an IDS:

•   *Fault tolerance* This property measures the capability of the IDS to work properly in the presence of attacks and failures. After an attack or a failure occurrence, the IDS must be able to recover from its previous state and resume its operation unaffected.

- *Resistance to subversion* This criterion requires that the IDS must be able to monitor its activity, detect attacks made by intruders, and protect itself.
- *Ease of deployment and operation* The IDS must be configurable to accurately implement the security policies of the system under protection, have a minimal overhead on the system that runs it, and be easy to deploy.
- *Adaptability* The IDS must be adaptable to changes in system and user behavior over time.
- *Timeliness* This characteristic measures how fast the IDS can perform and propagate its analysis.

A classification of intrusion detection systems can be based on five principles: (a) the intrusion detection methods, (b) the behavior on the detection of intrusions, (c) the source location of the input information, (d) the detection paradigm, and (e) the usage frequency. The detection methods describe the properties of the alert analyzer as being behavior-based or knowledge-based. The IDS is behavior-based when it uses information about the normal behavior of the system it protects in order to recognize malicious activities. It is knowledge-based when the IDS uses information about the intrusions it wants to detect. The behavior on intrusion detection describes the type of response of the IDS in response to an attack it detects, in the sense that an IDS can be passive alerting, or active responding. When it actively reacts to intrusions by taking corrective actions or pro-active actions, the IDS is said to be active. It is called passive if it simply generates alerts and sends them appropriately to the concerned entities. On the other hand, the event source location classifies the IDSs based on the location of the input information they analyze. The input information source can be the system logs on a host, application logs, network packets, or alerts generated by other intrusion detection systems.

Let us finally notice that three terms will have a particular role in this chapter. They are: *system, alert*, and *audit*. The term *system* is used in this chapter to denote the information system, the network, or a host connected to the network being monitored by the intrusion detection system. The term *audit* denotes the information provided by a system concerning its activity, internal state, and behavior. The term *alert* designates a message reporting on occurring actions that are observed through log files, traffic flows, or entity behavior. An intrusion detection system collects information about the system that is to be protected to perform a diagnosis on its security state. To achieve its objectives, the intrusion detection system uses three types of information: long-term information, which is related to the techniques used to detect intrusions, audit information, which reports information about the events that are happening in the system, and configuration information, which reports about the current state of the system (Undercoffer, 2003; Curry, 2003).

11.2   IDS architecture and classification

The detection paradigm in an IDS considers the nature of the detection mechanism used by the IDS and monitors the security state of the system to be protected. Based on the information it collects, the IDS can classify the states as *secure* or *insecure*. It can also classify the transitions between states as evolving from a secure state to an insecure state. While in

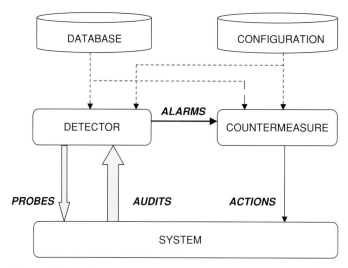

Figure 11.1   A simple IDS architecture.

the former case the detection mechanism is called state-based processing, the latter classifies the mechanism as transition-based. The usage frequency-based classification considers whether the IDS uses a real-time continuous monitoring of the system capabilities or runs only periodically. Major features of intrusion detection systems address the architecture and the location of information sources.

11.2.1   Generic IDS architecture

A simple IDS architecture is depicted in Figure 11.1. Detectors/sensors are components/devices used to monitor the network traffic or host activities for signs of intrusions or malicious behavior. The solid lines in Figure 11.1 represent actual communication links used within the IDS paradigm, while the dashed lines represent the secure communications used by the IDS to pass detection information between the IDS components.

The object *Probes* in Figure 11.1 represents a set of functions that look at the system activity and search for specific events or queries in particular situations. Entity *Audits* represent the input information that a detector is in charge of collecting. *Database* represents the location where useful information can be stored for the detector to perform its tasks. This may include a library of attacks and the archive of collected events. *Countermeasure* contains a set of decisions that the IDS can recommend to respond to a detected attack after the analysis of the related alarms, which are generated by the detectors. *Configuration* data represent the state that affects the operation of the IDS such as how and where to collect audit data and how to respond to intrusions.

A taxonomy of IDSs can be based on the following dimensions:

- *Time of detection* Two groups of intrusion detection systems can be considered: those that try to detect intrusions in a real-time manner, and those that analyze the data collected after some period of time.

- *Source of audit data* These sources belong mainly to three classes: network-based detectors, host-based detectors, and internal detectors. The latter class considers the possibility of inserting procedures, routines or sub-programs within the source code of an application allowing close observation of data and behavior in a program (Zamboni, 2001).
- *Granularity of data analysis* Audit information can be analyzed at different levels of detail. This ability may require tools for de-fragmentation, reassembly and decryption.

11.2.2   IDS location

The most typical way to classify intrusion detection systems is to consider the location of the sources of collected audits (or events, information about executions, packets, etc.). Typical audit sources are communication packets, operating systems, and sensitive files (e.g., logs). As the center of attention of computing moved during the last decade from single host environments to distributed networks of hosts, several products proposing IDSs were developed to accommodate network technologies and enterprises' policies. The first interest in that area was to get host-based IDSs to communicate and cooperate to achieve their objectives. In a distributed environment, users move from one host to another, possibly changing identities during their changes of location, and perform their attacks on several targets. Information exchanged between local intrusion detection systems takes place at several levels. An IDS can exchange simple audit information over the network, issue alerts generated based on a local analysis, and correlate its alerts with those received from its peers to detect distributed attacks.

Network attacks (such as port scanning, spoofing, and TCP-based attacks) cannot be detected easily by only examining audit information. Specific tools have been developed to give complementary mechanisms that use packet and flow control to process the transmitted data (e.g., packet payload content) and look for suspicious commands or patterns. More specialized intrusion detection tools have emerged to allow the capability of managing many components of the enterprise's presence on a communication network. These products can monitor firewalls and web servers, process host-based audits, and handle network-based information. Host audit sources help collecting information about the activity of the users on a given machine. This unfortunately makes them vulnerable to attacks and creates an important real-time constraint on host-based intrusion detection systems, which have to process the audit information and generate alerts before the intruders taking over the target can complete their attack and subvert the audit of the IDS itself.

Host-based sources allow the collection of audit information from multiple sources that are located in various computer systems, including: (a) operating system sources, which provide a snapshot of information about the activity taking place based on the system commands (such as the commands *ps, and pstat*, for Unix), (b) accounting information sources, which provide information on the use of shared resources (e.g., memory, process time, and archive), (c) system log information, which is provided and stored by the operating system based on a timestamp applied on a text string from the user, and (d) security audit information, which records all potentially security-significant events on the system (e.g., security audit defined by the well-known common criteria).

Network packets constitute another source of information that is largely used in detecting intrusion. Network-based sources are regarded as an efficient means for gathering information about the events that occur on the network and most accesses to resources that take place over a network. Therefore, capturing the packets before they enter a given server is probably a good way to monitor the security of this server. Network-based detection can perform the traffic analysis at two different levels. At a low level, the analysis is made by performing pattern matching, signature analysis, or some other kind of analysis of raw content of packets. The approach is called stateless, in the sense that it does not take into consideration session information and traffic flow information. It only considers the information that is contained in the packet under analysis. Useful information could, however, span over several network packets. At a high level, the intrusion detection acts as a gateway and analyzes each packet with respect to the application or protocol being followed by the packet.

Host-based intrusion detection systems (H-IDS) operate on information collected from host-based detectors in order to inspect auditable resources looking for unauthorized changes or suspicious patterns of host activity. H-IDSs can analyze activities with high reliability determining precisely which processes (and users) are involved in a particular attack and directly monitor the data files and system processes that are targeted by adversaries. H-IDSs can help in detecting attacks (or attempts of attacks) involving software integrity breaches, weak mis-configuration, or inappropriate access policies. However, H-IDSs present some disadvantages including:

- H-IDSs are difficult to manage, as audit information must be configured and managed for every host they monitor.
- An H-IDS can be disabled by various types of denial of service attacks.
- H-IDSs are unable to detect surveillance attacks (e.g., port scans) since they can only see those packets that are received by the host where the H-IDS is installed.

Network-based intrusion detection systems (referred to as N-IDS) address four important issues. First, an N-IDS attempts to detect network-specific intrusions, particularly the denial of service attacks that cannot be detected in a real-time manner by searching for audit information on the host, but only by analyzing network traffic. Second, it allows the study of the impact of auditing/collecting data on the performance of hosts, by entirely collecting the needed information on separate machines. Third, N-IDSs are very secure as they operate in a secret mode and cannot be easily located. Fourth, an N-IDS facilitates the acquisition and formatting of audit information. Finally, network intrusion detection can contribute to the detection of attacks against hosts by signature analysis that their components can perform on the traffic packets.

Network-based intrusion detection systems have, however, three major drawbacks: First, encryption makes it impossible to analyze the payload of a packet because encryption hides a considerable amount of the information carried by the packets that the detection process may need. Second, N-IDSs have problems in coping with high speed communication and high volumes of traffic. Third, systematic scanning operations are difficult to implement with network detection since they might constitute a source of traffic congestion. Finally, tools used in the network detection are inherently vulnerable to security attacks since they rely on off-the-shelf products to acquire network information.

In addition to host and network based IDSs, other locations were shown to be important for detection. Storage-based intrusion detection systems (S-IDS) can be valuable tools in monitoring and notifying administrators of malicious software executing on a host computer, including many common intrusion toolkits. An S-IDS runs inside a storage device, watching the sequence of requests for signs of intrusion and offering complementary views into system activity since it is not controlled by the host operating system or by the host's IDS. S-IDS can continue to operate even when the network-based IDS is circumvented and the host-based IDS is turned off (Pennington, 2003).

11.3   Detection techniques

There are several different approaches for the detection of malicious traffic or activity. Detection techniques depend on the data to monitor and involve various approaches. Pattern matching, signature-based, traffic anomaly based, heuristic-based, and protocol anomaly based analyses are among the major methods that are used by intrusion detection systems providers.

11.3.1   Detection methods

As intrusion detection becomes more accepted, organizations worldwide start deploying intrusion detection mechanisms on a large scale. Businesses who desire to have a presence on the Internet and to develop a security solution involving IDSs, should consider the nature of information useful for the detection analysis, the number of alert types to collect and process, and the location where to place the IDS devices. There are several trends in intrusion detection to use the knowledge built up about intrusions and look for evidences, which reveal the traces of these intrusions and build a reference model of the actual behavior of the system under protection. Detection methods represent the major components in detecting malicious activities. Methods typically include knowledge-based detection techniques, signature analysis, and anomaly-based analysis.

Knowledge-based intrusion detection techniques develop the knowledge collected about specific intrusions and system vulnerabilities and look for attempts that aim to exploit these vulnerabilities. When such attempts are detected, an alert is reported. This means that any action that is not explicitly recognized as intrusion would be considered as acceptable because it is assumed that it does not represent a situation where an alert is generated. Advantages of the knowledge-based approaches include the fact that they may have very low false alert rates, provided that the collected knowledge is sufficiently refined, reliable, and efficient to detect potential intrusions. Drawbacks, on the other hand, include the difficulty of collecting the required information on the known attacks, coping with the increasing complexity of attacks, and keeping the knowledge up to date with new vulnerabilities and the environment evolution. Moreover, the maintenance of knowledge-based IDSs is a time-consuming task because it requires careful analysis of all vulnerabilities, threats, and risks.

Expert systems and decision support systems represent the major tools to the efficiency and completeness of knowledge-based IDS techniques. The expert system manages a set of rules and axioms that describe intrusions, correlate alerts, and select responses. Audits are then translated into facts carrying their semantic meanings to the expert system. The inference engine establishes conclusions using these rules and facts. Various rule-based languages have been developed as natural tools for modeling the knowledge collected about intrusions and for allowing systematic browsing of the audit information in search of evidence of attempts to exploit known vulnerabilities.

Signature analysis, on the one hand, follows exactly the knowledge-based approach as expert systems do, but the collected knowledge is exploited in a different way. The semantic description of the intrusions is transformed into pieces of information that can be found directly in the audit information by searching for a signature (like virus scanners do). To be efficient, this method requires a complete database of signatures of all known attacks and their variants. This approach allows a very efficient implementation, but suffers from several shortcomings. The major drawbacks of this method include: (a) the need for frequent modifications to update the IDS with newly discovered vulnerabilities and the requirement to represent all facets of the intrusion signature, (b) the method is easy to elude by attacks such as the so-called zero-day-attack, and (c) the method does not cope with high speed communication since it induces unacceptable delays for traffic under monitoring.

Behavior-based intrusion detection techniques, which constitute another set of knowledge-based methods, assume that an intrusion can be detected by observing a deviation from the normal (or expected) behavior of the system. Typically, this method ignores every activity that is evaluated *normal* and generates an alert if the behavior deviates from normal. The model of normal behavior is extracted from the reference information collected by various means, including the construction of user profiles from historical data collected over a period of normal activity. The IDS compares this model with the current activity and when a deviation is noticed, an alert is generated. Therefore, any user activity that does not correspond to a previously learned behavior is considered abnormal. Advantages of behavior-based techniques include:

- They are effective in detecting unknown attacks. They can even contribute to automatic discovery of new attacks.
- The detectors used by these techniques can produce information that can be used to define signatures for signature-based analysis.

Shortcomings of the behavior-based methods are important. They often generate a large number of false alerts, while intrusion attempts that appear to be normal for the IDS may cause missed detection. Identification of attacks cannot be efficiently provided by behavior-based IDSs, nor can they report on the success state of attacks. Finally, one can notice that behavior-based IDSs require extensive training activity.

Another detection method of importance is used in different IDS products. It is called the *heuristic-based* detection. This method often uses some form of artificial intelligence techniques to detect the intrusions attempts. Such methods usually present statistical evaluations of the type of traffic being presented. It also uses neural networks, rule-based techniques, and heuristics and presents the advantage of detecting more complex forms of malicious activities. However, it presents some important drawbacks including the generation of too many false positives and the intensive frequency of changes.

11.3.2   Response generation

A response is defined as the set of actions that the IDS selects to respond to an intrusion once it arrives at a conclusion that the information source reveals a malicious activity. Response therefore defines the capability of the IDS to recognize a given alert (or a set of alerts) as representing an attack and then providing the countermeasures to react against the attack. Responses fall into two categories: passive responses and active responses. Active responses are enforced by an IDS to react to an attack. They are taken immediately after the attack is identified and aim to implement actions against the intrusion source and suspend the progress of the attack or mitigate its effects. Response may include collecting additional information about suspicious attacks to prevent the future success of the attacks, checking whether the attacks have succeeded, and supporting the forensic investigation of the attack. Stopping the progress of an attack is the other major form of an active response. It allows, for example, blocking the IP addresses, reconfiguring firewalls and routers, and modifying the access list. On the other hand, passive responses are normally taken by the security administrator after receiving the alerts revealing an attack. Passive responses involve two major activities as performed by the IDS: alarm notification generation and alarm reporting. Alarm generation is typically displayed on the IDS site or on any other system. Using protected common communication channels, alarm reports allow the entire network to respond to the identified attack.

A major concern in responding to attacks is the rapidity of reaction, which refers to the elapsed time between the occurrence of the events and the identification of the true attack related to those events. This is mainly related to the event collection process and the time spent in analyzing the collected events. Post-event audit analysis is made on a periodic scheme basis to send the collected audits from the monitoring points to the analysis modules. This mode of analysis can address studying a large size of information and refine the collected data. It also can reduce the costs incurred with the auditing process. Real-time audit analysis operates on continuous information feedings.

11.3.3   Forensic analysis

Digital computer forensic analysis can be defined as the process of collecting information sources in order to obtain evidence against a completed attack and compel legal actions against the author(s) of intrusion. IDSs can assist forensic investigation in collecting the needed information. Typically, forensic analysis involves three essential processes:

- The evidence identification, which collects all the information needed to provide useful results in analysis. Evidences are then stored to protect them from security attacks.
- The evidence analysis, which is the process of exploring the collected evidences to build the proof that an adversary has performed actions.

- The evidence presentation, which provides clearly and perceivably the evidences to the ad-hoc authorities. The forensic analysis provides complete technical knowledge on how an attacker has performed his/her attack, how the related actions have been relayed from source to target, and what has been compromised.

Intrusion detection systems contribute extensively to the forensic process through host-log monitoring and network activity monitoring. Host logs are defined as the high-level information built by integrating audit logs, system logs, application logs, and storage logs. Logs generated by an H-IDS can often summarize many lower level events about intrusions such as a login failure or a storage of a piece of information in a specific disk track. Application log files take a greater importance as a data source for intrusion detection proofs. Compared to system audits or network packet capture, the use of application log files has the following useful features for investigation:

- The application log contains all relevant information, even if the application is distributed over a set of sites. It also can provide internal data that cannot be made available in audit information or in network packets.
- The application log provides the guarantee that the information contained in the log file is accurate, since it does not require processing before the intrusion detection system extracts it. This is true when the log files are protected appropriately.

Application logs, however, have some limitations. Intrusions can be only detected after the application logs are written. Therefore, if a malicious adversary can prevent the writing operation, application log based IDS would not be able to detect the intrusion, and hence no evidence would be found within the application log file. Moreover, low-level attacks (such as some denial of service attacks) would not be detected by just application log based IDSs, except when the log is designed to include an exhaustive set of information. This unfortunately would impose a large activity of maintenance (need for logging new facts for new attacks) and contribute to excessive loads in processing the files.

11.4   Modeling the intrusion process

The model, which is proposed in this section to describe events, alerts and correlation, defines the fundamentals of most intrusion detection methodologies currently under utilization. Because of the applicability of the concepts involved in the model to most common intrusion detection systems, an overview of the primary concepts of the model is discussed in the following subsections in order to provide a basis for understanding the core technology.

11.4.1   Foundation of ID

Practically, an intrusion, whether it is centralized or distributed, is defined as a sequence of actions executed by one or a group of malicious adversaries that results in the achievement of intrusion objective (e.g., the occurrence of unauthorized access or damage on distributed computing or networking domains). Actions are generally represented by *pre-* and

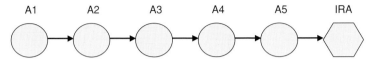

Figure 11.2 A simple intrusion scenario.

post-conditions. The *pre-conditions* of an action correspond to conditions the system's state must satisfy to carry out the action successfully and may require additional information provided by the adversary (e.g., password, privilege). *Post-conditions* correspond to the effects of executing the action on the system state and the conditions that the new state must satisfy after the attack had succeeded. This may be an information that the system outputs. A formal language can be used to specify actions and prove/check properties of the actions. It can be based on a first-order logic using predicates, logical connectives, and invariants. The language can be completed to include a large set of useful meta-predicates such as the binary meta-predicates *knows* and *exploits*. Expressions *knows(A,π)* and *exploits(X,v)* are binary expressions stating that user A knows that property π and attack X exploits vulnerability v, respectively.

The actions composing an intrusion are performed in a specific order, in a way that the pre-conditions of an action may need that the post-conditions of other actions should be satisfied. This allows the representation of an intrusion as a directed graph where nodes are actions and links are dependencies between post-conditions and pre-conditions needed to achieve the objective of the intrusion. The graph is called intrusion scenario. An example of intrusion scenario that allows an intruder to obtain a root access on a target UNIX workstation (as an objective) can be described by the sequence of the following actions (Cuppens, 2002a):

- A_1 Execute command *rpcinfo* to check whether *portmapper* is running on the target.
- A_2 Execute the *showmount* command to see the target exportable partitions.
- A_3 Execute *mount* command to enable the intruder to mount one of these partitions, say the root partition.
- A_4 Modify the *.rhost* of this partition and get a root access on the target system.
- A_5 Execute *rlogin* command to access the target system with root privileges.

Figure 11.2 depicts the intrusion scenario of the illustrative example and a simple linear scenario with an objective called "*Illegal root access*" as denoted by IRA. The pre-conditions and post-conditions of the five attacks are easy to deduce.

An attack is assumed to be revealed by an event (or a set of events) that can be observed by an IDS detector (or a set of detectors). Therefore, each event characterizes the occurrence of an action. It can be the observation of the occurrence of a signature (i.e., a sequence of consecutive bits, that form an address for example) on a packet or an anomaly incident (such as those reported by the use of unusual high bandwidth, abnormal use of computation time, or a high number of processes). At the occurrence of an action related event(s), the intrusion detection system generates an alert if it estimates that action is critical. The set of related alerts can compose an alert scenario, which may reveal a possible intrusion scenario. Formally, an event in a computing network can be defined by a 4-tuple of the form:

$$(Act, In, Out, Eff),$$

where *Act* is an operation request made by a user (or even an intruder), *In* represents all input information required to execute action *Act*, *Out* represents all the information obtained after the execution of *Act*, and *Eff* represents the damages caused by Act on some network resources (*Eff* may be empty in some cases). An alert can be defined any time a critical event is observed. An alert is formally represented by a 3-tuple of the form:

$$(ev, , tm, , \pi)$$

where *ev* is an event, *tm* is the time of occurrence of event *ev*, and $\pi \in [0, 1]$ is a real value measuring the criticality of event *ev*. The criticality value estimates the degree of damages caused by event *ev* on system resources. A scenario of alerts links a set of alerts together in order to identify scenarios of intrusions. It is formally represented by a 3-tuple $((A, F), Rel, \pi_{si})$ where:

- (A, F) is a serializable directed graph (meaning that no oriented loops can be found on the graph).
- *Rel is* a mapping that associates to each arrow $(n, n') \in F$ a relation between the values involved in alert *n* and those occurring in *n'*. Relations are assumed to be implementations of pre-condition/post-condition links between actions observed through the events.
- The timestamps $t(n)$ and $t(n')$ should satisfy: $t(n) \leq t(n')$, for all $(n, n') \in F$.
- For all node $n \in A$, the criticality $\pi(n)$ is higher than a predefined threshold T.

The process of intrusion detection can then be decomposed into four major tasks: alert generation, alert correlation, scenario management, and scenario correlation (or intrusion identification). The first task aims at detecting the events that report on actions capable of being part of intrusion scenarios, analyzing them, computing their criticality, and raising alerts when the events are shown to be critical. The second task attempts to build scenarios of alerts and link the newly generated alerts with ongoing intrusion scenarios (i.e., uncompleted scenarios with respect to known scenarios of attacks), if any. The third task manages the generated ongoing scenarios while simplifying and refining them. The fourth task checks whether a scenario of alerts matches an intrusion scenario that is already available within an ad-hoc library of intrusions.

11.4.2 *Intrusion correlation*

Three types of correlation can be considered in conducting the process of intrusion detection: (a) alert correlation, which attempts to relate a detected alert to the alerts (or nodes) occurring in a given scenario of alerts, (b) alert/action correlation, which aims at correlating the alerts involved in a scenario of alerts that the intrusion detection process has produced, with nodes (or actions) occurring in a scenario of intrusion, and (c) scenario correlation, which attempts to check whether a given intrusion scenario is correctly reported by a scenario of alerts, assuming that all the actions involved in the scenario of intrusions are correlated with reported alerts (or nodes) in the given scenario of alerts.

Alert correlation process

Let *SA* be an ongoing scenario of alerts (also called scenario under construction since it has not completed the detection of intrusion) and let α be an alert. The alert correlation process first retrieves all the alerts, say $\beta_1, \ldots, \beta_n$, occurring in *SA* that can be correlated with alert α, meaning that a set of generic form of relations is made available for this purpose. Second, the process updates *SA* by adding the arrows (β_i, α), $i \leq n$, to it. The formal definition of an alert states that it has the form:

$$((Act, \; In, \; Out, \; Eff), \; tm, \; \pi).$$

Relation (β_i, α) is stated provided that the *Out* part of β_i is related to the *In* part of α with respect to some correlation relation (e.g., matching of attribute values) and that $\pi(\beta_i)$ and $\pi(\alpha)$ are higher than a given threshold value, which is pre-defined by the security administrator.

Action/ correlation process

Let *SA* be a scenario of alerts produced during intrusion detection and *SI* be a given scenario of intrusions. The objective of the node correlation process, when applied to *SA* and *SI*, is to check that every action in *SI* is reported by an alert in *SA*. Formally, this means that if:

$$a = (action\_name, \; pre\text{-}condition, \; post\text{-}condition)$$

is an action in *SI*, then there is an alert in $\alpha = ((Act, \; In, \; Out, \; Eff, \; tm, \; \pi)$ such that *pre-conditions* and *In* are correlated and *post-conditions* and *Out* are correlated with respect to predefined correlation relations (we say in this case that alert α is an alert observing action a). When the action/alert correlation process is successful, it outputs a tuple of the form $(\alpha, \; a, \; R_1, \; R_2)$, where R_1 is a correlation between $In(\alpha)$ and the *pre-conditions* of a and R_2 is a correlation between $Out(\alpha)$ and the post-conditions of *a*. The process outputs a "null" value when it fails to find correlation relations.

Scenario correlation process

Assume that *SA* is a scenario of alerts and *SI* is an intrusion scenario, and that all nodes in *SI* are correlated with nodes in *SA*. The scenario correlation process, when applied to *SA* and *SI*, attempts to find for each arrow (a, b) in *SI* a path $(c(a) = \alpha_1, \ldots, \alpha_s = c(b))$ in *SA* between the alert $c(a)$ reporting on action a and the alert $c(b)$ reporting on action b such that the chain of correlation relations between the *Out* component of α_1 and the *In* component of α_s matches the link between the *pre-conditions* of b and the *post-conditions* of a. This means that association between actions a and b (in attack *SA*) can be observed through the sequence of s alerts.

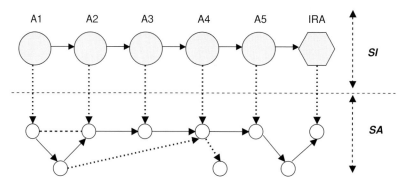

Figure 11.3   A scenario of correlation.

The scenario correlation process fails whenever no path can be found for a pair of consequent actions in *SI*. In this case, one can conclude that *SI* and *SA* could not be correlated since *SI* is a connected graph, and so should be any accurate scenario that reports effectively on *SI*. The failure of the path retrieving occurs when no path involving critical values higher than the predefined threshold can be found. To overcome this and provide a candidate, a path with lower critical values is needed (showing a pessimistic point of view in monitoring the scenario of alerts). A second approach to solve this attempts to retrieve a sub-graph linking alert $c(a)$ to alert $c(b)$ instead of looking for paths. This means that the detection process has to use a large set event detectors to find the correlation between $c(a)$ and $c(b)$. In optimized situations, this case does not occur.

Illustrative example

Let us consider the aforementioned attack (Figure 11.2). Assume that a scenario of alerts involving 9 alerts, which is depicted in Figure 11.3, shows the action/alert correlation process. One can see six alerts observing actions $A1, \ldots, A5$, and *IRA*. Three links in *SA* do not contribute to the observation of the attack (as depicted by dashed arrows). In addition, the scenario correlation shows that two paths of length 2 report on links $(A1, A2)$ and $(A5, IRA)$.

Finally, let us notice that different operations can be formally defined on the set of scenarios of alerts. Among the most important operations, one can mention: (a) the concatenation (or union operation), which constructs a scenario of alerts from two given scenarios, provided that the scenarios have a common node, (b) the full correlation of scenarios of alerts, which builds on the union of two scenarios the arrows that can be added to link the alerts involved in one scenario to the alerts belong to others, and (c) a refinement operation, which attempts to reduce the number of nodes whenever the following situations occur:

- If there are two arrows (α, β) and (β, χ) such that a relation between $Out(\alpha)$ and $In(\chi)$ can be clearly deduced and that no other arrows are involving β, then node β can be deleted and the link between α and χ is deduced accordingly.

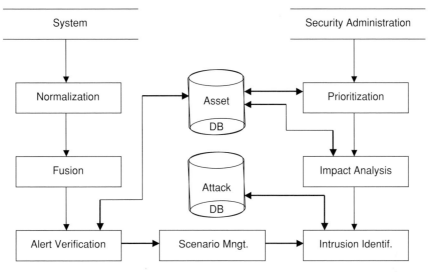

Figure 11.4 Typical fusion/correlation architecture.

- A path $(\alpha = \alpha_1, \ldots, \alpha_n = \beta)$ can be deleted if there is another $(\alpha = \beta_1, \ldots, \beta_s = \beta)$ with higher criticality, meaning that: $\pi(\alpha_i) > \pi(\beta_j)$ for all $n > i > 1$ and $s > j > 1$.

11.5   Correlation in practice

The correlation process can be decomposed into a collection of modules that transform detected events into intrusion reports. Because alerts can refer to different kinds of attacks at different levels of granularity, the correlation process cannot treat all alerts equally and should contain modules that can operate on all alerts, regardless of their types. These modules are used in the initial and final phases of the correlation process to implement a general functionality that is applicable to all alerts. The other modules can only process certain classes of alerts since they are responsible for performing specific correlation tasks that cannot be generalized for arbitrary alerts. Figure 11.4 depicts a generic architecture involving the major general modules.

Typical tasks that are performed on all alerts include a normalization module that transforms every alert that is received into a standardized format that can be understood by all the correlation modules. Normalization helps to hide the differences in alert specification provided by detectors and sensors and may use a specific language for alert description. It also aims at preparing cooperation between correlation processes. The normalized alerts are then augmented so that all required attributes are assigned useful values. Specific correlation functions operate on different spatial and temporal parameters, meaning that while some functions can correlate alerts that occurred within a small period of time, other functions operate on alerts targeting the same address or look for similarities. The fusion module is responsible for combining alerts that represent independent detections of the same attack instance (eventually using different ID sources). The verification module is responsible for taking a single alert and determining the success of the intrusion related to that alert. The task of the scenario construction module is to combine the alerts that may refer to intrusions launched (by a single or multiple intruder) against a single target, while associating the host-based alerts with network-based alerts that are involved in the same scenario of intrusion.

The next two tasks depicted by Figure 11.4 are the intrusion identification and the focus identification. They operate on the scenario of alerts under construction and identify common intrusion patterns, which are composed of a sequence of individual actions (scenario of intrusion). Moreover, they can occur at different points in the network and identify hosts that are involved in the scenarios of intrusions. The final two tasks are the impact analysis and the prioritization. While the first task determines the impact of the detected scenario of intrusions on the system being monitored and the assets that are targeted by the intrusions, the prioritization task assigns an appropriate priority level to each detected scenario of alerts and correlated scenarios of intrusions that measure the likelihood of the correlation to reduce the number of false positives (Haines, 2003).

In the following subsections we consider the fusion and impact analysis with more detailed description since they are highly important in the correlation process.

11.5.1 Alert fusion

Fusion methods have been useful in reducing false positives in different domains such as diagnosis and signal processing. In intrusion detection, the process of the alert fusion aims at combining alerts that represent the detection of the same occurrence by independent intrusion detection systems. The decision to combine two alerts is based on different criteria including the time of alert occurrence and the information contained in the alerts. For example, the decision to combine considers that two alerts are candidates for fusion if their start-times and end-times differ only by less than a predefined period of time. Since the fusion only considers alerts detected by different detectors, the attributes contained in the alerts should be used to check for criteria such as the identity of values (e.g., equality of source addresses).

When two alerts are fused, they are assumed to be related to the same intrusion. A so-called meta-alert is therefore constructed to include the parameters of the alerts in a normalized way. The start-time and end-time of the meta-alert are assigned the earliest start-time and end-time, respectively. This is done because alerts are assumed to be duplicate and that a later time stamp is the result of a processing delay at a detector. Typically the attributes defined for the meta-alert are set to the union of the values of the related attributes in the fused alerts. The alert fusion is an important alert reduction process. It is particularly significant if the existing intrusion detection platform contains a substantial level of redundancy. A meta-alert can be fused with other alerts and other meta-alerts. The fusion process therefore results in a hierarchical structure. Often a tree structure is used, where the leaves are simple alerts and intermediate nodes are meta-alerts.

During the fusion of a set of alerts, each individual detector computes its estimation of the criticality value (or intrusion decision) on whether an intrusion has taken place. The fuser combines the intrusion decisions to generate a fused decision to be added to the meta-alert deduced by the fuser. The output of the fuser, π, is determined using various approaches.

In a first method, the fuser can just compute the mean of all decision values of the provided criticality values $(\pi_1, \ldots, \pi_s)$. In another approach, the output π is given by the result of the following function expression (which is given for two detectors):

$$P(\pi_1, \ldots, \pi_s | Int) / P(\pi_1, \ldots, \pi_s | \neg Int)$$

where $P(\pi_1, \ldots, \pi_s | Int)$ is the probability that all the detectors produce decisions, $\pi_1, \pi_2, \ldots$, and π_s show that an intrusion is detected when there is really an intrusion and $P(\pi_1, \ldots, \pi_s | \neg Int)$ is the probability that the detectors produce decisions $\pi_1, \pi_2, \ldots$, and π_s for an intrusion when there is no intrusion (Shankar, 2003).

11.5.2   Alert verification

As it may be easily understood, reducing efficiently the number of false alerts would provide a high-level view of malicious activities on the network and help to obtain an efficient correlation process. For this, the correlation process needs to differentiate between the following cases when a detector outputs an alert: (a) the detector has correctly identified a successful intrusion (meaning that this is a true positive), (b) the detector has correctly identified an intrusion, but the intrusion failed to reach its objectives (i.e., non-relevant positive alert), and (c) the detector incorrectly identified an event as an alert (i.e., false positive). The objective of alert verification is to differentiate between successful and failed intrusion attempts. This process is relevant to correlation because it does not provide any additional information that helps the correlation with other detected alerts. While it does provide information about vulnerabilities, it adds more complexity to the correlation process.

The verification of a reported alert can be achieved by extending intrusion detection signatures with an expected *effect* of the intrusion, if it would succeed. The effect specifies the observable and verifiable traces that the succeeding intrusion may leave at a host or a communication node (on a log file or a resource, for example). This is what we assumed when we gave the definition of alerts in Subsection 11.4.2. Different techniques can be used to perform verification. Passive techniques require initial knowledge about network topology, hosts, and networked services in order to verify whether: (a) the target of an intrusion exists, (b) a malicious packet can reach the target, and (c) a potentially vulnerable service is running. Passive techniques present, however, different drawbacks including the limitation of the type of information that has to be collected initially and the fact that the state under monitoring is often different from the state initially stored (in a knowledge base, for example).

Active verification techniques dynamically check for a proof of the success of an intrusion associated with a gathered alert. This often requires a communication connection be established with the vulnerable target. The connection allows operations of scanning the intrusion target and assessing whether the target application is still available or whether it has failed. Active techniques, however, present some disadvantages. Active actions are observable on the network and scanning operations could have an adverse effect on the scanned systems. To overcome these shortcomings, the solution designer can use techniques that are able to collect information using authenticated accesses to hosts. The alert verification process can remotely log and execute system commands by using such techniques.

However, one can argue that an adversary can act on the compromised system to hide or delete suspicious traces or to hide his/her malicious activity from the auditing system and digital investigation capabilities. One solution to protect against such attacks is to run audit tools that require root privileges to be turned off. The sensors can operate in a best-effort mode and deliver accurate and protected results, while operating autonomously.

11.5.3   Intrusion identification

The identification of the scenario of intrusion attempts to link network-based alerts to host-based alerts by finding the relations between these two different types of alerts. While network-based detectors provide information about the content of packets that contain evidence of attacks, the host-based detectors include information about the process that is attacked and the user on whose behalf the process is running. Relations use spatial and temporal dependencies based on a priori collected data about the port(s) used by network services and other useful information. This allows a network-based alert related to a traffic containing a known attack to be associated with an alert generated by a host-based detector with respect to an application listening to the port(s) addressed by the traffic. Other possible relations specify certain network-based alerts that report on intrusions, which are known to prepare for other attacks (Ning, 2002).

The second step in intrusion identification is to identify exactly the attack performed and the hosts that are actors (i.e., the source or the target of the intrusion). In particular, this step looks for distributed denial of service attacks (DDOS) by aggregating the alerts associated with single sources attempting to undermine a single target; and alerts that associate single victims attacked by multiple adversaries. Meta-alerts can be fused to report on DDOSs integrating sliding structures, which store the parameters needed to identify each attack and link alerts (e.g., number of different targets, number of occurrences of corresponding events during a period of time, etc.) and modify their content with time.

11.6   IDS products

Different classes of IDS products can be found commercially. They use various methods for detecting intrusions. Selecting and deploying the right IDS for the needs of a company is a difficult task. However, three major issues can be considered within this process: the placement of detectors, the selection of the IDS components, and the integration of the existing security solution. Different other requirements may be useful for particular needs including the capability of performing investigations, structured reporting, and system reliability. Assessing a product and comparing IDS products, with respect to the needs of a company, is also important for the selection of efficient solutions.

11.6.1   IDS requirements

Choosing and deploying the right IDS product is an important activity. Without a clearly defined set of requirements, the selected IDS product may generate more problems than it can help to solve. The main requirements for an intrusion detection system include, but are not limited to, the effectiveness, protection against zero-day intrusions, management capabilities of encrypted traffic, availability of specific security features, and maintainability.

Effectiveness

The capability of detecting intrusions is generally the first criterion to address when assessing an IDS product (or comparing two IDS products). Effectiveness can be addressed by monitoring, controlling, and analyzing various metrics including detection rate, false alarm rate, and detection duration. For this, an IDS product should be able to recognize a large number of types of attacks and should at least recognize typical attacks against the applications it is protecting. It should not alarm about normal behavior. It also should be able to stop attacks before they terminate or shortly after they end instead of only detecting them. In addition to these requirements, an IDS product should be able to fulfill its role correctly under heavy load or high-speed communication. It also should impose minimal overhead on the system it is protecting and use separate resources.

Ability to handle encrypted traffic

This is the second major requirement for IDS products. Since a large part of the authenticated and/or protected traffic involves encryption, processing the encrypted traffic and analyzing it to find signs of malicious activity is very important because encrypted traffic may hide the intrusion attempts that the traffic is transporting. A good IDS product should be able to inspect the encrypted traffic generated by the most used protocols such as the SSH, SSL, and IPSec protocol without any need for additional resources. The assessment of IDSs should check whether they can support this minimal list. It should also consider how the IDS products perform when decrypting the traffic and whether they create security holes.

Ability to secure IDS components

An IDS product should not be easily attacked. If an adversary succeeds to have an effect on the IDS security, he/she may be able to modify the alerts the IDS generates and thus he/she is able to control the decisions of the IDS. Therefore, nothing reported by the IDS can be trusted. An IDS product should have implemented mechanisms to protect its components and the communication it generates. The IDS product must be able to monitor its own security status to ensure that it has not been compromised. Particularly, sensitive traffic between the sensors and the analysis centers should be protected against confidentiality and replay attacks.

Any detector needs a mechanism for generating reports when it detects an intrusion. Therefore, there are some desirable characteristics in the reporting mechanisms that should be met:

- The reporting mechanisms used by a detector should not be used by any other sensor (or another application) at the host where the detector is implemented.
- If a number of detectors have to coexist in a host, the reporting mechanisms need to use a minimum amount of resources. Reports need to be available as soon as they are generated and may need to be guaranteed confidentiality and integrity.

- It should be difficult for an attacker to disrupt the reporting mechanism, either by inserting invalid messages, or by intercepting or modifying the messages that the detectors generate. The messages should be stored inside kernel memory, so that they cannot be modified by an attacker unless he has root privileges.

Protection against zero-day intrusions

It is commonly understood that zero-day attacks represent a major challenge that the IDSs have to fight against. A zero-day attack can be defined as any exploit that is written for a previously unknown vulnerability (i.e., a new undocumented attack for which a signature or definition has not been written or posted yet). A zero-day attack can exploit a known vulnerability with a new pattern of malicious activity. Zero-day attacks can cause more damage to the signature-based IDS tools compared to that of anomaly-based detectors. Since signatures would not be available at the occurrence of a zero-day intrusion in the system to protect, signature-based IDSs can easily be got around and the related attack would not be detected until its damages are observed.

IDS products should protect against zero-day intrusions. They also should be protected against these intrusions. Since signature-based systems depend on the availability of a signature database, the database itself is a major victim of zero-day attacks. In order to protect against this threat, signature-based IDSs must be updated accordingly. Assessment of IDSs with respect to zero-day intrusions should consider the frequency of updates, the time spent by the product provider to release a new signature for an announced vulnerability and the capabilities offered by the anomaly-based IDS components to take care (partly or totally) of the task of protection against zero-day attacks.

Maintainability

An efficient maintainability is able to reduce the problems encountered by security administrators and provides a better protection of the IDS products. To promote maintainability the IDS manufacturer should first make easy the installation and deployment of the IDS components. Automated install scripts and centralized management for the deployment of IDS instances represent examples of tools that help dealing with installation and deployment. Second, the manufacturer should provide an easy management of the IDS product. A graphic user interface allowing the administrator to remotely control the components of the IDS will help the maintainability of the IDS. The system should be easily customized to support the creation of user profiles. Finally, the databases used for the need of detection should be easy to maintain and interoperate with.

11.6.2   Product survey

Nowadays, the number of available commercial products exceeds 200. Based on requirements, definition, architecture, and function of an IDS, this number can be reduced to less then 20 (Anttila, 2004; Raju, 2005). The authors in (Anttila, 2004) have conducted an evaluation of about 217 IDS products in order to find solutions which are able to secure the

Table 11.1 *Protection level against attacks for products unselected by (J. Anttila, 2004).*

	Storm Watch	Secure IIS	Applock Web	STAT neutralizer
Application buffer overflow	medium	medium	medium	non
Back door	non	low	non	non
Cookie poisoning	non	non	non	non
Cross side scripting	non	non	non	non
Forceful browsing	medium	medium	non	medium
Hidden field manipulation	non	non	non	non
Parameter tempering	non	low	medium	non
Known vulnerabilities	high	high	medium	high
Third party misconfigurations	medium	medium	non	medium

so-called "*Microsoft-based reference system*," while considering the reduction of the load of the security administrator in this process. The purpose of the reference system is to provide web-based services for users accessing the system from the Internet. The evaluation process involved three phases. During the first phase, the products are checked based on the following criteria:

- Vendor and product availability This is decided based on the information available at the vendor's homepage.
- Compatibility with the reference system This imposes the operating platform for the IDS components, which must work with Windows.
- Product type This imposes that the communication to servers be encrypted, which limits the use of network-based components.
- Integration This measures the ability of the solution to integrate smoothly host-based solutions.

Among the 217 evaluated solutions, only ten solutions were selected, meaning that only ten IDS products can smoothly integrate host-based IDSs, have an encrypted communication with servers, work with the reference model, and have vendor available on the Internet. The second phase attempted to define how the ten solutions protect web applications against major attacks. For this, the products were installed, configured, and tested with a list of attacks including: the application buffer overflow attack, back door and debug options, modification of the content of cookies, hidden field manipulation, parameter tampering, and other important attacks. Table 11.1 shows the level of protection provided by four non selected solutions (Storm Watch, Secure IIS, Applock Web, and STAT neutralizer) against nine attacks.

During the second phase, four products are selected (i.e., SecureWeb, TerosAPS, InterDo, and AppShield). They all protect against the aforementioned attacks at high level. During the third phase, the four solutions were checked against some of the criteria developed in the previous subsection, say effectiveness, performance security, reliability, and maintainability. The evaluation finally shows that TerosAPS presents a better solution compared to the others. Even though the evaluation has defined a constraining reference system and cut down a lot of products without looking for how they protect against attack (at least when implemented in another reference model), the evaluation process made a good job with the selected ten products.

Another evaluation of IDS products has been discussed in Raju (2005), where six products have been assessed using the criteria developed in the previous subsection. Most of these products have been shown to support a large range of advanced features such as stateful detection of intrusions, signature management, deep packet inspection, and active responses. However, despite the quality of these evaluations, it is essential that a more elaborate evaluation process be performed on the IDS product. Moreover, provision of the methodology of IDS assessment, and the development of a set of assessment guidelines should be according to those provided by the American Information Security Commission for PKI in (ISC, 2001).

11.7   Advanced issues in intrusion detection

11.7.1   Distributed intrusion detection

Typically, real-time level intrusion detection sensors were stand-alone components which were able to collect events, process alerts, and take actions on their own. However, the deployment of these sensors in large-scale networks has led to new developments to improve these products, since they are expected to cooperate and share information about traffic and data. This also has led to an important effort in research and development activities on distributed intrusion detection systems. Distributing the location of the intrusion detection sensors across the network allows the security administrator to have a large view of the network activity, since he/she can identify intruders, which perform attacks against the network, correlate attack signatures among different segments on the network, coordinate counter actions by following the physical route taken by the packets to trace the attack sessions (e.g., identify the user location even if he provides a false IP address) and reduce the detection cost by sharing intrusion detection devices' resources.

Coordinating the interaction between these detectors has created a number of problems. Some of these problems were already known through the activity developed on network management. However, while the network management tends to be centralized and to have system states easy to define, intrusion detection takes this to a dynamic situation where the system monitors periods of time and series of activities, instead of comparing system snapshots. Some of the major issues that need to be addressed here include: event generation and storage, distributed state-space management and rule complexity, and knowledge repositories and distributed inference activities. Event generation and storage consider the locations where events are generated and stored in a distributed or centralized manner. Previous intrusion detection systems have been based on centralized generation, storage and processing of events. This approach does not cope with the scale of large networks. The huge number of collected events accompanied by the large number of detectors deployed across the network can generate too many alerts to be stored efficiently in one location. It also does not cover distributed services.

Distributed state-space management and rule complexity deal with the decision on how complex the rules should be and the demands they impose on processing the traffic through the network. Complex rule models are more effective, but require complex state management and analysis algorithms. The CPU requirements to process large audit logs will typically limit the real-time application of rules. Knowledge repositories are locations containing the rules. Since rules need to be updated and distributed to intrusion detection entities, the use of one central point of distribution would minimize management concerns. But, this model presents a scaling problem, since rule updates need to be distributed across the network in an ordered real-time manner.

Various practical issues need to be considered by the network designer to address the concern on how this distributed intrusion detection will impact the existing network infrastructure. Four issues need to be addressed:

- Traffic management The communication between intrusion detection components across the network can generate a large traffic on the network.
- Protection of log traffic The intruders may be able to compromise the flow of data across the network or obtain useful information on systems security by monitoring the log traffic.
- Event description The intrusion detection components need to be able to have a standard event description so that they can share information. The development of an event/alert model is essential. Events should not be distributed as audit logs across the network. The distributed sensors should process the logs and share, in a distributed manner, their events and/or alerts.
- Traffic pattern description The modeling of network traffic patterns needs to be developed for optimum placement of the intrusion detection components. Network implementations vary widely from company to company. Hence, determining traffic patterns for the need of intrusion detection is a non-trivial task.

11.7.2 Intrusion detection for high-speed network

As transmission speeds in networks are getting higher and higher, the emerging need is to get better performance of the network-based IDSs since it is very difficult for traditional N-IDS to capture all the packets and perform so complicated an intrusion analysis during short periods of time. When data processing cannot cope with the speed and the throughput of networks, the packets which are not analyzed in time are often dropped. In addition, some intruders can attempt to overload the network with a large number of packets in order, for example, to force a denial of service.

To solve this problem, various approaches have been made available. The first approach introduces distributed frameworks in data collection to enable parallel and real-time data processing (Kruegel, 2002). The overall network traffic can be divided into traffic of manageable size and accordingly the traffic load on a single machine can be reduced. A second approach considers the problem of improving the performance of data analysis and concentrates on the construction of fast and efficient parallel multi-pattern search algorithms

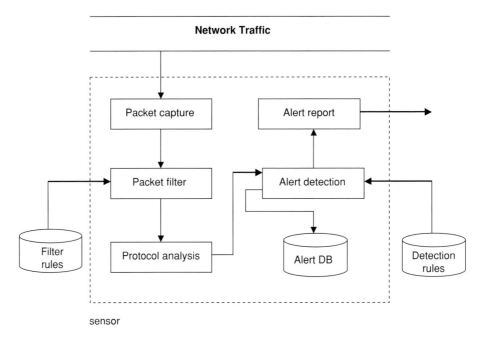

Figure 11.5   A generic sensor architecture.

(McAlemey, 2001). Figure 11.5 depicts a generic architecture of a sensor and shows the major functions that can be affected by the network performance and data load.

In a high-throughput network, the total amount of resources (e.g., CPU) devoted to system calls is not trivial. The memory bandwidth is another concern because too many operations will consume a large amount of CPU and memory resources. In order to improve the performance of the packet capture process, it is necessary to reduce the intermediate steps during packet transmission, attempt to eliminate kernel's memory, and use a zero-copy technique.

The packet filter is a function that aims to reduce the volume of data to be analyzed by the security analysis function. This is achieved by removing non-interesting network packets, while protecting the N-IDS itself from intrusions such as denial of service. The packet filter simply analyzes some fields of the packet header, such as the IP addresses and port numbers and applies filter rules to drop non-interesting packets. In practice, packet filters often use a large number of filter rules. To satisfy the requirement of performance, the packet filter should be designed with the following objectives in mind: the packet should be processed only once and all the rules applying to the packet should be found and the system running should be made insensitive to the number of rules.

Protocol analysis takes the application layer into consideration. It often includes an IP de-fragmentation and TCP stream reassembly sub functions. The analysis has to identify the application protocols of the packet and calls the appropriate application protocol decoding (or reassembly) function. Then the process decodes each packet and marks any anomalous packet appropriately. The application protocol analysis should concentrate on primitive packets. It also should be possible to define specific rules on applications, which direct the pattern matching to the interesting packets and skip non-interesting packets.

References

Anttila, J. (2004). Intrusion detection in critical e-business environment. Master Thesis, University of Helsinki.

Cuppens, F., F. Autrel, Al. Miege, and S. Benfarhat (2002a). Correlation in an intrusion detection process. SECI02, France.

Cuppens, F., F. Autrel, A. Miege, and S. Benferhat (2002b). Recognizing malicious intention in an intrusion detection process, Proceedings of Soft Computing Systems – Design, Management and Applications, HIS 2002, December 1–4 2002; *Frontiers in Artificial Intelligence and Applications*, Vol. **87**, 806–17.

Curry, D. and H. Debar (2003). Intrusion detection message exchange format: Extensible Markup Language (XML) document type definition. draft-ietf-idwg-idmef-xml-10.txt.

Haines, J., D. K. Ryder, L. Tinnel, and S. Taylor (2003). Validation of sensor alert correlators. *IEEE Security & Privacy Magazine*, **1**(1), 45–56.

Information Security Committee (2001). PKI Assessement Guidelines.

Kruegel, C., F. Valeurm, and G. Vigna (2002). Stateful intrusion detection for high-speed networks. 2002 IEEE Symposium on Security and Privacy, Berkeley, California, May 2002, pp. 266–274.

J. McAlemey, C. Colt, and S. Staniford (2001). Towards faster string matching for intrusion detection or exceeding the speed of snort, DARPA Information survivability conference and exposition, Anaheim, California, June 2001, pp. 367–373.

Ning, P., Y. Cui, and D. S. Reeves (2002). Constructing attack scenarios through correlation of intrusion alerts. In *Proceedings of the ACM Conference on Computer and Communication Security*, Washington DC, Nov. 2002, ACM, pp. 245–54.

Pennington, A. G., J. D. Strunk, J. L. Griffin, C. A. N. Soules, G. R. Goodson, and G. R. Ganger (2003). Storage-based intrusion detection: watching storage activity for suspicious behavior. Usenix Security Symposium, Washington, August 2003.

Porras, P. A. and A. Valdes (1998). Live traffic analysis of TCP/IP gateways. Proceedings of the 1998 ISOC symposium on Network and Distributed System Security (NDSS'98), San Diego, CA.

Naga Raju, P. (2005). State-of-the-art intrusion detection: technologies, challenges, and evaluation. Master Thesis, University of Helsinki.

Shankar, M., N. Rao, and S. Batsell (2003). Fusing intrusion data for detection and containment, Milcon 2003, *IEEE military communication conference*, Vol. **22**, No 1, 741–8.

Undercoffer, J., A. Joshi, and J. Pinkston (2003). Modeling computer attacks: an ontotlogy for intrusion detection. 6th International Symposium on Recent Advances in Intrusion Detection.

Zamboni, D. (2001). Using internal sensors for computer intrusion detection. PhD Thesis, TR CERIAS TR 2001–42, Purdue University.

12 Virtual private networks

This chapter deals with virtual private networks (VPNs), which have become more and more important for all kinds of businesses with a wide spectrum of applications and configurations. This chapter presents the basics and techniques of virtual private networks. We also review VPN services that include Intranet, Extranet and Remote Access VPNs. Security concerns that arise when transmitting data over shared networks using VPN technology are also addressed in detail. The fundamental VPN models, namely the peer and the overlay model are treated as well. The protocols employed in VPNs such as PPTP and L2TP as well as security aspects are also discussed. It is expected that VPNs will be in a position to support a set of QoS levels. We treat this subject in a dedicated section. We conclude this chapter by summarizing the main advantages and challenges of VPNs.

12.1 Introduction

A Virtual Private Network (VPN) is a private network connecting different sites or corporate offices by using public telecommunication infrastructure (Internet) using encryption and tunneling protocol procedures for secured and reliable connectivity. One other definition states that a VPN is a private data network that makes use of the public telecommunications, maintaining privacy through the use of tunneling protocol and security procedures. Others have defined a virtual private network as a network that allows two or more private networks to be connected over a publicly accessed network (Papadimitriou *et al.*, 2004; Metz, 2003; Ferguson and Huston, 1998; Hunt and Rodgers, 2004; Arora *et al.*, 2001). Ferguson and Huston came up with the following definition: A VPN is a communications environment in which access is controlled to permit peer connections only within a defined community of interest, and is constructed through some form of partitioning of a common underlying communications medium, where this underlying communications medium provides services to the network on a non-exclusive basis. VPNs use the concept of "virtual" connections (Ferguson and Huston, 1998). By routing through the Internet, a VPN connects the organization's private network and the remote users. Thus, a VPN can be viewed as an improved WAN (connecting multiple LANs). However, "using Internet as a medium" presents concerns for network security. Everyone who uses the Internet can see the traffic that passes between a remote office and main office over these insecure Internet connections. Clearly, efficient encryption schemes should be used for VPNs.

The mode of business and communication has changed a lot in the last few decades, where companies deal with the global community and have offices all around the world. Many companies have facilities spread out across the globe, and the major requirement is a fast, secure and reliable communication between their offices located all over the globe. Until fairly recently, companies used leased lines (T1), private lines, ATM or Frame Relay techniques to communicate between their offices.

Virtual Private Networks (VPNs) have evolved as a compromise for enterprises desiring the convenience and cost-effectiveness offered by shared networks, but requiring the strong security offered by private networks. Whereas closed WANs use isolation to ensure security of data, VPNs use a combination of encryption, authentication, access management and tunneling to provide access only to authorized parties, and to protect data while in transit (Strayer and Yuan, 2001; Brahim et al., 2003; Younglove, 2000; Gunter, 2001; Wright, 2000; Cohen, 2003; Braun et al., 1999; Harding, 2003; Ribeiro et al., 2004; Srisuresh and Holdrege, 1999; Rosenbaum et al., 2003; Gleeson et al., 2000; Pall et al., 1999; Hanks et al., 1994; Simpson, 1994; Townsley et al., 1999; Yuan, 2002; Rekhter et al., 1995; Tomsu, 2002; Zeng and Ansari, 2003; Braun et al., 2001; Papadimitriou and Obaidat, 2006; Bova, 2006; Nicopolitidis et al., 2003). To emulate a point-to-point link, data is encapsulated, or wrapped, with a header that provides the routing information allowing it to go across the shared or public internetworks to reach its destination. To emulate a private link, the data being sent is encrypted for privacy and confidentiality. Packets that are intercepted on the shared or public network are indecipherable without the encryption keys (Arora et al., 2001; Younglove, 2000).

In order to enable the use of VPN functions, special software can be added on the top of a general computing platform, such as a UNIX or Windows Operating System (OS). Alternatively, special hardware augmented with software can be used to provide VPN functions. Sometimes, VPN functions are added to a hardware-based network device such as a router or a firewall. In other cases, VPN functions are built from the ground up and routing and firewall capabilities are added. A corporation can either create and manage the VPN itself or purchase VPN services from a service provider (Strayer and Yuan, 2001).

VPNs came into existence around 1984, when the service was offered to US users by Sprint. Shortly after that, MCI and AT&T produced competing products. In their early form, VPNs were offered as a flexible, cost-effective scheme to the problem of connecting large, isolated groups of users. Private networks were the main alternative, and although equivalent functionality could be achieved via the Public Switched Telephone Network (PSTN), this was a restricted solution due to the length of full national telephone numbers and the lack of in-dialing capabilities from the PSTN. Such a VPN acted as a PSTN emulation of a dedicated private network, using the resources of the PSTN in a time-sharing arrangement with other traffic (Hunt and Rodgers, 2004).

There are many forms of VPNs that are available commercially despite the relatively short lifetime. The reason for these different forms is due to the distinct domains of expertise of the companies developing and marketing VPN solutions. For instance, hardware manufacturers may provide customer premise equipment-based solutions, whereas an Internet Service Provider (ISP) would provide a network-based solution.

The level of disorder and lack of inter-operability between rival products led to the submission of a VPN framework to the IETF in 1999, which defined VPNs as "the emulation of a private WAN facility using IP facilities." Various VPN implementations have been proposed to operate over a variety of underlying infrastructures and protocols, including Asynchronous Transfer Mode (ATM), Multi-Protocol Label Switching (MPLS), Ethernet, IP, and heterogeneous backbones (Hunt, 2004).

Conventional private networks assist connectivity among various network entities through a set of links. These are leased from public telecommunication carriers as well as privately installed wiring. The capacity of these links is available at all times, though fixed and inflexible. The traffic on these private networks belongs only to the enterprise or company deploying the network. Thus, there is an assured level of performance associated with the network. Such assurances come with a price. The drawbacks of this approach are (Papadimitriou *et al.*, 2004; Arora *et al.*, 2001):

1.   Conventional private networks are not cheap to plan and install. The costs associated with dedicated links are especially high when they involve international sites and security concerns are aggravated due to the international transport of information.
2.   The planning phase of such networks involves detailed estimates of the applications, their traffic patterns and their growth rates. Also, the planning periods are long because of the work involved in calculating these estimates (Papadimitriou *et al.*, 2004).
3.   Further, dedicated links take time to install. It is not unusual that telecommunication carriers take on the average about 75 days to install and activate a dedicated link, which negatively affects the company's ability to react to the quick changes in these areas (Arora *et al.*, 2001).

In order to support the upsurge in home offices, corporations have to provide a reliable Information Technology (IT) infrastructure so that staff can access a company's information from remote locations, which has resulted in large modem pools for employees to dial-in remotely. The cost keeps increasing due to the difficulty in managing the large modem pools. A supplementary charge in the case of mobile users is the long-distance calls or toll-free numbers paid for by the company. The costs are much higher if we consider international calling. For companies with a large mobile workforce, these expenses are significant (Papadimitriou *et al.*, 2004; Arora *et al.*, 2001).

Furthermore, in the case of the private network, the corporation has to manage the network and all its associated elements, invest capital in network switching infrastructure, hire trained staff, and assume complete responsibility for the provisioning and on-going maintenance of the network service. Such a dedicated use of transport services, equipment, and staff is often difficult to justify for many small-to-medium sized organizations, and while the functionality of a private network system is required, the expressed desire is to reduce the cost of the service through the use of shared transport services, equipment, and management (Ferguson and Huston, 1998; Papadimitriou and Obaidat, 2006).

In this way, VPNs have great potential to help organizations support sales over the Internet more economically, tying business partners and suppliers together, linking branch offices, and supporting telecommuter access to corporate network resources. A VPN can minimize costs by replacing multiple communication links and legacy equipment with a single connection and one piece of equipment for each location (Younglove, 2000). To extend the reach of a company's Intranet(s), a VPN over the Internet promises two benefits: (a) cost efficiency and (b) global reachability. On the other hand, there are three major concerns about VPN technology: security, manageability, and performance (Wright, 2000; Tyson, 2006).

1. Security In order for Virtual Private Networks to be private the transmitted data must be encrypted before entering the Internet, since the Internet is considered an untrusted network. Everybody can connect to the Internet and there is no warranty that participants stick to any policy or rule. However, protecting the traveling data will not protect the information inside the Intranet from unauthorized access (Wright, 2000).
2. Manageability The organizations' communication requirements grow at a high rate. VPN management has to deal with these rapid changes in order to avoid extra operating cost. Moreover, VPNs are connected to a lot of different entities that are difficult to manage as well. Such entities include the organization's physical network, security policy, electronic services and Internet Service Providers (ISPs).
3. Performance Because ISPs deliver IP packets on a "best effort" basis, the transport performance of a VPN over the Internet cannot be forecasted. Moreover, security measures can reduce performance considerably (Wright, 2000).

A VPN solution requires having multiple, appropriately configured VPN devices that are placed in the appropriate locations within the network. The network must be monitored and managed continuously (Strayer and Yuan, 2001).

The most familiar VPN device is the VPN gateway, which acts as the doorkeeper for network traffic to/from protected resources. Tunnels are established from the VPN gateway to other appropriate VPN devices acting as tunnel endpoints. A VPN gateway is usually located at the corporate network edge, and it works on behalf of the protected network resources within the corporate intranet to negotiate and provide security services. The gateway assembles the tunneling, authentication, access control, and data security functions into a single device. The particulars of how these functions are incorporated within a VPN gateway are unique to a vendor's implementation. Occasionally, these functions can be incorporated into existing router or firewall products. Occasionally, a VPN gateway can be a separate device that executes pure VPN work, without firewall or dynamic routing exchange means (Strayer and Yuan, 2001).

The inbound traffic is thoroughly checked by the gateway according to the security policies. Usually, only the traffic from the already existing secured tunnels must be processed by the VPN gateway. If no secured tunnel is established, the traffic should be dropped instantly, except if its purpose is to negotiate and establish a secured tunnel. Conditional on the realization, an alert can be generated to inform the network management station.

The outbound traffic is considered secure a priori, even though many network attacks are created within a corporate network. The outbound traffic is examined based on a set of policies on the gateway. If secured tunneling is needed for the traffic, then the VPN gateway first finds out whether such a tunnel is already in position. Otherwise, the gateway tries to establish a new tunnel with the anticipated device. As soon as the tunnel is made, the traffic is dealt with according to the tunnel strategy and is sent into the tunnel. The traffic from the private interface can also be dropped if a policy cannot be found. Based on how quickly a secured tunnel can be established, VPN gateway may buffer the outbound packet before the secure tunnel is in place (Strayer and Yuan, 2001; Papadimitriou and Obaidat, 2006).

The VPN client is basically a program used for remote VPN access for a single computer or user. In contrast to VPN gateway, which is a specific device and can guard multiple network resources at the same time, VPN client software is typically mounted on an individual computer and serves that specific computer only. In general, VPN client software creates a secure path from the client computer to a designated VPN gateway. The secure tunnel allows the client computer to acquire IP connectivity to get into the network resources protected by that specific VPN gateway.

VPN client software also must realize the same tasks as VPN gateways – tunneling, authentication, access control, and data security – although these implementations may be simpler or have less options. For instance, VPN software typically realizes only one of the tunneling protocols. Since the remote computer does not act on behalf of any other users or resources, the access control can also be less complex.

In contrast to VPN gateway, in which all of the gateway's hardware and software is directed toward the VPN functionality, VPN client software is typically an application that runs on a general-purpose operating system on the remote machine. Therefore, the client software should thoroughly consider its interactions with the OS.

One main worry regarding VPN client software is the ease of installation and operation. Since client software is likely to be deployed widely on end users' machines, it must be easily installed and operated by regular computer users who may not have much knowledge about the OS, software compatibility, remote access, or VPNs. Conversely, a VPN gateway is usually installed on the company's corporate network site and is managed by information technology specialists. VPN client software should operate within the limits of hosts' operating systems, regardless of whether the software is deeply integrated with the OS or runs as an application. Since networking is a foremost function of today's computers, nearly all state-of-the-art operating systems have fixed networking functionality. Nevertheless, a few have keen support for VPN capabilities integrated into the OS. As a result, separate VPN client software must be written for those operating systems that lack the VPN functionality. Furthermore, the same VPN client software may need to be ported to different operating systems, even for different release versions of the same OS.

There exist two chief architectures for realizing network-based VPNs: virtual routers (VR) and piggybacking schemes. The major distinction between them is basically the model used to attain VPN reachability and association functions. In the VR scheme, each VR in the VPN domain runs a routing protocol in charge of distributing VPN reachability information between VRs. Thus, VPN association and VPN reachability are dealt with as distinct operations, and separate mechanisms are used to realize these functions. VPN reachability is performed by a per-VPN type of routing, and a range of schemes is likely to determine membership. In the case of the piggyback model, the VPN network layer is ended at the edge of the backbone, and a backbone routing protocol such as the Extended Border Gateway Protocol-4, EBGP-4, is responsible for distributing the VPN membership and reachability information between provider edge (PE) routers for all the VPNs configured on the PE (Brahim *et al.*, 2003; Younglove, 2000; Gunter, 2001; Wright, 2000; Cohen, 2003).

12.2   Elements of VPNs

Some general elements are identified and described in order to familiarize readers with basic terminology used in the VPN world:

* Secure firewall
* VPN client software on user desktop
* Dedicated VPN server
* Network Access Server (NAS)

A brief description of these is given below.

Secure Firewall/VPN Router

A firewall is a program or hardware device that guards the virtual private network from potential hackers and offensive web sites. It filters (blocking or letting through) the information coming from the Internet connection to the network. The firewalls can be configured to restrict the number of open ports, types of packets and protocols that are allowed to pass through. Thus, if an incoming packet is flagged by the filter, the firewall blocks and does not allow the packet to go through.

A VPN router usually includes firewall capability and provides routing and Internet connection sharing.

VPN Client Software

The organization provides this client software on the user's computer for VPN access to a private network. The user runs the VPN Client software to establish the connection to the VPN server to keep connection and data secure.

VPN Server

A VPN server can be a piece of hardware or software that acts as a gateway to a network or a single computer. In general, it waits for a VPN client to connect to it and processes requests from the VPN client.

Network Access Server (NAS)

A network access server is a network device that authenticates clients and associates clients with specific VPN servers. Usually, it is used by Internet Service Providers (ISPs) for remote-access VPN users.

12.3　Types of virtual private networks

There are varieties of VPN implementations and configurations that exist to meet the needs of different users. Organizations may require their VPN to provide dial-up access, or permit third parties such as clients or suppliers to access specific components of their VPN. VPNs can be categorized into three wide categories (Hunt and Rodgers, 2004): (a) Intranet, (b) Extranet, and (c) Remote Access VPNs.

Intranet VPNs

VPNs based on the Internet have become attractive for a wide range of applications. In general, when we speak about VPNs, we mean Internet-based networks as an alternative to private networks based on public network services such as leased lines or Frame Relay. Internet service providers (ISPs) have increased their services tremendously. It is now possible to obtain connections from anywhere including remote locations. The majority of countries worldwide now have ISPs offering connections to the Internet. Thus, it is possible for many organizations, both large and small, to consider the Internet not just for external communications with customers, business partners and suppliers, but also for internal communications by using VPN technology. Whereas, Internet VPNs are appropriate for remote access needs, there are still problems that have to be overcome before moving to a full Intranet VPN solution.

An Intranet VPN connects a number of local area networks (Intranets) located in multiple geographic areas over the shared network infrastructure. Typically, this service is used to connect multiple geographic locations of a single company (Arora *et al.*, 2001). In this scheme, the goal is to share information and resources amongst dispersed employees. For instance, branch offices can access the network at the headquarters, normally including major resources such as product or customer databases. Intranet access is limited to these networks, and connections are authenticated. Various levels of access may be allocated to different sites on the Intranet based on their goal (Papadimitriou and Obaidat, 2006). Because an Intranet VPN is formed by linking two or more trusted sites such as corporate LANs that are certainly protected by firewalls, most security fears are eased.

Extranet VPNs

In this type of VPN, limited access of corporate resources is given to business partners, such as customers or suppliers, enabling them to access shared information (Wright, 2000). These users are allowed to access specific areas of the Intranet that are referred to as the De-Militarized Zone (DMZ). The firewall and access management facilities are responsible for differentiating between the company's employees and other users as well as each group's privileges. The requests for connection by the company's employees must be directed to the company Intranet, while these by a third party must be directed to the DMZ (Hunt and Rodgers, 2004; Papadimitriou and Obaidat, 2006).

There are many possible Extranet configurations that vary in their degree of security and access. The major schemes are reviewed briefly below in a decreasing order of security (Hunt and Rodgers, 2004; Papadimitriou and Obaidat, 2006):

- Private Extranet Here, access to a private Extranet is only limited to members with no use made of shared networks. This configuration cannot be considered a VPN, as it is physically private.
- Hybrid Extranet A hybrid Extranet corresponds to a private Extranet with the exception that it exploits one or more shared networks in order to give connectivity. Connection is limited, and access to private resources is restricted to pertinent resources.
- Extranet Service Provider Here, the service is offered by an ISP that develops Extranet services based on its backbone network. Clearly, this is a type of provider-provisioned VPN.
- Public Extranet This configuration provides data that is globally available. One good example is an organization that provides a public Web site and possibly a public free FTP site that Web users can use free of charge. These facilities are normally separate from private file servers; therefore, public servers cannot be used as platforms for compromising the classified component of the Extranet.

Remote Access VPNs

This type of VPN connects telecommuters and mobile users to corporate networks. Ideal VPNs enable remote users to work as if they are at their workstations in their offices. Installing remote access VPNs can result in considerable cost savings, eliminating the need for organizations to manage large modem pools, and substituting the need for toll-calls to these modems by calls to local ISPs. By using a high-speed access infrastructure, some of the performance limitations typically associated with remote access can be alleviated (Hunt and Rodgers, 2004; Papadimitriou and Obaidat, 2006).

Many organizations have allowed more employees to telecommute due to their business requirements. Examples include sales representatives on the road or software developers who work at home. VPNs do not only support individual remote users, but also support Intranet or Extranet services as an enhancement to conventional WANs. Obviously, VPNs can be categorized into two main classes: (a) remote-access VPN or (b) site-to-site VPN based on the organization's need.

- Remote-Access VPN In a remote-access VPN, if a remote user desires to log into an organization's VPN in order to use its internal network resources, he only has to link up to a local service provider or an ISP to get to the VPN server. The VPN client software is in charge of the establishment of a connection to the VPN server. The user is permitted to access the resources using tunneling if he is on the internal LAN. Layer 2 Tunneling (L2T) Protocol is used to support the tunneling of Point-to-Point Protocol (PPP) sessions containing user data through the Internet. An illustration of a VPN Remote Access Architecture is depicted in Figure 12.1 (Ferguson and Huston, 1998; Papadimitriou et al., 2004).

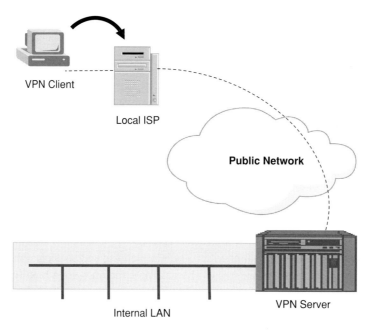

Figure 12.1 An example of a VPN remote access architecture.

- Site-to-Site VPN In this configuration, a company can link multiple fixed sites using the Internet. There are two types of Site-to-Site VPN (Metz, 2003):

  1. Intranet-based In this configuration, a company uses an Intranet VPN to connect LAN to LAN in order to make a single private network connecting one or more remote locations.
  2. Extranet-based Here, a company can employ Extranet VPN to connect LAN to LAN from other organizations such as business associate, dealer, or client.

Figure 12.2 shows the types of VPNs: remote-access VPN and site-to-site VPN (Intranet VPN and Extranet VPN). As shown in the figure, a VPN extends the main office LAN across the un-trusted Internet to the remote office, where both end points consist of corporate VPN gateways. The VPN gateways take away the encapsulation of protected Internet traffic and present it to the local network as LAN traffic.

12.4   VPN considerations

Before an organization plans to implement a Virtual Private Network (VPN), there are certain factors that should be considered. Among these are:

- Is it needed to connect sites together or individual remote users to central LAN?
- What are the types of network equipment and operating systems?
- Are there any existing devices that have VPN functionality?
- What are the main features of software and hardware solutions?

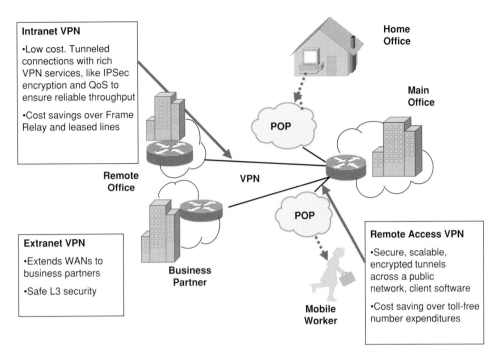

Figure 12.2   Examples of VPN types.

- How many users are supposed to be in the VPN environment?
- What tunneling scheme is used?
- What speed is to be used? Speed is a key criterion in choosing a VPN. For large ventures, there are VPN solutions that run at speeds up to 2 Gbit/sec and offer 100 to in excess of 40 000 VPN tunnels. The significant component of utilizing the speed is the technology's capacity to scale over a line of products.
- What degree of flexibility is to be considered? Proper VPN infrastructure equipment should provide a scalable architecture that can grow with business. VPN equipment should provide multiple network interfaces, high reliability and availability, and inter-operability with existing infrastructure equipment and network management. VLANs and virtual routers enable multiple virtual networks to be hosted on one physical infrastructure. Realization flexibility should include a management interface that per-mits customization of the communication platform as well as rock solid fail-over for nearly every hardware and software component as is necessary within the network infrastructure.
- What degree of security and cost reduction? VPNs are considered a convincing business solution since they offer the highest end-to-end security at wire-speed at a substantial cost reduction compared to a private network. VPNs offer strong security for users and managers. A VPN solution, an IPSec client, and a firewall unite to control access to information, while protecting against Internet incursion. By implementing a VPN system, which incorporates additional security applications such as intrusion detection, digital certificate support, DOS and client authentication, an efficient and powerful

communications platform is developed using the Internet for the transfer of company business information.

- How secure the data needs to be during transmission?
- What approach will be used to manage the VPN? VPNs include a total management answer that offers network professionals easy, integrated access tools for global, site and unit management. Only a sole point of control is required for monitoring and provisioning the entire network by sustaining widely used enterprise-class management tools. Recent VPN products can integrate many VPN requirements into one machine, including firewalling, load balancing, content checking, intrusion detection, denial-of-service (DoS) detection, policy routing and management, among others. Centralization of important means is essential for managing mission-critical VPN applications trusted with the digital assets of an organization. Since VPN requirements have become more and more exact and integral to an organization's overall IT success, devices exist now to meet or surpass such expectations.

Such considerations need to be addressed before invoking the process of implementing a VPN system.

12.5   VPN implementations

There are two main VPN models: (a) the peer model and (b) the overlay model. In the peer model, the network layer forwarding path computation is realized on a hop-by-hop basis, where each node in the intermediate data passage path is a peer with a next-hop node. On the contrary, in the overlay model, the network layer forwarding path computation is not realized on a hop-by-hop basis. Here, the intermediate link layer network is used as a "cut-through" to another edge node on the other side of a large cloud. The overlay model presents some serious scaling worry in cases where large numbers of egress peers are needed. The reason for this is because the number of adjacencies increases in direct relationship with the number of peers and consequently the amount of computational and performance overhead needed to maintain routing state, adjacency information, and other detailed packet forwarding and routing information for each peer turns out to be a big responsibility in huge networks. In the case that all egress nodes in a cut-through network become peers in order to make all egress nodes one "Layer 3" hop away from one another, then there will be a great limitation on the scalability of the VPN overlay model (Papadimitriou et al., 2004; Metz, 2003; Ferguson and Huston, 1998; Hunt and Rodgers, 2004; Arora et al., 2001, Strayer and Yuan, 2001; Brahim et al., 2003; Younglove, 2000; Gunter, 2001).

A number of different approaches to the problem of providing VPN links and services may be taken. Particularly, a VPN may be implemented and secured at a number of different layers of the TCP/IP protocol stack. The chief approaches currently available are briefly described below.

- Network Layer VPNs (Papadimitriou and Obaidat, 2006) These VPNs are mainly based on IP and implemented using network layer encryption, and perhaps tunneling. All packets that enter the shared network are added with an extra IP header that contains a destination address, which marks the other end of the tunnel. After this node gets the packet, the header is detached and the original packet, which is destined to the given network, is retrieved. Because of this encapsulation, the original packets could be based on any network layer protocol without disturbing their transport across the shared network.

- Data Link Layer VPNs (Ferguson and Huston, 1998; Papadimitriou and Obaidat, 2006) This type of VPN uses a shared backbone network founded on a switched link layer technology like the Frame Relay (FR) or ATM. Connections among VPN nodes are realized using virtual circuits, which are usually economical, and adaptable with some level of guaranteed performance. These VPNs are very suitable for providing Intranet services; dial-up access is not well supported since the majority of ISPs offer connectivity via IP. The overwhelming cost savings related to VPNs are due to the use of ISPs. Clearly, IP-based network layer VPNs are more appealing than link layer VPNs if dial-up access is needed. Virtual circuit-based VPNs face the same scalability problems as those of node-to-node tunneled VPNs. Therefore, full-mesh architecture may not be likely. Other possible choices such as partial meshes or hub-and-spoke organizations address this restriction to a certain degree; however, such solutions may produce sub-optimal performance.

- Application Layer VPNs (Ferguson and Huston, 1998; Papadimitriou and Obaidat, 2006; Bova, 2003) These VPNs are realized in software, whereby workstations and servers are needed to carry out tasks like encryption, rather than postponing these tasks to dedicated hardware. Consequently, software VPNs are economical to implement; however, they can have a considerable influence on performance by limiting the throughput of the network, increasing the CPU usage, especially over high-bandwidth connections.

- Non-IP VPNs (Papadimitriou and Obaidat, 2006) Multiprotocol networks might also have necessities for VPNs. The most favorable method for implementing VPNs in multiprotocol systems is to rely on the encryption of the application layer.

12.5.1   Hardware components

In order to implement any VPNs, we need a number of hardware devices. The majority of these devices are common to customary networks; however, some of them have added burdens and tasks placed on them when applied to VPNs and their particular necessities. Below is a brief description of the common hardware devices employed by VPNs (Ferguson and Huston, 1998; Papadimitriou and Obaidat, 2006).

Firewalls

VPNs should be secluded from other users of the backbone network. This can be achieved by a tool called "Firewall." The latter provides significant services such as tunneling, cryptography, and route and content filtering. A firewall is a set of related programs, located at a network gateway server, which protects the resources of a private network from users from other networks. Essentially, a firewall, working directly with a router program, filters all network packets to find out whether to forward them toward their destination or not. A firewall is often installed away from the remainder of the network in order to avoid incoming requests from getting directly to the private network resources. Many firewall screening schemes are available. The simplest one basically screens requests to make sure they come from acceptable domain names and IP addresses (Papadimitriou, 2004; Bova, 2003).

Routers

Adding a VPN functionality to existing routers may degrade the performance, particularly at network critical points. Particularly, Multiple Protocol Label Switching (MPLS) VPNs tackle this problem by making only the perimeter (PE) routers VPN-aware. Thus, the core routers do not need to maintain the multiple routing tables which introduce a huge overhead on PE routers (Ferguson and Huston, 1998; Hunt and Rodgers, 2004; Arora *et al.*, 2001; Strayer and Yuan, 2001; Brahim *et al.*, 2003; Younglove, 2000; Gunter, 2001; Papadimitriou and Obaidat, 2006).

Switches

A number of switches provide facilities for improved separation of traffic by permitting a physical network to be divided into a number of Virtual LANs (V-LANs). In a typical switch, all ports are part of the same network, while a V-LAN switch can deal with different ports as parts of different networks if preferred.

Tunnel Servers

This can be offered by a VPN router or a firewall. If this additional responsibility is assigned to an already existing network's component, then a performance degrade may result (Hunt and Rodgers, 2004; Strayer and Yuan, 2001; Papadimitriou *et al.*, 2004).

Cryptocards

All effective encryption scheme algorithms, such as Triple-DES, are computationally costly and can limit the effective bandwidth to about 100Mbps if a dedicated cryptographic hardware is not used. In the case of a workstation, this dedicated hardware is offered as an extension card that could be integrated as part of the network interface card or as a separate card. Furthermore, some types of firewalls can offer hardware support for a range of encryption schemes (Hunt and Rodgers, 2004).

12.6   Protocols used by VPNs

There are several protocols that are employed by virtual private networks (VPNs). The popular ones are described below.

12.6.1   Point-to-point tunneling protocol (PPTP)

The PPTP (Pall *et al.*, 1999) allows the Point-to-Point Protocol (PPP) (Network Working Group, 1994), to be tunneled all the way through an IP network. The Point-to-Point Tunneling Protocol (PPTP) does not state any alteration to the PPP protocol, but rather explains a new vehicle for carrying PPP. The network server for PPTP is envisaged to run on a universal operating system whilst the client, referred to as a PPTP Access Concentrator (PAC), works on a dial access platform. PPTP states a call-control and management protocol that permits the server to control access for dial-in circuit switched calls initiating from an ISDN or PSTN to initiate outbound circuit-switched connections. A Network Access Server (NAS) affords temporary, on-demand network access to users that are point-to-point using PSTN or ISDN lines.

PPTP utilizes an extended version of the Generic Routing Encapsulation mechanism (Hanks *et al.*, 1994) to transmit user PPP packets. These improvements permit low-level congestion and flow control on the tunnels used to carry user data between PAC and PNS. Clearly, this scheme allows for effective use of the bandwidth offered for the tunnels and prevents needless retransmissions and buffer overruns. Moreover, PPTP does not decide the specific algorithms to be used for this low-level control, but it does describe the parameters that must be communicated so as to permit such schemes to work. PPTP permits available Network Access Server (NAS) functions to be divided using client-server architecture. Usually, the following functions are realized by a NAS (Pall *et al.*, 1999; Papadimitriou and Obaidat, 2006):

*   Physical local interfacing to PSTN or ISDN and control of external modems or terminal adapters.
*   Membership in PPP authentication protocols.
*   Channel aggregation and bundle management for PPP Multilink Protocol.
*   Logical termination of a Point-to-Point-Protocol (PPP) Link Control Protocol (LCP) session.
*   Logical termination of various PPP network control protocols (NCP).
*   Multiprotocol routing and bridging between NAS interfaces.

The PPTP splits these functions between the PPTP Access Concentrator (PAC) and PPTP Network Server (PNS). The decoupling of NAS functions provides the following benefits (Pall *et al.*, 1999):

*   A resolution to the "multilink hunt-group splitting" problem. Multilink PPP, normally used to combine ISDN B channels, needs all of the channels composing a multilink package (bundle) be grouped at a single NAS. Since a multilink PPP bundle can be handled by a single PNS, the channels comprising the bundle may be distributed across multiple PACs.
*   Adaptable IP address management. Users that employ dial-in service may maintain a single IP address as they dial into different PACs provided that they are served from a common PNS.
*   Provision of non-IP protocols for dial networks behind IP networks.

The PPTP protocol is realized only by the PAC and PNS. It is not required that systems be aware of PPTP. Also, dial networks may be linked to a PAC without being aware of PPTP. Typical PPP client software should remain operating on the tunneled PPP links. Moreover, PPTP can be used to tunnel a PPP session over an IP network. In such a configuration, the PPTP tunnel and the PPP session run between the same two machines with the caller acting as a PNS (Pall *et al.*, 1999).

Each tunnel is delineated by a PNS-PAC pair and it carries PPP datagrams between the PAC and the PNS. The protocol used in a tunnel is defined by a modified version of GRE. We can multiplex several sessions on a single tunnel. A control connection operating over TCP manages the founding, release, and maintenance of sessions and of the tunnel itself.

The PPP takes care of the security of user data passed over the tunneled PPP connection. Since the PPTP control channel messages are neither authenticated nor integrity protected, it is possible for an attacker to hijack the underlying TCP connection. It is also possible to manufacture false control channel messages and alter genuine messages in transit without detection. The GRE packets forming the tunnel itself are not cryptographically protected. Because the PPP negotiations are carried out over the tunnel, it may be possible for an attacker to eavesdrop on and modify those negotiations. Unless the PPP payload data is cryptographically protected, it can be captured and read or modified.

PPTP provides remote connections to a single point. It cannot support multiple connections and cannot easily support network-to-network connections. Furthermore, its security is also limited; it does not provide defense from substitution or playback attacks. It is worth mentioning that PPTP does not have a clear mechanism for renegotiation if connectivity to the server is lost (Gunter, 2001; Pall *et al.*, 1999).

12.6.2 *Layer-2 tunneling protocol (L2TP)*

The Layer-2 Tunneling Protocol (L2TP) helps the tunneling of PPP packets across a superseding network in a way that is as transparent as possible to both end-users (Hanks *et al.*, 1994). PPP describes an encapsulation scheme for transporting multiprotocol packets across layer-2 point-to-point links. Normally, a user obtains a layer-2 connection to a Network Access Server (NAS) using schemes such as dialup through the telephone system, ISDN, ADSL, and then runs PPP over that connection. In such a configuration, the layer-2 termination point and PPP session endpoint reside on the same physical device (Townsley *et al.*, 1999).

L2TP expands the PPP model by letting the L2 and PPP endpoints be located on different devices unified by a packet-switched network. A user that employs a L2TP has an L2 connection to an access concentrator such as a DSL Access Multiplexer. Then the concentrator tunnels individual PPP frames to the NAS. This way the actual processing of PPP packets is able to be separated from the termination of the L2 circuit. One advantage of such a division is that rather than needing the L2 connection to come to an end at the Network Access Server (NAS) the connection may end at a local circuit concentrator that then increases the logical PPP session over a shared infrastructure like a frame relay circuit or the Internet. From the point of view of the user, there is no practical distinction between having the L2 circuit end in a NAS exactly or via the L2TP (Townsley *et al.*, 1999; Papadimitriou and Obaidat, 2006).

In L2TP, it is easy to multiplex various calls from numerous users over a single link. Many L2TP tunnels can exist between the same two IP endpoints. These are usually identified by a tunnel-id, and a session-id. Signaling is supported via the inbuilt control connection protocol, allowing both tunnels and sessions to be established dynamically (Pall *et al.*, 1999).

L2TP can guarantee the sequenced delivery of packets. Such capability can be negotiated when the session is initiated. After this it can be switched on and off. As far as the tunnel maintenance is concerned, L2TP employs a keep-alive protocol so as to differentiate between a tunnel outage and lengthened periods of tunnel inactivity (Papadimitriou and Obaidat, 2006).

The tunnel endpoints in L2TP may carry out an authentication process of one another during tunnel initiation. Such a process offers realistic protection against replay and snooping during the tunnel initiation procedure. This scheme is not intended to provide any authentication away from tunnel founding. Clearly, it is not difficult for a cruel user to insert packets when an authenticated tunnel establishment has been finalized (Townsley *et al.*, 1999). In order to secure L2TP, the underlying transport should provide encryption, integrity and authentication services for all L2TP traffic. Such a secure transport runs on the whole L2TP packet and is functionally autonomous of PPP and the protocol used by PPP. Hence, L2TP only deals with confidentiality, authenticity, and integrity of the L2TP packets between the tunnel endpoints. Both L2TP and PPTP target the remote access scenario. It is worth mentioning that L2TP designates security features toward IP Security (IPSec). Moreover, it suffers from the same problems as PPTP (Gunter, 2001; Townsley *et al.*, 1999).

The decision on whether to choose PPTP or L2TP for deployment in a VPN depends on whether the control has to be with the service provider or with the subscriber (Papadimitriou and Obaidat, 2006). In reality, the dissimilarity can be distinguished based on the client of the VPN. The L2TP scheme is one of a "wholesale" access provider that has a number of configured client service providers, which appear as VPNs on the common dial access system, whereas the PPTP scheme is one of distributed private access where the client is a separate end-user and the VPN structure is that of back-to-back tunnels. The difference can also be based on economics where the L2TP model permits the service provider to essentially provide a "value added" service, away from basic IP-level connectivity, and charge their subscribers based on the privilege of using it, hence creating new sources of income. Conversely, the PPTP model facilitates distributed reach of the VPN at a higher atomic level, allowing corporate VPNs to broaden access capacities without the need for unambiguous service contracts with a multitude of network access providers (Papadimitriou, 2006).

12.6.3   IP Security (IPSec)

The Internet Protocol Security (IPSec) is an open standard that is based on the network layer (layer 3) security protocol. IPSec adds extra headers/trailers to an IP packet and can tunnel IP packets in new packets (Nicopolitidis *et al.*, 2003). There are several security services that are provided by IPSec, such as connectionless integrity, access control, non-repudiation, protection against replay attacks, confidentiality, and limited traffic flow confidentiality, that are offered at the transport layer, providing protection for IP and upper layer protocols. The main functions of IPSec can be divided into three protocols: (a) the first is the authentication using an Authentication Header (AH); (b) the second is encryption using an Encapsulating Security Payload (ESP); and (c) lastly, automated key management using the Internet Key Exchange (IKE) protocol. IPSec offers a design with solutions for management, encryption, authentication and tunneling (Cohen, 2003). It was meant to be an algorithm-independent scheme that can support several encryption and authentication paradigms, which enable companies using VPN to go for the preferred security degree for each VPN (Papadimitriou *et al.*, 2004). Moreover, IPSec is an optimal solution for trusted LAN-to-LAN VPNs (Youn-glove, 2000). It is flexible to accommodate a wide diversity of encryption schemes. It is considered to be an application transparent and a natural IP expansion, which ensures interoperability between VPNs over the Internet. Nevertheless, there are weaknesses in IPSec. Among these are: (1) It is based on the TCP/IP stack and IP addressing is part of IPSec's authentication scheme. It is less secure than higher layered methods and it is a trouble in dynamic address situations that are regular to ISPs. (2) It entails a public key infrastructure, and it does not state a technique for access control other than simple packet filtering (Gunter, 2001).

12.6.4   Encapsulating security payload

The Encapsulating Security Payload (ESP) protocol is responsible for packet encryption. An ESP header is added into the packet between the IP header and the remaining packet contents. As for the header, it includes a Security Parameter Index (SPI) that indicates to the receiver the appropriate SA for dealing with the packet. Moreover, a sequence number, which is a counter that increases each time a packet is sent to the same address using the same SPI and indicates the number of packets that have been sent with the same group of parameters, is included in the ESP header. This sequence number offers protection against replay attacks. In this latter attack, an attacker copies a packet and sends it out of sequence to perplex communicating stations. ESP permits multiple schemes of encryption, with DES as its default scheme. Moreover, it can be used for data authentication. An optional authentication field is included in the ESP header that contains a cryptographic checksum, which is computed over the ESP packet when encryption is achieved. It is worth mentioning here that the ESP's authentication services do not guard the IP header that precedes the ESP packet.

There are two modes for ESP: transport and tunnel modes. In the former mode, the packet's data (not including the IP header) are encrypted and optionally authenticated. Despite the fact that the transport mode ESP is enough for defending the contents of a packet against eavesdropping, it leaves the source and destination IP addresses vulnerable to alteration if the packet is captured. On the other hand, the packet's contents and the IP header in tunnel mode are encrypted and may be authenticated if desired. Even though the tunnel mode ESP offers better security than transport mode ESP, traffic analysis is still possible. For instance, the IP addresses of the transmitting and receiving gateways could still be determined by checking the packet headers.

12.6.5 Management of keys

In order to handle key management in IPSec architecture, we can rely on either manual keying or Internet Key Exchange. In manual keying, face-to-face key exchanges are made (e.g. exchanging keys on paper or magnetic disk) or transferring keys using an e-mail or a bonded courier. Even though manual keying is proper for a small number of sites, automated key management is needed in order to accommodate on-demand establishment of keys for security associations (SAs). The Internet Key Exchange (IKE) is the IPSec's default automated key management protocol. It was created by merging the Internet Security Association Key Management Protocol (ISAKMP) that describes procedures and packet formats to establish, negotiate, modify and delete SAs, with the Oakley Key Determination protocol, which is a key exchange protocol based on the Diffie–Hellman algorithm (Ferguson and Huston, 1998).

Because key exchange is very much related to the management of security associations, part of the key management procedure is handled by SAs, and the security parameter index (SPI) values that refer to the SAs, in each IPSec packet. Every time an SA is formed, keys must be exchanged. The structure of IKEs unites these functions by requiring the starting node to define: (a) a scheme for encryption to look after data, (b) a hash algorithm to minimize data for signing, (c) a scheme for authentication, and (d) information about the material required to produce the keys, on top of which a Diffie–Hellman conversation is carried out.

12.6.6 Packet authentication

The data integrity feature in the Authentication Header (AH) of IPSec avoids the hidden modification of a packet's contents at the same time as the packet is in transit, and the authentication characteristic thwarts address spoofing and replay attacks, and allows the receiver to validate the application and filter traffic consequently (Ferguson and Huston, 1998; Wright, 2000). Similar to the ESP header, the AH is placed into the packet between the IP header and the remaining part of the packet. Moreover, the AH holds an SPI that states to the receiver the security protocols that the sender is utilizing for connection. There is a difference between the authentication provided by the AH and that offered by the ESP. The external IP header that precedes the ESP header, along with the entire contents of the ESP packet is protected by the AH services. It is worth mentioning that the AH holds authentication data, acquired by executing the hash algorithm specified by the SPI to the packet's contents. There are two schemes that can be used: (1) Hash-based Message Authentication Code with Message Digest version 5 (HMAC-MD5), and (2) Secure Hash Algorithm version 1 (SHA-1) (Wright, 2000; Nicopolitidis *et al.*, 2003).

Similar to the ESP, the authentication header can be used in the same two modes: the transport and tunnel modes. In the first mode, the data of the packet, and the original IP header are validated (authenticated). On the other hand, in the tunnel mode, the entire IP packet fields are authenticated. Besides applying either AH or ESP to an IP packet in the transport or tunnel modes, IPSec supports blends of these modes. Some combinations are likely, among these are (Wright, 2000):

1. It is possible to use the ESP in the transport mode without the authentication option. Then it is possible to authenticate the ESP header, the packet's data, and the original IP header.
2. It is possible to use the AH in the transport mode in order to authenticate the packet's data and the original IP header. After that the ESP could be applied in the tunnel mode in order to encrypt the entire authenticated inner packet and to append a new IP header for routing the packet from source to destination (Tomsu, 2006).
3. It is possible to use the ESP in the tunnel mode so as to encrypt and, maybe, to validate a packet and its unique header. After that the AH could be applied in the transport mode in order to validate the data of the packet, original IP header, and outer IP header.

12.6.7   Authentication (validation) of users

In order to have effective control access to the venture resources and keep illicit users out of a company's networks, a virtual private network should be able to reliably authenticate users (Wright, 2000). In this context we can identify three protocols: (a) Password Authentication Protocol (PAP), (b) Challenge Handshake Authentication Protocol (CHAP), and (c) Remote Authentication Dial-In User Service (RADIUS). A brief description of these protocols is given below.

Password authentication protocol (PAP)

This scheme was initially intended as a simple way for one computer to validate itself to one more computer when PPP is employed as the communication's protocol. Password Authentication Protocol (PAP) is a two-way handshaking scheme. Whilst the PPP link is created, the client system launches the plaintext user ID and password pair to the receiver. The latter either accepts the pair, or the connection is ended. The PAP is not considered a secure way of user authentication. There is nothing to protect the authentication information from playback attacks or intense endeavors by invaders to deduce valid user ID/password pairs (Papadimitriou, 2004; Wright, 2000).

Challenge handshake authentication protocol (CHAP)

This scheme is more secure for authenticating users. It is a 3-way handshaking protocol; it uses a three-step process to produce a verified PPP link as described briefly below (Wright, 2000):

(a) The authenticator launches a challenge message to the client.
(b) The client system determines a value using a one-way hash function and returns the value back to the authenticator.
(c) The authenticator grants authentication if the reply corresponds to the expected value.

CHAP takes away the likelihood that an attacker can try continually to log in over the same connection, which is considered a weakness inherent in PAP. Nevertheless, both PAP and CHAP have common drawbacks. Among these are: (1) They rely on a confidential password that should be saved on the remote user's computer and the home computer. In

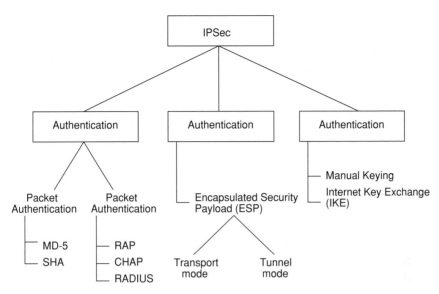

Figure 12.3   Main services of IPSec.

such a case, if either of these computers becomes under the control of a network assailant, the password will be compromised. (2) They both permit only one set of rights to be assigned to a particular computer, which avoids different network access privileges from being allocated to various remote users who use the same remote host (Wright, 2000; Papadimitriou and Obaidat, 2006).

Remote authentication dial-in user service (RADIUS)

The RADIUS protocol offers more flexibility for overseeing remote network connection users and sessions. It employs a client/server paradigm to validate system users, find out access levels and sustain all needed assessment and bookkeeping data. RADIUS protocol employs a network access server (NAS) to admit user connection requirements in order to get user ID and password information and deliver the encrypted information firmly to the RADIUS server. The latter returns the authentication standing in terms of approved/denied along with whichever pattern data are needed for the NAS to offer services to the user. Authentication of users and admission to services are administered by the RADIUS server, allowing a remote user to use the same services from any communications server that can connect to the RADIUS server (Wright, 2000; Papadimitriou and Obaidat, 2006).

Figure 12.3 summarizes the main services and protocols of IPSec.

SSL and SOCKS v5

The Internet Engineering Task Force (IETF) approved SOCKS v5 as a standard protocol for authenticated firewall traversal (Ferguson and Huston, 1998; Wright, 2000). If merged with the Secure Socket Layer (SSL) it can offer the groundwork for constructing well secure VPNs. SOCKS v5 is known for its good access control, where it controls the flow of data at layer 5 (the session layer). Moreover, SOCKS v5 sets up a circuit between a client and a host on a session-by-session basis. Hence, it can offer a better access control when compared to protocols in the lower layers without the necessity to rearrange each application. It is worth mentioning that SOCKS v5 and SSL can interoperate on top of IPSec, PPTP, L2TP, and IPv4, or any other lesser level VPN protocol.

12.6.8   MPLS (multiprotocol label switching)

Multiprotocol label switching is usually installed in an Internet Service Provider's IP backbone network that employs IP routers at the boundaries of the network (Wright, 2000; Papadimitriou and Obaidat, 2006). The fundamental scheme is to implement layer-3 routing at the edges of the backbone network and to implement layer-2 forwarding within the backbone network. Inward IP packets acquire an attached MPLS header and they are tagged based on the IP header information. For each route, a label switched path is set up. When this is completed, all successive nodes may well simply forward the packet alongside the label-switched path recognized by the label at the front of the packet. Labels are negotiated among nodes using the label distribution protocol of MPLS. For each label switched path, an ATM connection is set up. This means that MPLS can support quality of service, QoS, which is an important feature. Furthermore, MPLS makes the fundamental backbone infrastructure hidden to the layer-3 resources. This light-weighted tunneling offers a flexible foundation that provides VPN and other facilities. Additionally, the MPLS structural design allows network operators to describe explicit routes (Younglove, 2000). It is noteworthy to state that MPLS technologies are valuable for Internet Service Providers that desire to provide their clients a variety of services (Papadimitriou et al., 2004).

The Border Gateway Protocol (BGP-4), which is a scalable protocol, is very popular in industry. BGP-4 is excellent at applications that involve carrying additional information in its routing updates. This feature is necessary in the MPLS VPN schemes (Rekhter et al., 1995; Patel et al., 2001).

12.7   QoS provision

VPNs have the potential to support a collection of service levels in addition to establishing a separated address environment to permit private communications. It is possible to specify such per-VPN service levels in terms of a defined service level that the VPN can rely on at all times, or in terms of a level of differentiation that the VPN can draw upon the common platform resource with some level of priority of resource allocation. The Integrated Services Working Group within the IETF have produced a set of specifications to support the guaranteed and controlled load end-to-end traffic profiles using a mechanism which loads per-flow state into the switching elements of the network (Zeng and Ansari, 2003; Papadimitriou and Obaidat, 2006).

As for the Differentiated Services (DiffServ) approach, it tries to provide a solution for QoS support with more enhanced scalability than Integrated Services (IntServ). It is worth mentioning that differentiated services can offer two or more QoS levels without keeping per-flow state at every router. The concept of the DiffServ scheme is to employ a DiffServ field in the IP header to assign the appropriate DiffServ level that the packet must receive. It is possible to offer scalability using DiffServ by combining the flows into a small number of DiffServ classes and by realizing traffic conditioning at the boundary routers of a network (Cohen, 2003).

QoS and VPN techniques introduce new challenges: (a) they both require broad configurations in the routers, and (b) local configurations have to be reliable across the network. The majority of firms may not have enough expertise and resources to install and administer improved Internet services locally. Instead, they subcontract the service management to their Internet service provider (Braun *et al.*, 2001).

The traditional model for specifying QoS arrangements involves drawing a set of network service requirements between an ordered pair of customer sites. The pair describes a way of data flow, and the entire collection is usually called the "traffic matrices". There is a new QoS specification model called the hose model that has been devised recently. Such a model is expected to be desirable to VPN customers. In the hose model, two hoses are specified to the customer with one hose used for traffic distribution out of the network and the other for getting traffic from the network. In other words, the network provider provides collective traffic service to the customer where the traffic coming to the customer's site should be inside the receiving hose capability and all the traffic leaving the customer's site should be inside the sending hose capability. The specification of the hose includes outbound traffic and the in-bound traffic pair for each customer's site, which is involved in the VPN. The main advantages of this scheme from the user's point of view is the fact that the requirement is easy and there is a high possibility for multiplexing gain due to the use of collective needs. Such an easy requirement permits alterations such as changes in the VPN membership to be performed easily. Even so, the user (customer) in this scheme squanders monitoring of the data flow between sites and the service guarantee is on a combined basis, which means coarser service guarantee level. From the point of view of the network provider, the traffic matrices paradigm identifies heavy limitations, which restrict the network optimization that may be achieved (Harding, 2003; Papadimitriou and Obaidat, 2006).

12.8 Summary

In this chapter, we present the main concepts of virtual private networks (VPNs), which make use of the public communications infrastructure, maintaining privacy through the use of efficient protocols such as tunneling protocols and security procedures. The chief objective of VPNs is to provide the organization/company the same capabilities as private leased lines at reduced costs by using the shared infrastructure. VPNs are becoming appealing solutions to many organizations in order to create both Extranets and wide-area Intranets.

We have reviewed the main functions and structure of the basic building blocks of a VPN. The main blocks include the secure firewalls, VPN client software on user desktop, dedicated VPN server, and network access server (NAS). We also described the main attributes and functions of these blocks. We also reviewed the paradigms that can be used to implement VPNs including the Intranet, Extranet and remote access VPNs. Before an organization plans to implement a VPN, certain factors should be considered. These include whether to connect sites together or individual remote users to a central LAN, the types of network equipment and operating systems, existing devices that have VPN functionality, main features of software and hardware solutions, number of users that are supposed to be in the VPN environment, tunneling scheme to be used, speed to be used, degree of flexibility to be considered, degree of security and cost reduction, and management approach of VPN. Finally, we addressed the quality of service (QoS) issue in VPNs.

It is possible to specify per-VPN service levels in terms of a defined service level that the VPN can rely on at all times, or in terms of a level of differentiation that the VPN can draw upon the common platform resource with some level of priority of resource allocation. We reviewed the set of specifications devised by the Integrated Services Working Group within the IETF to support the guaranteed and controlled load end-to-end traffic profiles.

References

Arora, P., P. R. Vemuganti, and P. Allani (2001). *Comparison of VPN Protocols – IPSec, PPTP, and L2TP*, Project Report ECE 646, Fall 2001, available at http://ece.gmu.edu/courses/ECE543/reportsF01/arveal.pdf

Bova, R. (2001). VPNs: the time is now? available at: http://intranetjournal.com/articles/200110/vpn_10_03_01a.html

Brahim, H. O., G. Wright, B. Gleeson, R. Bach, T. Sloane, A. Young, R. Bubenik, L. Fang, C. Sargor, C. Weber, I. Negusse, and J. J. Yu (2003). Network based IP VPN architecture using virtual routers, Internet draft <draft-ietf-l3vpn-vpn-vr-00.txt>.

Braun, T., M. Günter, M. Kasumi, and I. Khalil (1999). *Virtual Private Network Architecture*, CATI Project Deliverable, January 1999, available at http://www.tik.ee.ethz.ch/~cati/deliverables.html

Braun, T., M. Guenter, and I. Khalil (2001). Management of quality of service enabled VPNs. *IEEE Communications Magazine*, May 2001, 90–98.

Cohen, R. (2003). On the establishment of an access VPN in broadband access networks. *IEEE Communications Magazine*, 156–63, February 2003.

Ferguson, P. and G. Huston (1998). What is a vpn? White paper; available online at http://www.employees.org/~ferguson

Gleeson, B., A. Lin, J. Heinanen, G. Armitage, and A. Malis (2000). A Framework for IP Based Virtual Private Networks. RFC 2764, February 2000.

Günter, M. (2001). Virtual private networks over the Internet, available at http://citeseer.nj.nec.com/480338.html

Hanks, S., T. Li, D. Farinacci, and P. Traina (1994). Generic Routing Encapsulation (GRE). RFC 1701, October 1994.

Harding, A. (2003). SSL virtual private networks. *Computers & Security*, Volume **22**, Issue 5, 416–20.

Hunt, R. and C. Rodgers (2004). Virtual private networks: strong security at what cost? available at http://citeseer.nj.nec.com/555428.html

Metz, C. (2003). The latest in virtual private networks: part I. *IEEE Internet Computing*, 87–91, January/February 2003.

Nicopolitidis, P., M. S. Obaidat, and G. I. Papadimitriou (2003). *Wireless Networks*. Wiley.

Papadimitriou, G. I., M. S. Obaidat, C. Papazoglou, and A. S. Pomportis (2004). Design alternative for virtual private networks. Proceedings of the 2004 Workshop on Electronic Government and Commerce: Design, Modeling, Analysis and Security, CDMA 2004, pp. 35–45, Setubal, Portugal, August 2004.

Papadimitriou, G. I. and M. S. Obaidat (2006). Virtual private networks (VPNs) basics. In *Handbook on Information Security*, Vol. 3, Wiley, pp. 596–611.

Pall, G., W. Verthein, J. Taarud, W. Little, and G. Zorn (1999). Point-to-Point Tunneling Protocol (PPTP), RFC 2637, July 1999.

Patel, B., B. Aboba, W. Dixon, G. Zorn, and S. Booth (2001). Securing L2TP using IPSec. RFC 3193, November 2001.

Rekhter, Y., T. J. Watson, and T. Li (1995). A Border Gateway Protocol 4 (BGP-4). RFC 1771, March 1995.

Ribeiro, S., F. Silva, and A. Zuquete (2004). A Roaming Authentication Solution for WiFi using IPSec VPNs with client certificates. Proceedings of TERENA Networking Conference, June 2004.

Rosenbaum, G., W. Lau, and S. Jha (2003). Recent directions in virtual private network solutions. Proceedings of the IEEE International Conference on Networks (ICON 2003), September 2003.

Simpson, W. (ed.) (1994). Network Working Group. The Point-to-Point Protocol (PPP), 1661, July 1994.

Srisuresh, P. and M. Holdrege (1999). IP Network Address Translator (NAT) Terminology and Considerations, RFC 2663, August 1999.

Strayer, W. T. and R. Yuan (2001). Introduction to virtual private networks, available online at http://165.193.123.40/isapi/page~1/sort~6/dir~0/st~%7B62A1DC08-8A24-47CD-B772-E55E08C2D481%7D/articles/index.asp

Tomsu, P. and G. Wieser (2002). *MPLS-Based VPNs – Designing Advanced Virtual Networks*. Prentice-Hall.

Townsley, W., A. Valencia, A. Rubens, G. Pall, G. Zorn, and B. Palter (1999). Layer Two Tunneling Protocols. RFC 2661, August 1999.

Tyson, J. (2006). How virtual private networks work, available at: http://computer.howstuffworks.com/vpn.html.

Wright, M. A. (2000). Virtual private network security. *Network Security*, volume **2000**, Number 7, 11–14.

Younglove, R. (2000). Virtual private networks: secure access for e-business. *IEEE Internet Computing*, Volume **4**, Number 4, 96.

Yuan, R. (2002). The VPN client and the windows operating system, available online at: http://165.193.123.40/isapi/page~1/sort~6/dir~0/st~%7B62A1DC08-8A2447CD-B772- E55E08C2D481%7D/articles/index.asp.

Zeng, J. and N. Ansari (2003). Toward IP virtual private network quality of service: a service provider perspective. *IEEE Communications Magazine*, April 2003, 113–119.

13 Protecting against malware

During the past few decades, there has been a significant interest in computer malicious programs. As the number of these programs keeps on increasing, efficient software solutions are needed to protect the enterprise from other living software without excessive requirement of user intervention. This chapter discusses malware definition and classification. It describes the ways that major classes of malware (e.g., viruses, worms, and Trojans) are built and propagated. It finally discusses the protection measures that an enterprise needs to develop to protect against such malware destructions. It also develops a non exhaustive set of guidelines to be followed.

13.1 Introduction to malware

Malicious software, often referred to as malware, is defined as a program or part of a program that executes unauthorized commands, generally with some malicious intention. Types of malware can be classified based on how they execute their malicious actions and propagate themselves. Viruses, worms, Trojan horses, and backdoors are the major examples of malware (Garetto *et al.*, 2003). Other malware related terms include malcode and malware payload. Malcode refers to the programming code that contains the malware logic, while the malware payload represents the malicious action it is designed to realize (Briesemeister *et al.*, 2003; Anagnostakis *et al.*, 2003).

A malware can damage the host on which it is running by corrupting files and programs or over-consuming resources. Typically, this is done while the malware is avoiding the complete devastation of the host because a system failure would prevent the ability of the malware to propagate further. However, some new malware types have been created with the blameworthy intention to destroy the hosts they infect. In addition, the malware is designed to gain control of a large set of computers quasi-concurrently. Malware can be classified into two main categories (as depicted by Figure 13.1): the class of malware that cannot exist independently of some specific program and those that can be scheduled to run by the operating system. The malware classes are described below.

Viruses

A virus is a sequence of code that is inserted into the executable code of a normal program so that, when the program is executed, the code composing the virus is also executed. This

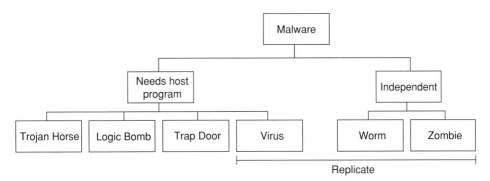

Figure 13.1   Malware classification.

code causes a copy of itself to be inserted in one or more other normal programs (this is called code propagation). Therefore, viruses are not distinct programs and cannot run autonomously. They need to have some host program, in which they are inserted. They are active when the host program is running. Programs to which the virus copies or attaches itself are said to be infected with the virus.

Computer viruses are the most pervasive threat to the security of computers and networks. An example of a virus among the huge set of available viruses that can be considered is W97M.Melissa.A (also known simply as Melissa) which is a typical Microsoft Word macro virus with a relatively simple code, which was first mailed to some newsgroups on March 26, 1999. Although there is nothing special in the infection routine of this virus, it has a payload that utilizes MS Outlook to send an attachment of the infected document being opened. It uses a self-check method to verify a setting in the registry to test whether the system has already been infected. This virus also sets macro security level to low security in Office2000. Melissa creates an Outlook object using Visual Basic instructions and accesses the list of members from the address book in the host. An e-mail message is then created and sent to the first 50 recipients in the list. The message is created with the subject:

"Important Message From"

and a body text containing:

"Here is that document you asked for . . . don't show anyone else."

The active infected document is attached to the message and the email is sent. Often, the content of the document is a list of web sites.

Worms

A worm is a program that makes copies of itself; for example, from one disk drive to another, or by copying itself using email or another transport mechanism. The worm may damage and compromise the security of the computer. It may arrive in the form of a joke program or software of any sort.

SQLSlammer is a well-known example of a worm. It causes a denial of service on many Internet hosts. It also dramatically reduces the performance of the Internet traffic. SQLSlammer worm aims simply at spreading from one host to another and does not convey a destructive payload. This worm exists only in the memory of unpatched Microsoft SQL servers. It causes increased traffic on UDP port 1434 and propagates between SQL servers. SQLSlammer exploits two buffer overflow bugs in Microsoft's database product SQL Server. It doubles in size periodically (every period of about 8.5 seconds). Therefore, it can infect very quickly a large percentage of vulnerable hosts. The malformed packet (which is the full worm) is only 376 bytes long and simply transports the following strings:

"h.dllhel32hkernQhounthickChGetTf", "hws2", "Qhsockf" and "toQhsend".

Zombies

A zombie is a program that secretly takes control of another Internet-connected computer and uses the infected computer to start attacks that are difficult to trace to the zombie's creator. Typically, zombies are used in denial-of-service attacks against targeted Web sites. A zombie can be placed on hundreds of computers belonging to unsuspecting third parties. The computers are then used to overload the target by sending an overwhelming attack on Internet traffic.

Trinoo is an example of a zombie. This is not a virus, but an attack tool that performs a distributed denial-of-service attack. It allows an attacker to flood a remote computer with a very heavy network traffic performing a distributed denial-of-service attack. It can work in conjunction with other infected computers. It also allows an attacker to perform unauthorized access to the computer.

Trojans

A Trojan horse is a malware that performs unauthorized actions. The main difference between a Trojan and a virus, worm, and zombie is the inability to replicate itself. Trojans cause damage, induce unexpected system behavior, and compromise the security of systems, but do not replicate. A Trojan typically looks like any normal program, but it has some hidden malicious code within it. When a Trojan is executed, users are likely to experience system problems and loss of valuable data.

A known example of a Trojan is the so-called *Back Orifice* (also referred to as *Remote Access Trojan*, or *RAT*). It enables a user to operate remotely a computer running the Windows operating system. *Back Orifice* was designed based on a client-server architecture. In a typical attack, the intruder sends the Back Orifice Trojan horse to the victim target as a program attached to an e-mail. When the e-mail recipient executes the program attachment, the Trojan horse opens connections from the computer to another entity. The two entities communicate with one another using the TCP and/or UDP network protocol, via port 31337. This allows the intruder to control the computer. Back Orifice can stay invisible and can restart itself automatically even if Windows is re-booted. Also Back Orifice allows a hacker to view and modify any file on the infected computer. It can create a log file of user's actions and take screen images and send them to the hacker. Or it can simply force a computer crash.

Logic bombs

A logic bomb is programming code, which can be inserted deliberately or unpurposely. It is designed to execute under special circumstances, such as the completion of a certain amount of time. In fact, it is an event-based (e.g., delayed-action) computer virus or Trojan. A logic bomb may be designed so that, when executed (or "exploded"), it displays a false message, deletes or corrupts data, or has other undesirable effects on some resources or applications in the host. Time bombs represent a large subclass of logic bombs that execute at a certain time. An example of time bombs is represented by the so-called "*Friday the 13th virus*." It duplicated itself every Friday and on the 13th of the month, causing system slowdown. Another example of a time bomb is called the "*Millennium Time Bomb*," which was designed to take benefit of the arrival of the year 2000.

Trap doors

A trap door, also referred to as *back door*, is a secret entry point into a program that allows a malicious adversary that is aware of the trap door to gain access without going through the usual security access procedures. The difference between a trap door and a remote access Trojan is characterized by the fact that the trap door only opens a port, often with a shell, while the RAT is built with a client server architecture.

The problem of malware is escalating to epidemic proportions. The scale of the threats is made dangerous by repeated increases in various major malware attributes including: frequency of incidents, speed of propagation, and damage achieved with the new malicious codes. Persistence of this escalation threatens all systems connected to the communication networks. Technical protections against malware have been long awaited to be proposed and are likely to be vulnerable to human interaction. In fact, as long as computer operation is a mixture of technical mechanisms and human control, a defence against malware needs to deal with this combination and therefore may pose more problems than it can solve. Unfortunately, the computer security practices, as a whole, have made little effort to modify the risky behaviors adopted by the average computer user. This is one among the reasons, which explain why viruses remain the major source of loss.

The following sections will explain how the major malware classes work and propagate. The sections also describe some of the possible ways to get infected by malware. They also discuss the limits of an enterprise's protection mechanisms and address the development of protection policy against malware.

13.2   Virus analysis

A virus is conceptually a simple program. In its easiest form, a virus can be modeled under the following form:

begin
   Look for (one or more infectable objects)
   if (no infectable entity is found)
   then
      exit
   else (infect the object(s) found).
   endif
end

Typically, viruses do not remain in memory. They execute all their code at once and then give the control to the host program they have infected. However, some viruses install themselves into memory after the host program is executed, so that they can infect objects accessed after the infected application has terminated. They are called *memory resident*. The term *hybrid* is sometimes used for viruses that stay active as long as the host system is running.

Practically, some viruses perform more than just replicate themselves. For this, a virus is sometimes described as having three components: an infective routine, a trigger condition, and a payload. The trigger condition might, for instance, be the execution of a program, an access to a file, a particular date or time, or a particular accessible value. One can consider, for example, that finding an infectable object (e.g., program or file) can be the trigger for the infective routine. The payload is, in general, any operation that any other program can perform. A generic virus architecture looks more like the following:

begin
   (Get a resident status)
   if (infectable object exists)
 then
      if (object is not already infected)
      then
      (infect object)   <----------- Infective routine
      endif
   endif
   if (trigger condition exists)   < -----------Trigger condition
   then
      (deliver payload)   <----------- Payload content
   endif
end

13.2.1   Viruses classification

Typically, viruses can be classified conveniently into five main classes: boot sector infectors, file infectors, multipartite viruses, macro viruses, and scripting viruses. The main characteristics of the five classes are described as follows:

Boot sector infectors (BSIs)

These PC-specific viruses infect the Master Boot Record and/or DOS Boot Record. In the past, these viruses represented the majority of reported incidents. Nowadays, BSIs constitute a declining proportion of the total number of threats that are reported as damaging sources. In addition, new BSIs are becoming very uncommon. This might be explained by mainly two facts: (a) the exchange of files is made through e-mail and network tools instead of floppy disks, and (b) BSIs are harder to write than the other classes of viruses such as the macro viruses, scripting viruses, and file viruses. A BSI works based on the three following characteristics:

- When a PC starts, it goes through a process called *Power-on-self-test* (or POST). This process includes checking hardware components and getting useful information related to disk and memory type and configuration to start. Some of its information comes from data stored in CMOS. If the CMOS settings do not match the actual drive geometry, the machine will not be able to find the system areas and files at the location where they should be. Therefore, the PC will fail to finish the POST process.
- The Master Boot Record (MBR), called also the *Partition Sector*, is found only on hard disks, where it is always the first physical sector. MBR contains the information about the disk, including the starting address of the partition(s) into which it is divided. Conversely, a floppy does not contain a MBR because it cannot be partitioned. Like a partition in a hard disk, the first physical sector in a floppy is the boot record (or *DBR*). On a floppy, the DBR contains a program whose objective is to check that the floppy is bootable; and, if that is the case, it hands over the control to the operating system.
- If a bootable floppy is accessible on the PC, then the PC is likely to boot from the drive containing the floppy, rather than from the hard drive. This is actually a default configuration, which is unfortunate because this is the normal entry point for a boot sector virus.

If the PC attempts to start booting (from a floppy or the hard disk) with an infected boot sector, the system will be infected: even when a DBR infected floppy does not contain the necessary files to load an operating system, the MBR in a PC's hard drive will be infected. Additionally, when the MBR of the first hard drive of a PC is infected, the virus will infect all write-enabled floppies inserted to the PC.

File viruses (or *parasitic viruses*)

File viruses infect executable files. Historically, most file viruses have not been particularly successful in terms of propagation. Thousands of such viruses have been written. However, not many viruses have been seen in networked systems. Nonetheless, those that have survived have often surprising propagation activity (such as the *CIH virus*). Some of the most dominant present-day file viruses, however, are generally described as worms.

After a virus infects an executable file (by direct attachment), the file will infect other files when it executed. Each time the virus infective routine is executed, it may infect a whole directory, all directories on the infected program current path, a whole volume, and even all currently mounted volumes. Moreover, file infectors can spread quickly across systems and networks in an information system, where multiple binary executables can be open frequently. Every time an application is open, at least one executable file is loaded. Some applications can open multiple files when they start up, whereas others periodically open files when performing their operations. Examples of file infectors are the *sparse infectors*, which infect occasionally (only on specified occasions) or only files whose length falls within a given range. They may not infect every time the virus is executed, but only under very specific conditions, even when an infectable object is there to infect. By infecting occasionally, the sparse infector attempts to minimize the probability to be detected.

Multipartite viruses

Multipartite viruses are viruses that use more than one infection mechanism. File and boot viruses are the most common examples of multipartite viruses, for which both boot-sectors and binary executable files can be infected and used as the means of disseminating the virus. However, it is commonly agreed that there will be an increase in the development of new multipartite viruses integrating different combinations of other virus types.

Macro viruses

Macro viruses infect macro programming environments rather than specific operating systems. Macro viruses are viruses that execute when a macro is executed. They usually infect the global template and often modify commands within the application's menu system. Environments such as Microsoft Office allow users to write macros. Macros are programs that bind multiple commands to a single keystroke or action. Macros are very effective virus distribution tools since they are interpreted instead of compiled, can contain entire programs, and can address any function. These viruses are typically cross platform because the environments they are used in are multi-platform executable. They are particularly successful against Microsoft applications because they allow executable code (or macros) to exist in the same file as data (Microsoft Office or compatible suite, for example, can execute on DOS, Mac and Linux environments).

Script viruses

These viruses represent malicious codes that can be embedded in HTML scripts and executed by HTML-aware e-mail clients through the Windows Scripting. VBscript (visual basic scripting edition) and Jscript have been used for the creation of a large set of script viruses. The view of script viruses is, however, rather restrictive. A broader definition might include HyperCard infectors, batch file infectors, UNIX shell script infectors, and many more.

13.2.2   Defense against viruses

There are several methods of defense against viruses. Unfortunately, no defense is perfect. Defense against viruses generally takes one of three forms:

Activity monitors

Activity monitors are programs that are resident in the operating system. They monitor activity and generate alerts warning or execute specific actions, in the event of suspicious activity. Thus, attempts which aim, for example, at altering the interrupt tables in memory, modify files, or rewrite the master boot sector would be detected by the monitor. This form of defense can be thwarted (if implemented in software) by viruses which activate earlier than the monitor code in the boot sequence. They are highly vulnerable to virus alteration if they are used on computers without hardware memory protection. Unfortunately, this is the case with all personal computers.

Another form of monitoring is the execution tracing of suspect applications. To this end, the monitor evaluates the actions taken by the application's code and determines whether the activity is similar to what a virus would undertake. Appropriate alerts are generated and reported, if suspicious activity is identified.

Scanners

Scanners have been the most popular and prevalent form of protection against viruses. A scanner operates by reading data from disk and applying pattern matching operations using a list of known virus patterns. If a match is found for a given pattern, a virus instance is decided. Scanners are fast and easy to use, but they suffer from many disadvantages. Among the major disadvantages, one can consider that the list of patterns should be updated. This is difficult to guarantee since new viruses are appearing in dozens each week. Keeping a pattern file updated in this rapidly changing environment is hard to achieve. Another drawback in using scanners is characterized by the large number of false positive reports. As more patterns are added to the current list of patterns serving the monitoring, it becomes more likely that one of them will match some otherwise legitimate code. A third disadvantage is represented by polymorphic viruses, which cannot be detected with scanners. Scanners, however, have multiple advantages including their rapidity of reaction. Scanning can be made to work faster. It can also be done remotely and across platforms (Kumar, 1992) because pattern files are easy to distribute and update. Additionally, among the new viruses that are developed continuously, only a few will ever become pervasive. Hence, out-of-date pattern files may still remain adequate for most environments. Scanners may also find almost all polymorphic viruses, when they integrate algorithmic or heuristic-based checking processes. For all these reasons, scanners are the most widely-used form among anti-virus tools.

Integrity checkers

These checkers are programs that generate check codes (e.g., checksums, cyclic redundancy codes (CRCs), secure hashes, or message digests) for files under watch. Periodically or on the occurrence of specific events, these checking codes are computed and compared with the saved versions. If the comparison shows difference, a change is known to have been made to the file, and it is labelled for further investigation. Integrity monitors are integrity checkers that run continuously and check the integrity of files on a regular basis. As viruses attempt to alter files to attach themselves, integrity checking will find those changes. Furthermore, the integrity check will discover the change no matter what causes it and whether the virus is known or not. Integrity checking also may find other changes caused by buggy software and system operation errors.

Integrity checking also has several drawbacks. In fact, one can notice the following limits:

- Repeated false positive alarms may lead the user to ignore future reports or disable the checker. For, example, executable files may change whenever the user runs the file, or when a new set of preferences is recorded.
- A change may not be observed until after an altered file has been run and a virus installed (or after a damage is produced).
- The initial computation of a file check code must be performed on a known-trusted version of the file. Otherwise, the checker will compare with an unreliable file. This probably will lead the user to accept that the system is uninfected.

13.3   Worm analysis

Worms represent another form of malware. Unlike viruses, worms are programs that can run independently and move from one computer to another through the communication networks (Kienzle *et al.*, 2003). Worms may have various components running on many networked computers. Worms do not change other programs and may transport codes that operate like a true virus, which are composed of two parts: the part that handles the spread of the virus and the payload. Unlike true viruses, a worm is characterized by two processes:

- The worm distribution, which is achieved through the execution of two processes. This process attempts to discover new targets to infect.
- The worm activation, which represents the mechanisms by which the worm's code begins operating on the target and the payloads, which are the non-propagating routines that the worm may use to accomplish the objectives of the worm creator (Weaver *et al.*, 2003).

In order to understand the worm threat, it is necessary to discuss its various types.

13.3.1   Target discovery

To achieve its objectives on a set of targets, a worm must first discover that the targets exist and then it has to transport itself to those targets. There are a number of techniques by which a worm can discover the targets to infect. Propagation mechanisms allow the worm to proliferate from computer to computer, or transport itself along the network as part of a normal communication. These techniques include network processing, but are not limited to operations such as: (a) scanning networks using target lists, which are pre-generated, externally generated or internally built and (b) passive monitoring.

In the following, we discuss the features of the major mechanisms used for target discovery: scanning, and target lists.

Scanning

This mechanism requires the search for a set of addresses to identify vulnerable targets. Two simple types of scanning can be utilized: (a) a sequential scanning, which assumes that the searching is made using an ordered set of addresses and (b) random scanning, which tries destinations out of a block of addresses in a pseudo random manner. In general, the performance of the scanning process is limited by several factors including: the size of the network, the density of vulnerable targets, and the performance of the edge routers to transfer a significant increase in traffic (as generated by the scanning).

Several optimization techniques can be applied to increase the performance of the scanning process including:

- The preference for local addresses, which enables the worm to exploit a simple firewall breach to scan the entire local network.
- The permutation. This allows a worm to utilize distributed coordination to more effectively scan the net and determine when a group of nodes in the network is infected.
- The bandwidth-limited scanning, which uses scanning algorithms that are limited by the throughput of the sender rather than by latency to the target.

Many worms, such as Code-red (Zou *et al.*, 2002), used scanning routines. The Code-red worm exploits the Index Server (.ida) buffer overflow vulnerability in Microsoft information system (IIS). The buffer overflow allows the worm to execute code within the IIS server to spread itself, disfigure the server's home page, and perform a denial-of-service attack on www.whitehouse.gov. The worm code executes only in memory and is not written to disk. The worm attacks other systems by randomly generating an IP address and then exploring that address to check whether it can connect to port 80. If it can, it sends a copy of the buffer overflow attack to that machine. The random IP address generator is likely to start with the same random seed; thus, the list of machines attacked by the worm is identical for every copy of the worm. The result is that machines whose IP addresses are on the list of random IP addresses will be accessed by the worm in every infected machine, creating an effective denial-of-service attack. The first version of Code-red required approximately 12 hours to reach endemic (widespread) levels, but could have required less than 15 minutes if it utilized a bandwidth-limited scanner (Moore *et al.*, 2003).

Scanning is highly anomalous behavior, so devices can effectively detect scanning worms as being very different from normal traffic. For example, as the Code-red worm runs only in memory, one cannot find evidences on the disk that indicates the worm presence. However, when the worm is active in a server, the server will show a heavy load and large number of external connections to port 80 on other machines.

Target lists

An attacker can get in advance a target list containing probable victims. Three types of lists can be considered: The pre-generated target list, the externally generated target list, and the internal target list.

Pre-generated Target Lists A small list can be used to accelerate a scanning worm, while a complete target list creates a worm capable of infecting all targets extremely rapidly. The biggest problem with the usage of target lists is the effort to create the target list itself. For a small target list, public sources or open access points can be used to perform small-scale scans. Comprehensive lists require more effort and can require tools such as a distributed scan or the availability of a compromised database.

Externally Generated Target List This is an external target list that is maintained by a separate server, such as a meta-server, which keeps a list of all the servers that are currently active in a system. Externally generated target lists exploit the fact that some servers maintain lists of active servers (e.g., games, peer-to-peer applications). A meta-server worm first requests the meta-server in order to discover new targets. Such a worm can quickly spread, even when the set of targets is relatively small.

Internal Target Lists Many applications contain information about other hosts providing vulnerable services. Such target lists can be used to create topological worms, where the worm searches for local information to find new victims by trying to discover the local communication topology. Topological worms make use of the representation of vulnerable machines by a graph directed $G = (N, A)$, where a node $(n \in N)$ is a machine and an arrow (n,n') $(n \in A)$ represents information that n has about n'. The time it takes a topological worm to infect the entire graph is a function of the depth of the graph with respect to the initial point of infection (i.e., the distance from the initial point to the fastest node using shortest paths).

Although topological worms may present a global anomaly, the local traffic may appear extremely normal since each infected machine only needs to contact a few neighbors. This means that highly distributed detectors may be needed to detect topological worms.

Passive worms

Such worms do not search for victim targets. As an alternative, they either wait for possible victims to contact the machine where the worm is installed, or they rely on user behavior to discover new targets. Passive worms produce no anomalous traffic patterns during target discovery, which potentially makes them quite silent. Passive worms require at least some interaction from victims in order to spread. In the sense that the user has to manually open an e-mail attachment or, otherwise, click something before the worm can propagate. An example of a passive worm is the Love worm. This worm is distributed as an attachment to an e-mail message entitled, "*I Love You.*" If its attachment is open, the worm sends copies of the same e-mail to everyone listed in the local address book. It then looks for files with extensions in {jpeg, mp3, mp2, jpg, js, jse, css, wsh, sct, and hta} and overwrites them with its content, changing the extensions to .vbs or .vbe. The new files cannot be used.

13.3.2   *Worm activation*

The means by which a worm is activated in a host severely affects its rapidity to spread, because some worms can plan to be activated immediately after their propagation, whereas other worms may stay inactive for days or weeks waiting to be activated. However, the slowest activation approach requires a worm to influence a local user to execute the local copy of the worm. Since most users do not accept to have a worm executing on their system, these worms rely on a variety of social engineering techniques to provide situations where the user executes them. Some worms indicate urgency or need or show attractive features. For example, the Melissa e-mail-worm (CERT, 1999) shows urgency from someone the target user knows (the sentence "*Attached is an important message for you*" is used), while the Benjamin worm attempts to exploit the insatiability (greed) character of the user (by sending him "*Download this file to get copyrighted material for free*") to tempt him into opening the infected e-mail attachment or file.

Furthermore, while some worms require that a user initiate the execution of a program, other worms attempt to exploit bugs in the software that brings data onto the local workstation so that simply viewing the data would request the execution of the program. The continued spread of these worms is disturbing, as they can be effectively used as secondary vectors of propagation (e.g., the case for Nimda). The Nimda worm appears to spread by multiple mechanisms: (a) from client to client via e-mail, (b) from client to client via open network shares, (c) from Web server to client via browsing of compromised Web sites, (d) from client to Web server via active scanning for and exploitation of various Microsoft IIS (4.0/5.0) directory traversal vulnerabilities and (e) from client to Web server via scanning for the back doors left behind by the "Code Red II," and "sadmind/IIS" worms.

The Nimda worm propagates through e-mail arriving as a MIME (Multipurpose Internet Mail Extensions) message and it consists of two parts. The first part is defined as MIME type "text/html," but it contains no text, so the e-mail appears to have no content. The second section is defined as MIME type "audio/x-wav" containing an attachment named "*readme.exe*," which is a binary executable. Due to an IIS vulnerability called *Automatic Execution of Embedded MIME Types*, any mail software running on an x86 platform that uses Microsoft Internet Explorer to render the HTML mail automatically runs the enclosed attachment and, as result, infects the machine with the worm. Hence, the worm payload will automatically be triggered by simply previewing the mail message. As an executable binary, the payload can simply be triggered running the attachment.

Finally, to further expose a vulnerable machine, the Nimda worm can enable the sharing of the c drive, create a "Guest" account, and add this account to the "Administrator" group. Furthermore, the Nimda worm infects existing binaries on the system by creating Trojan horse copies of legitimate applications. These Trojan horse versions will first execute the Nimda code (further infecting the system and potentially propagating the worm), and then complete their intended function.

Worm activation can be classified mainly into three classes based on how fast they can be activated:

Self Activation The worms that are the fastest activated are able to initiate their own execution by exploiting vulnerabilities in services that are always on and available or in the libraries that the services use. Such worms either attach themselves to running applications or execute other commands using the permissions associated with the attacked services. Activation occurs as soon as the worm can locate a copy of the vulnerable service and transmit the exploit code.

Scheduled Process Activation Worms among this class get activated using scheduled system processes. Such worms can propagate through mirror sites, or directly to the target machines. Many operating systems, packages, and applications integrate automatic updaters that periodically download and install software related updates. Some among the updaters do not (or did not) use server authentication to avoid propagating worms and not allowing attackers to serve a file to infect the target. Therefore, a simple protection is to provide updaters with authentication capabilities. Another way to propagate worms with scheduled activation is done through period system backup or other network software that includes vulnerabilities. An attacker who desires to exploit these vulnerabilities should depend on the scheduled process's design and implementation. In this case, if the attacked tool does not include authentication, the attacker can place its attack. If authentication is available, then the attacker might need to acquire the private keys for the update server and code signing, as well.

Human Activity-Based Activation A worm in this class is activated when the user performs a specific activity of interest to a worm, such as resetting the machine, logging in, executing login scripts, or opening a remotely infected file. The activation mechanism is commonly seen in open shared windows worms (such as one of Nimda's secondary propagation techniques), which will begin execution on the target machine either when the machine is reset or the user logs in, as these worms write data to the target disk without being able to directly trigger execution.

Currently, preventing self activation worms relies on various actions including:

- Run software that is not vulnerable, limit the access of services to reduce the effect of worms and keep patching from appropriate vendors.
- Ingress filtering This manages the flow of traffic as it enters a network under control. Servers are typically the only machines that need to accept inbound connections from the public Internet. Thus, ingress filtering should be performed at the border to prohibit externally initiated inbound connections to non-authorized services. With Nimda, for example, ingress filtering can prevent instances of the worm from infecting vulnerable servers in the local network that are not explicitly authorized to provide public web services.
- Egress filtering This manages the flow of traffic as it leaves a network under control. There is typically limited need for machines providing public services to initiate outbound connections to the Internet. In the case of Nimda, employing egress filtering at the network border will prevent certain aspects of the worm propagation.
- Recovering from the system compromise by formatting the system drive(s) and reinstalling the system software from trusted media.

13.3.3 Worm propagation

The means by which the propagation of a worm occurs can affect the propagation speed and secrecy. A worm can either actively spread itself from one machine to the other, or it can be carried along as part of normal communication. These means are called *self-carried* and *secondary channel*, respectively. A self-carried worm actively transmits itself as part of the infection process. This mechanism is commonly employed in self-activating scanning or topological worms, as the act of transmitting the worm is part of the infection process. Some passive worms also use self-carried propagation. Some other worms, such as the Blaster worm, require a secondary communication channel to complete the infection. The victim machine is required to connect back to the infecting machine using TFTP to download the worm body and complete the propagation process.

An embedded worm transmits itself as part of a normal communication channel, either as appended to a message or by replacing normal messages. As a result, the propagation does not appear as anomalous when viewed as a pattern of communication. The contagion (infection) strategy is an example of a passive worm that uses embedded propagation. Although relatively secret, an embedded strategy can succeed when the target selection strategy is also secret; otherwise, the worm will give itself away by its target selection traffic, and acquires little gain from the secrecy that the embedded propagation provides. Therefore, a scanning worm would not use an embedded distribution strategy, while passive worms can benefit significantly by ensuring that distribution is as silent as target selection.

The speed at which embedded worms spread is highly dependent on how the application is used and how far the worms could deviate from the natural patterns of communication in order to accelerate its propagation without compromising its secrecy. Similarly, the distribution of the worm body (or associated payloads) can either be: (a) one-to-many, which is the case when a single site provides a worm or module to other sites once they have been initially infected, (b) many-to-many, which is the case when multiple copies propagate the malicious code, (c) a hybrid approach, where the worm propagates in a many-to-many mode with updates received via a single host. Typically, many-to-many distribution can be significantly more rapid, if a limiting factor is the time it takes to perform the distribution. Many-to-many distribution also eliminates the ability for others to block further distribution by removing the originator of the malicious code from the network.

13.4 Trojan analysis

Typically, Trojans are built with two parts, a client part and a server part. When the infected machine runs a Trojan server, the attacker (or Trojan originator) can use the related Trojan client to link to the server and start using it. Both TCP and UDP based communication protocols can be used between server and client. When a Trojan server runs on a victim's computer, it tries to hide somewhere on the computer and then starts listening for incoming connections from the attacker on one or several ports. It then attempts to modify the registry and/or use some other auto-starting methods. For the Trojan to achieve its objectives, it is necessary for the attacker to know the victim's IP address to connect to his/her machine.

Many Trojans include the ability to mail the victim's IP and/or message the attacker via ICQ or IRC. This is used when the victim has a dynamic IP, that is, every time he connects to the Internet, he is assigned a different IP (most dial-up users have this). Many users with broadband connections have static IPs, meaning that in this case the infected IP is known to the attacker. This makes it considerably easier for an attacker to connect to your machine. Nearly all Trojans use an auto-starting procedure that permits them to restart and allow an attacker access to the target machine. That is why Trojan code creators are constantly searching for new auto-starting methods. As a rule, attackers start by "attaching" the Trojan to some executable file that is used very often on the target host, such as explorer.exe. Then they proceed to use known methods to modify system files or registries. Other ways to get infected include, but are not limited to, the protocol Internet Relay Chat, file sharing, and software download from non-trusted Web sites.

13.4.1 Types of Trojan horses

Generally, Trojans can be grouped into six major categories. Note, however, that it is usually difficult to classify a Trojan into a single group. They often have traits that place them in multiple categories. The categories below outline the main Trojan categories:

Remote access Trojans (RAT)

These are probably the most common Trojans. They provide the attacker with complete control of the victim's machine. This means that they can have full access to files, applications, communications (such as private conversations), accounting data and sensitive information. A RAT acts as a server and typically listens for a connection on a TCP/UDP port. A computer network attached to a security domain can be protected by a firewall placed at boundary since it is unlikely that a remote (outside the domain) hacker would be able to connect to the Trojan (assuming that the firewall has blocked the related ports). However, an internal hacker (attached to the domain from inside) can easily connect to the Trojan. Firewalls usually stop incoming connections and allow inside users to connect freely to the Internet, usually on the most common ports (such as FTP, HTTP, SNMP and ICQ).

Anti-protection Trojans

Generally called security software disablers, these special Trojans are designed to disable protecting programs such as the antivirus software and filters. Once these programs are disabled on a machine, the hacker is able to attack the victim's machine more effortlessly. The Bugbear virus (that hit the Internet in September 2002) installed a Trojan on the machines of the infected users that could give the remote attacker access to sensitive data and be capable of disabling popular antivirus and firewall software. The destructive Goner worm (revealed on December 2001) is another anti-protection Trojan that included a program capable of deleting anti-virus files.

Destructive Trojan horses

The unique function performed by the destructive Trojans is to destroy and delete files. They can automatically delete all the core system files in an operating system (e.g., files such as ".dll," ".exe," and ".ini") on the user's machine. The destructive Trojans can either be activated by the attacker or work like a logic bomb that starts on the occurrence of a specific event (e.g., date). This makes them very simple to use, harming any computer network. In many ways, they act similarly to viruses, but they are unlikely to be detected by the antivirus software.

Data sending Trojan horses

The data sending Trojans aim at transmitting data back to the attacker including information such as passwords or other confidential information such as credit card details, address lists, and private information. The Trojan can search for particular information in specific places or it may set up a key-logger and simply transmit all stored keystrokes to the attacker (who can obtain, with this in hand, the relevant passwords). Captured data can be sent back to the attacker's e-mail address, which in most cases is located at some free Web-based e-mail provider. Alternatively, captured data can be sent by connecting to a hacker's Web site – probably using a free Web page provider – and submitting data via a Web-form.

Internal and external attackers to a security domain can use data sending Trojans to obtain access to confidential information about the company hosting the domain. An example of data-sending Trojans is represented by the *Badtrans. B e-mail virus*, which was released in December 2001 and could log users' keystrokes on the victim machine.

Denial of service (DoS) attack Trojan horses

The DoS Trojans give the attacker the ability to initiate a distributed denial of service (DDoS) attack if there are a large number of victims. The major approach behind DDoS uses a 3-task process: (a) use of hundreds of infected zombies, (b) schedule them to attack a victim simultaneously, and (c) the zombies will then generate heavy traffic (higher than the victim's bandwidth can support), causing its access to the Internet to fail.

An example of DDoS is represented by WinTrinoo. Using this Trojan, an attacker who has infected many machines can cause major Internet sites to shut down. Another variation of a DoS Trojan is the mail-bomb Trojan whose aim is to infect as many machines as possible and simultaneously attack specific e-mail addresses with random subjects and contents that cannot be filtered. Like a destructive Trojan, a DoS Trojan acts similarly to a virus, but it can be created purposely to attack a specific target and therefore is unlikely to be detected by general purpose anti-virus software.

Proxy Trojan horses

These Trojans transform the victim's computer into a proxy server, making it available to all Internet users or only to the attacker. It is used for creating entities, referred to as *anonymizers*, which can be used for illegal actions, such as buying goods with stolen credit cards. This gives the attacker complete anonymity and the opportunity to perform any action from the victim's machine, including initiating attacks. If the attacker's actions are detected and tracked, however, the detected source would be the victim's machine and not the actual attacker. Legally speaking, the network or computer where the attack is launched is legally responsible for the damage the attack can induce.

13.4.2   Protection against Trojans

The current solutions implemented by enterprises to protect the integrity of their networked computers using anti-virus software and firewalls are in general inefficient and do not provide a good protection anti-virus. In addition, the firewalls (and particularly personal firewalls) cannot make many decisions by their own because they only apply rules decided by security administrators. If the administrator decides an application can be authorized to use the network, then the firewall cannot do anything against this. Thus, virus definition files must be updated continuously to protect against the known malware and deal with recently known malware.

In addition, the study conducted in Qattan (2004) proved that it was possible to disable protection software by using Remote Access Trojans. In fact, it is clearly stated that all products could be disabled, even though the attacker has limited experience with the malware used because he cannot experience serious problems in doing so. It is also shown that, if a Trojan server is revealed by the protection software, another Trojan can achieve its objectives by simply altering the original Trojan and distributing it to another application. The study showed that, when the Trojans were altered, it is difficult to discover them. It was also shown that the anti-virus programs perform better than the personal firewalls at discovering the Trojans. The firewall can only do what the user tells it. Even though the anti-virus programs are better at discovering Trojans, they are not faultless.

Solutions that are complementary to the aforementioned ones are developed to achieve more secure computer systems. Among these solutions is the next generation secure computing base (called NGSCB), which can help protect the host programs and data against malicious actions by dividing the computer environment into two areas. One area only contains secured applications and protected data, while the second area defines an insecure environment where anything can be run. Nevertheless, in order to enable the NGSCB to function in the way proposed, changes need to be made in computer hardware, operating system and applications. It has not yet been decided when the NGSCB will be available for the users.

The software part of the NGSCB is formed by two components: (a) the Nexus, which is a software kernel that manages the security of hardware and protected operating environment and provides system services to the applications that the user wants to run in the protected environments and (b) the Nexus Computing Agents (or NCAs), which are trusted user mode applications that run in the protected operating environment. They communicate directly with the nexus. The NCAs have four important security characteristics:

- *Strong isolation* Each NCA is isolated from the software that it does not trust, including the common operating systems.
- *Protected storage* Information can be stored in such a way that only the application from which data is saved can open it.
- *Secure I/O* A secure path is created to and from the user. Trusted devices (e.g., keyboard) and trusted video output so that I/O operations cannot be intercepted or modified by any software.
- *Digital signature* The NCAs have the ability to generate a digital signature and prove to other entities that data transmitted by the NCA was issued by an identified and unforgeable trusted software.

13.5 Protection techniques against malware

Typically, most enterprises use the network to exchange information with their suppliers and customers. The network interconnects many servers and deploys various applications. Strategies to protect the enterprise information systems can be classified into mechanisms for the protection of the network, anti-virus software, and e-mail filters.

13.5.1 Firewall-based protection

Firewall is significantly useful for the enterprises to protect their security, but it also has some disadvantages. A firewall is a tool that implements an access control policy used to protect a security domain (e.g., the enterprise's Intranet) from an untrusted network (such as the Internet). Designed to implement the enterprise's security policy, the firewall can improve network security and reduce risks to hosts in the domain it protects by filtering the insecure services. It can also prevent outside access to its components. Despite its many advantages, the firewalls present some drawbacks. For example, a firewall does not protect against back doors and often it provides weak protection for inside threats.

There are three types of firewalls: packet filtering firewalls, stateful inspection firewalls, and application-proxy firewalls. Packet filtering concentrates mainly on whether to accept or deny packets. Packet filtering deals not only with a packet's header, but also with its payload. The main strengths of packet filter firewalls are: (a) their rapid reaction and flexibility, (b) they can be deployed easily, and (c) they can be used to secure nearly any kind of communication protocol. However, because they do not inspect upper-layer information, . firewalls cannot prevent the network from complex viruses. Stateful inspection firewalls add awareness to the standard packet filter architecture by allowing observation at the fourth layer of communication. They share the robustness and weaknesses of packet filter firewalls. In addition, they are very costly as the state of connection is monitored continuously. On the other hand, application-proxy firewalls have larger logging capabilities. They are capable of authenticating users directly and can be made less vulnerable to address attacks. However, these systems do not cope with high-bandwidth and real time applications.

The basic components of a firewall include the network policy interpreter, authentication mechanisms, packet verifier, and packet-based classification; see Figure 13.2. The network policy defines the services that will be allowed or denied from outside and how these services will be used. It also specifies how the firewall restricts the access and filters the services that are allowed. The network policy focuses on external network access. The rules used to implement the service access policy usually implement two basic policies: (a) allow all services except the services occurring in a list that the service access policy has defined as forbidden to pass through and (b) deny all services except a set of services that the firewall declares allowed. Finally, one can say that there are two categories of network policy: service access policy and firewall design policy. While the former category defines the services that will be allowed or denied from outside, the latter describes how the firewall restricts the access and filters the services defined in the service access policy.

The major role of a packet verifier is to detect protocol anomalies and verify the compliance to standards. It also inspects the protocol header part of the packets, and verifies the

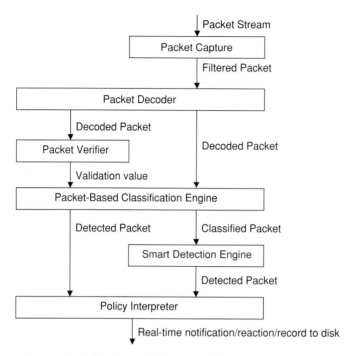

Figure 13.2   Architecture of a firewall model.

packet size and all packet parameters. The packet-based classification process tells which packets could be malicious (by computing an appropriate probability, for example). The smart detection tool will recognize and detect/check data packets that are considered malicious by the packet verifier. These detection entities cooperate to detect not only anomalous traffic, but also detect packet potentially belonging to malware. Authentication mechanisms are intended to counter the weaknesses of traditional passwords. A packet verifier is a useful place where the authentication mechanisms (software or hardware) can be implemented with appropriate strengths. The packet filtering process uses a packet filtering router to implement the filtering of packets that they pass between the router's interfaces. A packet filtering router typically filters IP packets based on the content of packet fields such as the IP address or port number of the source (or the destination). Finally, the smart detection module manages filtered packets, which have a high probability of being malicious. Unlike anti-malware software, the module does not have to match the infected part exactly.

13.5.2   Preventing malware by using anti-malware software

The enterprises that want to take part in worldwide communication need to connect their business network to the Internet. This unfortunately would expose the enterprises to harming risks. To operate a primary countermeasure in protecting the enterprise's network, the enterprise can deploy an anti-malware. There are several anti-virus techniques that can be used to detect and eliminate viruses, worms and Trojans. Each technique has its strong points and weaknesses. The major anti-malware software can be classified into three major categories:

- *Anti-malware gateway* An anti-malware occurring in this category can be an anti-virus for mail server that is in charge of providing comprehensive virus protection and checking all the file downloads using HTTP. It also can protect file transfer (FTP) and email (SMTP) before they reach the internal systems (or the security domain protected by the gateway.
- *Anti-malware for file servers* Software in this category checks continuously every new file on the server system to prevent the viruses and malicious codes from being added to servers. The checking process can use pattern matching and learning procedures to perform its duties.
- *Anti-malware for hosts* Such an anti-malware protects personal computers and portable devices against viruses, worms and malicious codes. It protects the fixed and mobile workers, ensuring the system availability and data integrity. However, commercial products for workstations need frequent updating to face the increasing number of new malicious codes and ensure the system availability and data integrity.

As explained in the previous sections, malwares hide themselves using different methods and locate themselves in different places. Scanners remain among the most popular anti-malware software used today, while checksum-based software systems are very common. Scanners contain detection or infection information for all known viruses. They are easy to use and are capable of identifying a large spectrum of malicious codes. However, the main disadvantage of scanners is that they also need to be updated with the latest malware information.

Checksum-based programs rely on detecting changes by computing checksum functions. When a malware infects a file, the file will change. The change will be picked up by the checksum-based software. They can detect known malware as well as unknown viruses, provided that the malware changes the file monitored by the checksum. The main weaknesses witnessed in checksum-based software include: (a) distinguishing legitimate from viral changes is hard to achieve since the results computed by the checksum function need a large effort of interpretation and (b) the checksum-based program is only able to detect a malware when the infection happens and cannot prevent the infection.

Finally, let us point out that a number of professional anti-malware software packages were made available in the marketplace. As attacks keep increasing in volume and severity, so the anti-malware manufacturers have to continuously update their products and signature databases. Among the major products one can consider the following:

- Sophos anti-virus (Sophos, 2005) Sophos uses a series of methods to rapidly identify new viruses, including code emulation, online decompression for scanning. It has an engine for detecting and disabling macro viruses. It scans incoming documents by analyzing their format.
- Internet security suite (McAfee, 2005) McAfee integrates elements of intrusion prevention and firewall technology into a single solution for computers and file servers. Therefore, it can deliver proactive protection against the newest threats and present

advanced management responses to reduce costs. It can protect files, block traffic, and filter e-mails.

- Anti-virus Corporate Enterprise Edition (Symantec, 2006) This product combines real-time malware protection for enterprise workstations and network servers with graphical Web-based reporting and centralized management and administration capabilities. The solution automatically detects and repairs the effects of spyware, viruses, and other malicious intrusions to enable enterprise-wide system uptime.

- *Norton AntiVirus* (Norton, 2005) This solution automatically keeps virus definitions updated to provide continuous protection of Internet and e-mail. It combines virus-detection capabilities with updating and scanning technologies to make it easy for users to secure their systems against malware. It also protects against malicious code in ActiveX controls and Java applets as well as against worms, Trojan horses and password stealers. Norton AntiVirus adds expanded threat detection alerts which alert you to non-virus threats, such as spyware and keystroke logging programs, and scans compressed file archives before they are opened.

13.5.3   *Invasion protection using IPS*

Intrusion detection systems (as described in Chapter 11) are often used to monitor the communication traffic in the enterprise's information systems and networks in order to detect, analyze, and identify malicious actions. Another technique has been developed to actively prevent those attacks at their beginning. This technique is referred to as *intrusion protection system* (IPS). Like IDSs, two main classes of IPSs can be used: the Host IPS (H-IPS) and Network IPS (N-IPS). An H-IPS relies on agents installed directly on the system and is used closely with the operating system and the services to prevent attacks and manage log files about them. H-IPS also monitors the data stream in order to protect the services from general attacks for which no signature is made available. Because it verifies all requests to the system it protects, it must be very reliable, has no negative impact and cannot block legitimate traffic; otherwise it should not be installed on a host.

An N-IPS combines the features of a regular IDS and a sophisticated firewall system. Performing like a firewall, the N-IPS has at least two network interfaces: the internal interface and the external interface. When a packet arrives at an interface, it is submitted to a detection engine. The IPS device behaves, in this case, like an IDS in determining whether the arriving packet poses a threat. However, if the N-IPS detects a malicious packet, it not only generates an alert, but also discards the packet and marks "untrusted" the flow to which the packet belongs. The following packets in the traffic flow will also be discarded upon their arrival.

Protection techniques used in an IPS include protocol identification and traffic analysis. A protocol can be identified using port assignment, heuristics, port following, protocol tunneling recognition, while traffic analysis takes place after the protocol has been correctly identified. It includes protocol analysis, packet reassembly, pattern matching, and statistical threshold analysis. IPS technology combines the filtering capabilities of a firewall and packet analysis in an IDS. The major drawback of IPS is the unreliability on frequent update to the signatures that must be applied to the operating system.

13.6   Protection guidelines

Generally, viruses propagate simply because of unsafe usage practices. Reducing some of these practices would hugely reduce the risk of infection. Guidelines for safe computer usage from a perspective of infection can be divided into six classes: security settings, signature updates, disk management, scanning processes, training skills, and user policies. Guideline rules are solely described for each class as follow:

Security settings

Various rules can apply for system settings. These include, but are not limited to, the following seven rules:

- Schedule periodic backups of system and data files. Protect all servers (including e-mail and firewall servers) with anti-virus software. Deploy anti-virus software on all hosts.
- If a firewall is installed, then only necessary ports can be kept open. The critical systems files (such as sys.ini, win.ini, autoexec.bat, and config.sys for Microsoft Windows) should be set to read-only mode, so that they cannot be written.
- Servers and clients should be set to scan incoming and outgoing files of all types (particularly compressed files).
- Any external drive of high-risk workstations should be disabled. If this is not feasible, then the option of booting from external drive must be disabled.
- Prevent users from gaining access to the infected files and perpetuating the infecting viruses. This can be done using a software package that allows files to be quarantined.
- All macro virus protection within software packages (such as Word Office and Open Office) must be enabled.
- Operating system administrators must set permissions to the system files to prevent unauthorized changes.

Virus signature updates

Guidelines providing protecting rules related to the signature policy include the following five rules:

- Regular updates of virus signature files must be scheduled based on what manufacturers offer in terms of updates (regular updates are done hourly, daily, weekly, biweekly or monthly).
- Updates should be deployed on all hosts. If automatic distribution is not feasible, then sending the updates as an e-mail attachment must be considered or a competitive upgrade to an anti-virus solution which does so automatically must be obtained.
- Any update should be built into users' network login scripts.
- Whenever a new signature file is received by a host, it should update the write-protected emergency boot disk, if any.

- If a dedicated server is set up to retrieve regular updates in a security domain, then internal users should be allowed to connect to the server to update their hosts. For this, special care should be taken of the protocol used to connect (e.g., ftp) and the available anti-malware that do not allow ftp downloads.

Disk management

Guidelines integrating rules relevant to storage management policy (and particularly the floppy management) include the following list of rules:

- Users should avoid using external unit containing data and programs (e.g., floppies) received from unknown sources.
- A policy that enforces the scanning of all diskettes before they are used in a host must be established. An alternative solution might set up a remote dedicated PC for this purpose, or use anti-malware software which can be configured to automatically scan the external drive of all machines.
- All data and programs of external storages should be made write-protected.

Scanning

Guidelines including a policy to follow for secure scanning should involve the following rules:

- The enterprise should consider using a dedicated workstation that continually scans data directories on the network.
- Full host scans with minimal intrusion to the user should be scheduled on a regular basis (scheduling can take place after working hours). User intervention of scans should be stopped.
- Web browser should offer a plug-in to scan files prior to downloading. The plug-in should be used when available. If a plug-in is not an option, make sure all downloaded files are scanned prior to installation.
- Scan operation should be performed on new computers upon their reception from vendors since they may contain viruses out of the box.
- Background monitoring on the workstations must be allowed.

E-mail policies

Guidelines to protect e-mails define a policy based on a set of rules including the following:

- E-mail server filters should be set up to eliminate spam and undesired useless e-mails that could contain viruses, worms or malicious codes. All incoming and outgoing e-mail and attachments must be scanned.
- The e-mail server should be set up to immediately send a notification to the network's administrator and involved user on the occurrence of an infected message before it is opened.

- The enterprise may not allow the use of Internet e-mail services (e.g., Hotmail and Yahoo), where incoming e-mails do not pass through the enterprise's e-mail server. Anti-virus protection can be set as an additional measure to help monitor this access.
- Non-work-related downloading of attachments should not be allowed.

User policies

Guidelines governing the user activity include the following rules:

- The enterprise should consider limiting Internet access to approved sites. This can be done since most Internet solutions allow an administrator to create a password-protected list of approved Web sites.
- The enterprise should not allow remote-access users to upload files to the network unless a trusted procedure can verify the integrity of the computer being used for remote access.
- The enterprise has to deploy a set of trusted applications that employees (or users) have available to do their activities in the enterprise. It also should prohibit any software to be installed beyond those provided and impose that all software installations be performed only by authorized people.
- A system to educate all users about polices such as the "no download rule" should be set up.

Educate your users

Guidelines addressing awareness can include the following rules:

- The enterprise should consider developing an Intranet site dedicated to virus information with links to anti-virus sites. It can also develop an e-mail system that informs about the same type of information.
- The enterprise should train the users of its network to check the Intranet site (and other links to reliable virus information) when they suspect they have a virus or when they need useful information about malware.
- The enterprise should inform users on the proper use of macro virus protection, train them to disable all macros when prompted and encourage them to install an anti-virus software package on their home computers.
- The enterprise should persuade users to report on the malwares they find on their system so that it can track which viruses surfaced in the enterprise's network.

13.7 Polymorphism challenge

A *Metamorphic virus* rewrites its code with each attack, so that it does not leave any recognizable signature, which may be difficult to find by anti-virus software. It may achieve rewriting by permuting its commands and altering its code from one iteration to the other or simply perform substitutions to obscure recognition. Nowadays, there are more complicated metamorphism schemes that aim at more devastating effects. One of the interesting examples of metamorphic viruses is found in the virus *Zmist* (Ferrie and Ször, 2001), which attempts to decompile executable files and randomly combines its instructions throughout blocks of code, while keeping their order, so that they can execute and ensure that the location of blocks is not predictable. In this way, virus *Zmist* masks and randomizes the point at which its code starts to execute. In addition, it can vary its own instructions, reversing branch conditions, and substituting instructions.

Detection of the most complex metamorphic viruses is still an open challenging issue. In fact, if the metamorphic virus is written in a way that no malicious command takes more than a few instructions long, then it is almost impossible to predict where in the program the new lines will be. Moreover, if the virus overwrites existing code rather than adding code, it is practically impossible to detect. In addition, combining metamorphism with encrypting procedures (that are used when the data to hide cannot be altered) makes it impossible to search for strings of any length. Fortunately, this cannot happen because it is clear that: (a) there is a huge amount of overhead associated with writing malicious code that transforms completely with every execution to avoid detection and (b) the virus creator needs to write code to control the code transformation and needs more code to mask that code and encrypt the data that cannot be altered.

Despite the theoretical limitations on size and sophistication of metamorphic viruses, various implementations have been developed showing that metamorphic viruses will clearly constitute a large subclass of the future of malware. Probabilistic methods are currently under investigation, but they lead to frequent false positives.

Another class of challenging viruses contains the *polymorphic viruses*. A polymorphism is the application of a cryptographic procedure that is used to make the malicious code unreadable by any entity, but a computer that is involved in the process of execution (of the polymorphic virus) by the time any decryption takes place. The goal of polymorphism is to provide little to no constant virus contents for scanners to find. This is implemented by encrypting the virus code with a different key each time. Advanced polymorphism introduces small variations into the decrypted code, such as inserting a random number or do-nothing instructions. Inserting random numbers can be done using random number generators to establish initial indices, numbers of iterations, etc.

The obvious weaknesses of polymorphic viruses make them less sophisticated among those which are easily detectable by anti-malware software, whereas metamorphic viruses present a far greater challenge to those who would attempt to stop them. Captivatingly, an assumption about the future of cryptography procedures in malicious programs often carries up the vision of viruses that may be encrypted themselves, which would encrypt the data of the infected computer and threaten to damage it or block access to it, if the polymorphic virus is destroyed. As high-speed malware gains popularity because of the increase of user awareness, anti-malware generalization and the growth of faster connections on hosts. The priority for malware creators has moved away from avoiding detection toward proliferating fast enough that enterprises and hosts are infected before they have time to react.

13.8 Conclusion

This chapter has reviewed the chief basics of malware systems and ways to protect computer systems and networks against them. As the number of malicious programs has grown recently, an efficient software solution is needed in order to protect the organization's systems from other harming living software without excessive requirement of user intervention. We defined and classified malware systems and described the ways that major classes of malware, such as viruses, worms, and Trojans, are designed, developed and propagated. The chapter introduced the concept of malware, and then reviewed the schemes that can be used to analyze viruses including scanners and integrity checkers. Warm analysis was investigated with an eye towards target discovery, target lists, warm activation procedures, and warm propagation techniques. We also analyzed Trojans and shed some light on their types, and ways to protect against them. We dedicated a section on the ways to protect against malware using schemes such as firewalls and anti-virus software packages. The chapter presented recommended guidelines on how to protect against malware software. Finally, polymorphism challenges as related to malware have been investigated and discussed.

References

Anagnostakis, K. G., M. B. Greenwald, S. Ioannidis, A. D. Keromytis, and D. Li (2003). A cooperative immunization system for an untrusting internet. In *Proc 11th International Conference on Networks (ICON), 2003* (available at http://www1.cs.columbia.edu/angelos/Papers/icon03-**worm**.pdf).

Briesemeister, L., P. Lincoln, and P. Porras. Epidemic profiles and defence of scale-free networks. In *Proceedings of the 2003 ACM Workshop on Rapid Malcode, SESSION: Defensive Technology*, Washington, DC, ACM Press, pp. 67–75.

Ferrie, P. and P. Ször (2001). Hunting For Metamorphic. Symantec White Papers, http://www.symantec.com/ avcenter/reference/hunting.for. metamorphic.pdf.

Garetto, M., W. Gong, and D. Towsley (2003). Modeling malware spreading dynamics. In *Proceedings of INFOCOM*, April 2003. www.telematics.polito.it/garetto/papers/virus2003.pdf

Kienzle, D. M. and M. C. Elder. Recent worms: a survey and trends (2003). In *Proceedings of the 2003 ACM Workshop on Rapid Malcode, SESSION: Internet Worms: Past, Present, and Future*, Washington, DC, ACM Press, pp. 1–10.

Kumar, S. and E. H. Spafford (1992). A generic virus scanner in C++. In *Proceedings of the 8th Computer Security Applications Conference*, Los Alamitos CA, December 1992. ACM and IEEE, IEEE Press, pp. 210–19.

McAfee (2005). http://www.mcafeestore.com.

CERT (1999). CERT Advisory CA-1999–04 Melissa Macro Virus, http://www.cert.org/advisories/ca-1999-04.html

Moore, D., V. Paxson, S. Savage, C. Shannon, S. Staniford, and N. Weaver (2003). Inside the slammer worm. *IEEE Magazine of Security and Privacy*, 33–9, July/August 2003.

Norton (2005). Norton AntiVirus 2005, http://www.download.com/Norton-AntiVirus/

Qattan, F. and F. Thernelius (2004). Software protection mechanisms and alternatives for securing computer integrity, Master Thesis. CSS Department, Stockholm University-Royal Institute of Technology.

Sophos (2005). Sophos Enterprise solution, http://www.sophos.com/products/

Symantec (2006). Symantec Anti-virus Corporate Edition, http://www.symantec.com/ Products.

Weaver, N., V. Paxson, S. Staniford, and R. Cunningham (2003). A taxonomy of computer worms. In *Proceedings of the 2003 ACM Workshop on Rapid Malcode*, Washington, DC, 2003, 11–18. ACM Press. http://www.cs.berkeley.edu/nweaver/papers/taxonomy.pdf.

Zou, C. C., W. Gong, and D. Towsley (2002). Code red worm propagation modeling and analysis. In *9th ACM Conference on Computer and Communications Security*, 2002, ACM.

14 Computer and network security risk management

The use of communication technologies to conduct business has become a crucial factor that can significantly increase productivity. The need to secure information systems and networked infrastructures is now a common preoccupation in most enterprises. As a result, strong links are being established between security issues, communication technologies, an enterprise's security policy, and an enterprise's business activity. Risk management has become an important procedure for any enterprise that relies on the Internet and e-means in its daily work. Risk management determines the threats and vulnerabilities of any e-based system. It also integrates architectures, techniques, and models. This chapter attempts to deal with all of the above concepts and techniques.

14.1 Introduction

The development of information and communication technologies, especially the Internet, has prompted enterprises to redesign their communication infrastructure in order to take benefit of this visibility factor and re-engineer their business processes by implementing projects online, managing virtual enterprises, and externalizing their activities. Renovation and ICT use have contributed significantly to the success of many companies. Nevertheless, the current growth of digital attacks has caused decision makers in enterprises to doubt the confidence in information technology. In fact, security incidents that occurred recently (as discussed in the previous chapters) have emphasized three important facts: (a) computer network attacks can induce a huge damage on business activity, (b) many of the attacked enterprises have active security infrastructures at the moment the security incident occurred, and (c) the security infrastructure costs vary highly from one enterprise to the other based on the security policy adopted and the nature of the activity performed by the enterprise. While the first statement is a natural consequence of information technology involvement in the production process, the second fact is very interesting as it may suggest that security systems have no efficiency in thwarting sophisticated malicious actions, and the third statement suggests that much work needs to be performed before investing in a security solution. Therefore, structured methodologies that identify, analyze, monitor, and mitigate computer security risks have been developed to help enterprises integrate security in their strategic plans. In practice, experience has shown that the ad-hoc application of a Risk Management

(RM) approach would considerably limit the aforementioned damages and reduce the cost of solution development.

A basic definition of security RM includes: *the ideas, models, abstractions, methods and techniques that are employed to control security risks.* Typically, a RM approach is intended to: (a) identify the risks threatening the information system of an enterprise, (b) select the appropriate security measures/solutions to limit this risk to an acceptable level, and (c) continue to monitor the system in order to preserve its security level and detect deviations. The efficiency of RM relies closely on risk analysis, which aims at controlling the state of a system in order to select the appropriate actions that should be taken to enhance its defence means when an attack occurs. This increases the effectiveness of security solutions as they are selected according to a clear strategy and not on the basis of human intuition.

In spite of its importance at the enterprise level, very few studies have discussed RM in detail and presented efficient methodologies for their utilization. Existing work mainly focuse on several facets of RM without considering the whole problem. Abundant research and development work has addressed vulnerability analysis, risk analysis, and network state monitoring while RM approaches have been rarely cited in the literature. This chapter is an attempt to bridge the gap between enterprise decision making and RM benefits. A review of the different methods, techniques and mechanisms is performed in order to present an overview of the work performed to build RM systems. Of course, as some of the considered issues may be used in other fields, RM discussion will be restrained to the digital security framework (Hamdi and Boudriga, 2005).

14.2   Risk management requirements

Since the development of the first risk management method, many projects to develop new approaches and increase the existing methods have taken place. None of these approaches has experienced a wide interest and application. The major factors that limit the application of those methods in a real enterprise's environment are: (a) the lack of mature methodologies that handle risk and estimate the return of investment and (b) the lack of appropriate software tools. Effectively, risk analysis is so complicated that performing it manually would be almost unfeasible. Moreover, a glance at the existing automated risk management tools shows that they do not keep up with the evolution of the results derived from related theoretical models. More significantly, all software packages using the tools tend to be incomplete, in the sense that they do not support a complete risk management cycle. The most important issues that should be covered by risk management include the following:

Planning cost estimation

As security planning itself has an important cost to the enterprise that conducts it, the amount of effort related to a risk management project should be conveniently estimated. Generally, if the cost of security planning overtakes the allowed budget, risk management operations have no reason to start. The absence of techniques estimating the required effort with respect to the features of the analyzed environment is one of the significant reasons that can modify the confidence of IT users, high-level managers, and decision makers in risk management.

Attack modeling

Specifying attacks performed against computer networks is among the most challenging tasks the risk analyst should address. These are strongly linked to the intrinsic complexity of the performed attacks, which can be explained partly by the following major facts:

1. As network attacks are conducted by humans, a predictive model integrating human profiles would be hard to develop. In addition, human ideas cannot be easily translated into a formal representation. Most of the existing methods assume the existence of a static attack library without addressing this issue.
2. As attacks are often performed under the form of a structured hierarchy of actions, attackers often follow a structured methodology to achieve their malicious objectives. Hence, the risk model should provide various kinds of links between elementary threats, vulnerabilities, and actions (e.g., causality, time precedence) and integrate security countermeasures and enterprise resources as a component of the expert knowledge related to attack hierarchies.
3. As the efficiency of the response to attacks is based on the quality of links made between the attacks and the related alerts and the effectiveness of the alert correlation process, the correlation handling and response management are difficult to integrate.

Decision selection

In typical risk management methodologies, security decisions are associated statically to digital attacks. In other terms, a set of countermeasures corresponds to each attack, in the sense that they mitigate it. The mapping often may not satisfy totally the needs of the risk analyst because it may generate more problems than it can solve and does not allow one to assess the efficiency of security decisions. Consequently, several conditions should be integrated within the security solution selection module to make the choice of the optimal countermeasure alternative depend not only on the nature of the attacks, but also on the environment of the information system to be protected (e.g., network topology, networked applications, and server nature). As a result of this remark, the security solutions that are optimal for an enterprise may not be necessarily optimal for another enterprise even if both of them face the same threats. This highlights also the fact that portability of control systems should be carefully adapted.

Monitoring

This task is typically reduced to a continuous control of the state of information and communication system under protection. However, given that it constitutes a step in the risk management cycle, it seems unwise to restrict monitoring to its pure security aspects related to information collection and analysis. It should be fully integrated in the business activity of the enterprise. Incident response, for instance, should be enriched by a set of models allowing the accurate measurement of an incident impact, and the selection of the appropriate reactions using multi-criteria resolution. It also should integrate a set of roles to help achieve an efficient incident response.

When applied inside an organization, RM should be structured into a framework including a life cycle, a set of techniques, and a set of computer assisted design and analysis tools. Evidently, all these components should fully interoperate to guarantee the efficiency of the process. Throughout the following sections, we show that monitoring constitutes one of the most important weaknesses of the existing frameworks. We also show that developing a structured decision database would be very important to conduct responses to incidents.

14.3   Risk management methods

The rapid growth of attacks against computer network systems and the high cost induced by the available security countermeasures have favored the development of structured high-level methodologies aiming at: (a) evaluating the security state of such systems, (b) performing risk analysis (before a solution is developed and on the occurrence of a threat), and (c) selecting the most suitable defence measures when an attack is identified. Therefore, integrating security issues into the business activity of an enterprise has been an important question over the past three decades. One of the earliest risk management approaches has been proposed by (Campbell, 1979) who developed a structured methodology based on a set of concepts that continued to be used in later approaches such as vulnerability analysis, threat analysis, risk analysis, and state control.

All through the complexities that were faced when applying the first RM methods, it has been observed that quantitative risk assessment is difficult to conduct due to two major factors:

- As security models are still immature, building a representation of the analyzed environment is a very difficult task. The large amount of data collected from various components of the information and communication systems could not be treated accurately.
- The lack of automated tools supporting the risk management activity made its manual execution more difficult. Consequently, easier subjective methods have been introduced. The qualitative evaluation of the basic risk analysis parameters (such as threat probability and threat impact) is basically a simple scaling. For instance, the probability of occurrence for a specific threat can belong to a set of three values denoted by {*high*, *medium*, *low*}. It is obvious that in this case the collection of information becomes much easier than in the quantitative case. In fact, instead of relying directly on digital security data such as audit log records, interviews and questionnaires are used in the qualitative case.

RM methodologies can be divided into two categories: bottom-up approaches and top-down approaches. The bottom-up approach aims at selecting a priori the residual risk;

meaning that the degree of protection of the system is determined and the countermeasures that allow reaching it are implemented. Top-down risk estimation specifies programmed tasks that are planned to reduce the identified threats. Examples of these tasks include, but are not limited to, risk identification, risk mitigation, system state monitoring, and risk assessment. Another significant aspect that has occurred due to the need of cost-effectiveness is the development of automated software tools that support the risk management actions. In the following sections, we do not focus on such tools but we rather discuss the influence of the automation requirement on the content of several methodologies. In other terms, it should be underlined that the simplicity of implementation is among the most pertinent factors to assess a RM approach.

14.3.1 The OCTAVE method

The Operationally Critical Threat, Asset, and Vulnerability Evaluation, OCTAVE (Alberts and Dorofee, 2002) has been structured with respect to the various categories of results that could be reached by the developing team. There are three main types of outputs that have been addressed: (a) the organizational data, (b) the technological data, and (c) the risk analysis and mitigation data. Unlike the typical technology-focused assessment, OCTAVE is aimed at organizational risk. It focuses on strategic and practice-related issues. The OCTAVE approach is driven by two aspects: operational risk and security practices. Technology is examined in relation to security practices, enabling an organization to refine the view of its current security practices. OCTAVE is organized around three basic phases (as illustrated in Figure 14.1), enabling organizational personnel to assemble a comprehensive picture of the organization's information security needs. The phases are:

- Phase 1 (*Build Asset-Based Threat Profiles*) This is an organizational evaluation. The analysis team determines what assets are important to the organization (e.g., information-related assets and useful applications) and what is currently being made to protect these assets. The analysis team then selects those assets that are most important to the organization and describes security requirements for each critical asset. Finally, it identifies threats to each critical asset, creating a threat profile for that asset. It is worth mentioning that threat profiles are derived from ad-hoc event-trees.
- Phase 2 (*Identify Infrastructure Vulnerabilities*) This phase attempts to perform an evaluation of the information infrastructure. The analysis team examines all network access paths and identifies classes of information technology components related to each critical asset. The team then determines to what extent each class of components is resistant to network attacks.
- Phase 3 (*Develop Security Strategy and Plans*) During this phase, the analysis team identifies risks to the organization's critical assets and decides what to do about them. The team creates a protection strategy for the organization and mitigation plans to address the risks to the critical assets based on an analysis of the information gathered.

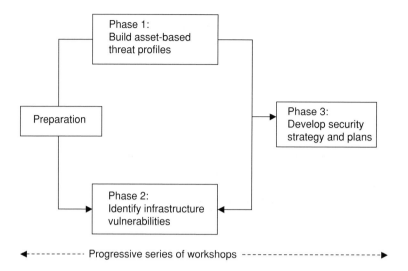

Figure 14.1 The OCTAVE method.

14.3.2 The CORAS framework

CORAS is a European-funded RM project that has been conducted between 2001 and 2003 (Stolen *et al.*, 2003). A large spectrum of risk-related matters has been addressed in this framework. This has resulted, among others, in the definition of a global model-based approach for security risk management relying on several concepts that have been taken from existing risk assessment techniques. One of the key components of the CORAS framework is an automated tool supporting the CORAS methodology. It integrates a UML-based specification language that can be used for three major purposes described as: (a) representing accurately the system state (from a security point of view) and the interaction between the various entities, (b) standardizing the communication between the risk assessment team members through the use of a uniform language, and (c) documenting the risk management activities. Moreover, the CORAS method highlights the role of security libraries.

The processes of CORAS risk management method are illustrated in Figure 14.2. In the following, these processes are briefly described:

1. *Context identification* This process aims at determining the security-critical assets as well as the related security requirements. CORAS relies on the initial phase of CRAMM (the UK Government's Risk Analysis and Management Method) to address this need. A questionnaire is submitted to system users to determine the essential *"data groups"* (i.e., the relevant assets that should be analyzed).

2. *Risk identification* This process aims at identifying the weaknesses of essential components and the potential threats that may harm them. Four different methods are being used to accomplish this process:
   - Fault-Tree Analysis (FTA) This method identifies the causes for an unwanted outcome event (in a top-down approach).

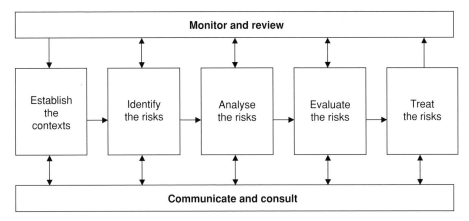

Figure 14.2 The Coras RM.

- Failure Mode and Effect Analysis (FMECA) It focuses on single failures of components (bottom-up approach).
- CCTA Risk Analysis and Management Methodology (CRAMM) It addresses threats/vulnerabilities of an asset by means of predefined questionnaires (Barber and Davey, 1992).
- Goal Means Task Analysis (GMTA) This method identifies tasks and preconditions needed to fulfil an identified security goal.

3. *Risk analysis* This process aims at investigating the possible consequences of unwanted outcomes, as well as the likelihood of their occurrences. FMECA supports well this activity if the system description is sufficiently detailed. FTA and Markov analysis can be used to support hazard and operation analysis as they assign probabilities to the basic events in the fault trees (to compute the ones on top of the events) and evaluate the likelihood of the sequences, respectively.

4. *Risk evaluation* It ranks the identified risk events on the basis of the likelihood of occurrence and on the impact they have. This process is based on FMECA to handle quantitative parameters and on CRAMM to combine qualitative weights. A cause-effect analysis is also performed to determine the relationships between risk events.

5. *Risk treatment* It defines the strategy that should be applied to thwart potential attacks. Multiple options are available at this level including risk avoidance, reduction of likelihood, reduction of consequences, risk transference and risk retention.

The execution of a CORAS-based RM project can be viewed as an iterative process where the initial input is specified using UML (or a similar object-oriented approach) state diagrams. Therefore, the iterations follow the RM-ODP process viewpoints. It is worth mentioning that two processes are run in parallel to the process depicted in Figure 14.2. *Monitoring and review* process re-evaluates the results of each iteration while the *Communication and consultation* process implies a strong cooperation inside the RM team.

14.4   Limits of the existing methodologies

The approaches that have been presented in the above section show several weaknesses that may affect the efficiency of the risk management process performed by the scheme. These shortcomings are categorized into architectural limits and technical limits.

14.4.1   Architectural limits

The following statements constitute, in our opinion, the most important architectural shortcuts of the existing RM methods:

There is no separation between preventive and reactive risk analysis A short look at Figures 14.1 and 14.2 shows that the related methodologies focus on preventive risk analysis and put as secondary the reactive risk analysis. As it will be explained in the next section, preventive and reactive risk analyses show many differences. The preventive analysis needs an a priori reasoning about security (before the occurrence of a security incident), while the reactive analysis requires *a later* intervention after an incident has occurred. Moreover, preventive risk assessment relies on probabilities of occurrence of the identified threats. Reactive risk assessment relies on alerts and therefore introduces probabilities which are principally related to the efficiency of the detection mechanism. As a result, it is suitable to give a larger importance to reactivity in the RM architecture. This is not addressed in existing RM approaches.

Real-time reactivity is relatively absent The time factor is basically important with reactive countermeasures. Depending on the nature of the attack, security solutions might be triggered immediately (this is called real-time response) or after a collaborative reasoning process involving both stakeholders and security specialists. This differentiation is not provided by the aforementioned methods. In fact, this depends essentially on how the attack impact changes with respect to time factor. For example, the decision to block a TCP port must be performed in a very short period of time. Other countermeasures may require the approval of high-level management before being implemented, especially when they may cause an important loss to the enterprise. This means that a compromise between security solutions cost and benefit should be guaranteed. With regard to the rapidity of most known attacks, a high proportion of this decision-making framework should be automated. Only critical decisions should be subjected to a human approval.

Libraries are not described as they should be Due to the large number of vulnerabilities, attacks and alerts, semantic links cannot be built manually (i.e., by the security analyst himself). Suppose that we just want to associate to every attack the vulnerabilities that it exploits. If N_A attacks and N_V vulnerabilities are considered; and giving that the risk analyst has to inspect the whole set of vulnerabilities, for every attack, in order to select those that can be exploited by the attack of interest, then the total number of semantic links that should be addressed equals $N_A \times N_V$. Obviously, one may argue that databases provided by official security organizations can be used to cope with this problem. Unfortunately, current databases can be used only in certain simple situations. In fact, they are restricted to system weaknesses and do not consider human or documentation-related vulnerabilities.

Consequently, one of the key components of a RM methodology is a language which uniformly handles vulnerabilities, attacks, and alerts. Obviously, this language should support several automated inference operations such as mechanically determining which vulnerabilities can be exploited by a given threat, which security alerts correspond to a specific attack, or which countermeasures can thwart a damaging action.

Absence of decision libraries A careful consideration of the current approaches shows that an accurate structure is needed to represent countermeasures, store them and retrieve them based on various criteria. This can substantially facilitate the search and selection of decision when responding to an attack. Furthermore, an appropriate syntax should be defined to allow the automated association with attacks, assets, and vulnerabilities. A query language may also be useful for exact search (or approximate search) of decisions to soften a situation.

14.4.2   Technical limits

The major technical limits that can be observed are:

Absence of homogeneity One of the most important limitations of OCTAVE and CORAS stems from their lack of uniformity. The techniques that have been proposed to implement RM processes seem different and difficult to integrate with each other. For instance, combining the myriad of risk analysis techniques proposed within the frame of the CORAS approach requires an important adaptation effort, even though a uniform language is used to specify security. Naturally, a RM cycle consists of various activities that can not be described nor implemented through the use of a single technique. Henceforth, what is required is to take into account the effort that should be made to set up a toolbox that adapts and integrates many techniques. In this context, the use of specification tools that provide an abstraction view of RM processes and tasks may be very useful.

Lack of powerful techniques As will be underlined in the following sections, many sophisticated tools and theories have been developed for most RM tasks, especially vulnerability detection and intrusion detection. Nonetheless, RM methodologies have not kept up with those research advances. Techniques that are proposed by OCTAVE or CORAS are certainly easy enough to be quickly understood and applied, but they also present severe limitations. For instance, HAZOP (used by CORAS to identify security threats) is not compatible with the automated vulnerability tools.

Lack of efficient models Some among the proposed RM processes are performed under constraining needs related to: (a) a decision structuring capable of optimizing search operations, (b) a correlation and fusion process that is able to reduce false positives while monitoring the system to protect, and (c) an attack representation allowing rich representation of the scenario of collected alerts. Models provided by the current RM are limited, where they exist.

14.4.3   The NetRAM framework

To overcome the aforementioned shortcomings, a risk management framework called NetRAM (Network Risk Analysis Method) has been developed (Hamdi *et al.*, 2003). A major characteristic of NetRAM is that it is heterogeneous, in the sense that it combines the use of three different components: (a) specific software tools and life cycle, (b) architectural design methodologies, and (c) theoretical models. NetRAM is characterized by five properties. First, NetRAM can be adapted to different types of enterprise structures and handle various network architectures and topologies. Moreover, NetRAM can be customized so that the business activity of the enterprise can be taken into account.

Second, NetRAM can be viewed as a collaborative and a collective framework that requires the contribution of several automated and human components. It guarantees optimal handling of the key risk management processes: analysis, decision and response. Many sources provide necessary data and multiple tools that can be used to execute various operations. For instance, information contained in Intrusion Detection Systems (IDSs) and automated vulnerability scanners can be imported and their outputs analyzed in a single uniform environment. Likewise, a group of experts might be involved at some stages of the decision-making process.

Third, NetRAM permits handling appropriately the uncertainty introduced by different factors. Both computer-aided risk analysis and expert-based decision making are performed through the use of structured methodologies that reduce the error-rate and prevent errors from propagating across the risk management steps. Obviously, this approach permits us to control efficiently the residual risk and make the enterprise aware of the actual threats related to its environment.

Fourth, monitoring the state security of the analyzed information and communication infrastructure is a critical task. A set of modules ensuring a continuous control of the system state has been integrated with the NetRAM framework. Detecting deviations of the system state from the normal behavior and conducting convenient reactions are the main objectives of these modules.

Finally, NetRAM is highly adapted to the rapid changes of ICT that might affect the analyzed system. A learning process is thus developed to limit human intervention during the update of several quantitative parameters and semantic links. In fact, automating the learning process makes its output more accurate because the sources from which the necessary data is gathered are tightly related to the system environment. For example, log files located in routers, firewalls, and intrusion detection systems are highly useful at this level. However, human intervention remains a necessary complement at this level to avoid.

From an architectural point of view, NetRAM can be represented by a ten-process life-cycle as depicted by Figure 14.3. The interested reader may say that it consists of two concurrent graphs. The graph designed with solid lines represents the workflow of the risk management activities. The arrows model the precedence relation between the ten processes of NetRAM, which follow the progress of the risk management activities. On the other hand, the second graph, represented by dashed lines, illustrates the retroactivity existing between some processes of NetRAM. It is worth noting that this characteristic is specific to NetRAM. In fact, existing risk assessment methods always ignore the importance of re-executing some risk management activities further to the emergence of vulnerabilities or threats.

A brief description of the constituency of each module is given below.

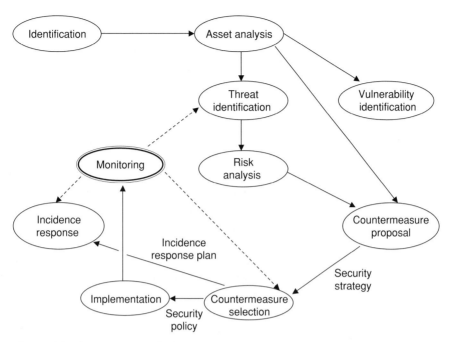

Figure 14.3 The NetRAM Method.

Initialization

This process aims at preparing the initial step of the risk management project by providing the managers with the cost of the project and the schedule of different risk management activities. Moreover, an estimation of the effort required to conduct the risk management project is performed. The underlying activities described next are then scheduled according to this estimation (Krichene *et al.*, 2003).

Asset analysis

The aim of this process is to gather information about the analyzed system. In addition to the evaluation of the inventory resources, some dependency links between them should be established. Such relationships might be needed when building attack scenarios.

Vulnerability identification

Weaknesses and security breaches of the analyzed system are identified at this level. A combination of automated tools (vulnerability scanners) and questionnaires is used for this purpose.

Threat identification

This process aims at identifying the attacks that endanger the analyzed system. A list of potential threats is deduced using an exploit relationship with the identified vulnerabilities.

Risk analysis

The aim of this process is to identify the risks that may threaten the assets of the analyzed system based on the identified vulnerabilities and threats. These risks are then ranked and prioritized with respect to their influence on the system.

Monitoring

The objective of this process is to maintain the analyzed system in an acceptable security level. A set of relevant security metrics should be continuously measured and addressed to detect security incidents as soon as they occur. Monitoring activity can result in the re-execution of some NetRAM processes if needed.

Countermeasure proposal

A security strategy mitigating the identified risks is proposed by this process. It consists basically of high-level statements that guarantee the respect of the security requirements.

Countermeasure selection

This process aims at defining the security policy that will be adapted by the analyzed system to mitigate security risks. A set of countermeasures is selected from a decision library according to the strategy established in the former step. A multi-attribute algorithm is set out to cope with this need.

Incident response

This process aims at reacting to security intrusions according to the incidence response plan. A risk analysis process is conducted to select the most cost-effective reactions based on so-called cognitive maps and to ensure the continuity of the business-critical processes (Hamdi *et al.*, 2004).

Implementation

At this level, selected security countermeasures are implemented according to the security policy. A schedule for the deployment of these solutions is performed depending on the available budget and the total period.

14.5 Management of risk libraries

Two libraries are important in risk management. These are: the vulnerability library and the attack library. We discuss in this section the use of these libraries.

14.5.1   The vulnerability library

A study of networked computer systems shows that three major factors need to be studied to determine whether a system is secure: (a) the state of hardware and software infrastructure, (b) the behaviour of the system's users, and (c) the correctness of the security documentation. This means that a security breach can be related to one or more of those factors. Conceptually, it is obvious that vulnerabilities corresponding to each of those classes have different natures. The hardware and software components of the IT system are by far the most complex entity. Assessing their security should take into account operation systems, applications, and network infrastructures. Subsequently, a large part of the data must be correctly collected and scanned. Practically, security questionnaires are often used to detect user-related vulnerabilities. Finally, analyzing security policies for breaches requires methods and mechanisms which state whether a given policy fulfils a set of specific requirements. To this end, a policy should be translated to formal expression in order to be validated rigorously.

Effectively, most of the existing RM methods reduce vulnerability analysis to vulnerability identification and thus subordinate many key issues. For example, OCTAVE defines this activity as *"running vulnerability evaluation tools on selected infrastructure components."* A similar definition has been given by Ozier in (Ozier, 2000). It states essentially that:

This task includes the qualitative identification of vulnerabilities that could increase the frequency or impact of threat event(s) affecting the target environment.

It appears from these definitions that the goal of vulnerability analysis has been commonly viewed as ascertaining the existence of a set of vulnerabilities in several components of a given system. In other terms, the risk analyst has to map some elements of the vulnerability library to each asset of the studied system. Nevertheless, this task can be more skilfully applied by considering the following concepts:

Vulnerability detection strategies Security specialists should be aware of the various testing methodologies that can be applied to detect vulnerabilities. The choice of a testing strategy depends upon the nature of the weakness and the characteristics of the target system.

Vulnerability identification techniques Many vulnerability detection mechanisms are available. Thus, the most convenient ones should be selected according to the need of the risk analyst.

Vulnerability library structure The vulnerability library may contain thousands of security weaknesses. Therefore, it must be structured in a manner to support various types of queries.

We take this approach to describe vulnerability analysis as we believe it is more amenable to reason about this task. In what follows, these ideas are detailed to show how the richness of the vulnerability library enhances the efficiency of the risk management process.

Vulnerability detection strategies

Automated software vulnerability testing strategies can be classified into two categories: black-box testing and white-box testing. The former consists of evaluating the behavior of the system without having a precise idea about its constituency while the latter preassumes that the source code of the tested entity (e.g., piece of software, OS, communication protocol) is available. Most of the existing vulnerability detection methods belong to the black-box class. Three primary sub-classes can be defined:

- *Rule-based testing* The vulnerability detection engine relies on a set of rules that represent known patterns corresponding to security breaches. For instance, a password can be checked to see if it fulfils the requirements given in FIPS Pub 112. The major problem with rule-based testing is its incompleteness, since rules only discuss limited facets of the vulnerability to test.
- *Penetration-based testing* Intrusion scenarios are actually conducted on the analyzed system to assert whether the corresponding vulnerabilities exist. In the case where only hashed passwords are available, the vulnerability detection procedure should obviously differ from the aforementioned one. Various character combinations have to be generated and their hash be computed sequentially and compared with the original hash. In this case, the robustness of the password is measured according to the period of time necessary to guess it. The longer is this period, the stronger is the password.
- *Inference-based testing* Many events that do not correspond to security weaknesses may lead, if combined, to an insecure state. Therefore, the vulnerability detection engine should take into consideration the relationship between the components of the target environment. Consider the case where the hash function of a strong password is stored in a file that does not have suitable access rights. For example, if a user has a write access to the password file, he can substitute the password of a given account with an arbitrary password that he would use for a more privileged access. It appears, thus, that neither the rule-based nor the penetration-based testing strategies permit the detection of such vulnerabilities as the detection engine should perform inferences on the state of the information system.

The second possibility to detect software security breaches is to perform white-box testing by analyzing a specific source code. This analysis can be either static or dynamic. Static security scanning encompasses the cases where the security analyst searches for pre-defined patterns in the scanned code. On the other hand, dynamic testing means that the code should be modified, compiled, and executed to know whether it is vulnerable. This vulnerability scanning strategy (i.e., white-box) is clearly more efficient than black-box testing, especially for identifying buffer overflow vulnerabilities, which are among the most numerous and harmful. However, they can not be always applied. They are particularly useful to scan system-specific software (i.e., software that is used within the frame of a particular application) or open-source applications that can be freely enriched and customized by users. In addition, white-box testing can be used to analyze the security of software during its development.

As a result, the risk analyst should identify the tests to run in order to build an accurate view of the security level of the analyzed infrastructure. Often, the testing strategy is hybrid in the sense that it is made up of several black-box techniques and some white-box methods.

Vulnerability library structure

Vulnerability information is usually structured in various manners including traditional (or relational) databases, mailing lists, and newsletters. The most important Vulnerability Databases (denoted by VDBs) are listed below:

- *The Computer Emergency Response Team* (*CERT*) *Advisories* CERTs publish warnings related to the most important and most recent vulnerabilities. They collect information about different types of security incidents. After analyzing these events, reports including the exploits and related countermeasures are generated and published.
- *The MITRE Common Vulnerability and Exposures (CVE) List* A CVE list is defined as a standardized list of names for vulnerabilities and other information security exposures. Practically, when multiple vulnerability scanners are used for a better completeness of the detected weaknesses, CVE can serve to fuse the data emanating from those tools into a unified list.
- *The NITS ICAT Metabase* ICAT is a CVE-based VDB which offers interesting search possibilities. It may link the requester to other vulnerability databases where he(she) can find more detailed information. It can even redirect the user towards Web resources where he can find accurate defence solutions. ICAT is probably the most complete VDB as it indexes the information available in CERT advisories, ISS X-Force, Security Focus, and a large variety of security and patch bulletins. For this, it can be considered as a metabase, which can be used as a search engine rather than a VDB itself.

These vulnerability databases often include advanced search and browsing functionalities. The security analyst often needs precise information to process accurately the security weaknesses, especially when associating vulnerabilities with threats, an appropriate data structure should be used. An approach, based on data-mining, can be proposed to build VDBs as described in (Schumacher *et al.*, 2000). The main focus of this work was the identification of the faulty patterns. It has been argued that the use of data-mining within this context has been motivated by the fact that this technique can be used to discover existing patterns in data sets or classify the new elements (i.e., when the base is being built). A large set of real cases can be used at the training phase so that the mining engine acquires enough knowledge to manage vulnerability description.

To increase the efficiency of the training process, unsupervised learning algorithms should be used because they are more suitable to build VDBs. In fact, the system does not initially have a classification scheme; vulnerability classes are rather defined progressively according to the similarities between input descriptions. Furthermore, one of the most important aspects is the language that describes the vulnerabilities. As the use of a natural vocabulary may, at the mining stage, lead to various problems (e.g., homonyms and identical words), logic-based languages are preferred because their properties can be more easily controlled.

In addition, the VDBs should be structured in such a way that they can be accessed easily and securely. Four models have been proposed to cope with different situations (Schumacher *et al.*, 2000). The *central* architecture, as its name indicates, defines a single VDB which is used by all the community. The *federated* model requires the participation of multiple operators that can manage distinct (in the case of a partitioned architecture) or redundant data (in the case of a replicated architecture). The *balkanized* organization corresponds to the context where no coordination is set up between the different operators. Finally, the *open-source* notion can be applied to VDB by making available full access to all data. Users can even get a copy of the whole VDB and adapt it with respect to their needs.

14.5.2 *The attack library*

Modeling of attacks is one of the most crucial tasks of the risk management cycle. In fact, representing the behavior of various types of adversaries before or after they performed damaging actions is often considered a challenging task. Generally, the risk analyst can never reach absolute statements about the probability or the impact of a given attack. Moreover, as attackers become more aggressive, analyzing their activities is increasingly complicated. To this purpose, attack libraries' structures should be sufficiently sophisticated to take into consideration both of those aspects.

Modeling the gradual evolution of the attacker's goal to reach the main objective is among the most important issues. It has been noticed that attackers often proceed through multiple tasks in order to fulfil their malicious intentions. Tree structures have therefore been introduced to capture those tasks. In fact, fault trees, which have been used for a while to analyze failure conditions of complex systems, have been adapted to the computer network security context. As has been previously underlined, the CORAS RM method relies on the concept Fault Tree Analysis (FTA) to evaluate and prioritize risks. Two concepts have been especially focused: minimal cut sets and top event frequencies. The former identifies attack scenarios whilst the latter allows computing of the frequency of the main attack with respect to the frequencies of the elementary threats.

Attack trees that were first introduced in (Schneier, 2001) can be seen as an adaptation of FTA to the computer security framework. Their purpose is to determine and assess the possible sequences of events that would lead to the occurrence of harmful attacks. The root of a tree represents the main objective of an attacker while the subordinate nodes, in the tree, represent the elementary attacks which are necessary to perform in order to achieve the global goal.

A *qualitative analysis* yields a logical representation of the tree by reducing it to the form:

$$t_0 = S_1 v S_2 v \ldots v S_N,$$

where t_0 is the root of the tree and S_i, for $1 \leq i \leq N$, is the ith attack scenario corresponding to t_0 and having the following structure:

$$S_i = t_1^{S_i} \wedge t_2^{S_i} \wedge \ldots \wedge t_{N_{S_i}}^{S_i},$$

where $t_i^{S_i}$ are the elementary threats belonging to S_i. For instance, the tree shown in Figure 14.4 can be reduced to the logical expression $(b \wedge e \wedge f) v c v (d \wedge g) v (d \wedge h)$.

Writing a global threat as the conjunction of multiple threat scenarios offers a useful tool to the risk analyst who can henceforth describe the various manners for achieving the main attack. Then, a *quantitative analysis* is carried out to compute the security attributes, such

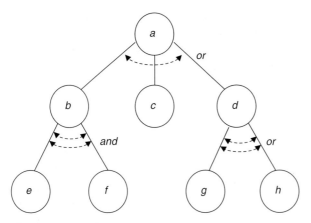

Figure 14.4 Attack tree.

as the probability of occurrence or the outcome of the composite attacks (i.e., scenarios). Often, the attribute of the main attack can be expressed as a function of the attributes of the elementary attacks through the use of simple algebraic operators. For example, the probability of the attack defined in the example depicted in Figure 14.4 follows the logical expression and is determined by:

$$P_a = [(P_a.P_e.P_f) + P_c + (P_d.P_g) + (P_d.P_h)]/4.$$

A substantial attack library has been proposed in (Moore *et al.*, 2001). This approach relies on attack patterns, which are essentially used to build generic representations of attack scenarios. The constituents of an attack pattern are mainly: (a) the goal of the attack, (b) a list of preconditions (i.e., hypotheses that should be verified by the system so that the attack succeeds), (c) the attack mechanism (i.e., the steps followed by the attacker to reach the objective), and (d) a list of post-conditions (changes to the system state resulting from the occurrence of the attack).

On the other hand, the concept of *Attack Net* was proposed in (McDermott, 2000). It defines an attack modeling approach relying on Petri nets. Obviously, a Petri net is composed of places, transitions, arcs, and tokens. Places are the equivalent of graph nodes while a transition is a token move between two places. According to this paradigm, attack steps are represented by places which are similar to nodes in attack trees. Formally, an attack net consists of a set of places, denoted by $P = \{p_0, \dots, p_n\}$, and a set of transitions, denoted by $T = \{t_0, \dots, t_m\}$. P represents the various states of the target system when the actions of the attack are conducted, whilst T models events that may trigger state changes. Additionally, tokens are used to evaluate the progression of a given attack through the study of a set of moves within the attack net.

Attack nets combine the advantages of hypothetical flaws and attack trees. The main difference with these trees is that attack events (i.e., transitions) are described separately from the system states (i.e., places). On the other hand, attack trees merge the preconditions, the post-conditions, and the running steps of a given attack into a single node. However, it is declared that attack nets are better than attack trees for top-down testing of poorly documented operational systems (McDermott, 2000). Furthermore, transitions can catch a vast variety of relationships between system states. As a passing remark, transitions turn out to be more efficient than AND/OR operators to build attack libraries. In fact, these traditional logical operators show some critical limits, especially when enriching an existing state. For example, adding a branch (or AND of nodes) to a set of alternatives (OR node) can not be achieved without restructuring the child nodes.

14.6   Risk analysis

Risk analysis basically stands for setting up relationships between the main risk attributes: assets, vulnerabilities, threats, and countermeasures. Obviously, the main objective of building this quad is to achieve a precise representation of events that can affect the security of the information system and to deduce the corresponding security countermeasures. In other terms, risk analysis aims at: (a) identifying the risk events threatening a particular system, (b) assessing their magnitudes, and (c) maintaining this magnitude below an upper threshold. Risk analysis is among the most complex components of the risk management framework. It generally involves many processes of the risk management cycle. In the following, we present the ingredients of a generic risk analysis activity and we describe several techniques that are commonly used in this context.

14.6.1   The risk analysis process

Generally, risk analysis is classified into three elementary activities which are: the threat analysis, business impact analysis, and cost benefit analysis. These activities can be described as follows:

- *Threat analysis* This activity consists of determining the potential attacks and modeling them accurately. It has been affirmed that threat analysis should involve the examination of three basic elements (Peltier, 2001): (a) the agent, which is the entity that carries out the attack, (b) the motive, which is the reason that causes the agent to act, and (c) the results, which are the outcome of the occurrence of the attack. Another element that would be helpful to add to the aforementioned elements is the attack mechanism. This key factor can serve to differentiate between threats that share the same agent, motives, and results. In addition, analyzing the steps to perform a harmful action is helpful for automating several tasks, such as the selection of appropriate defence solutions.
- *Business Impact Analysis* (BIA) This consists of evaluating the effect of the occurrence of an attack against a given asset (or a group of assets). The major outcome of CBA is the return on investment (RoI). This gives high-level managers the ability to define/approve a strategic view of the efficiency of the potential countermeasures. For instance, balancing the initial cost of a security decision against the value of the assets that it protects might be insufficient. An idea about the period necessary to get back the cost of the related safeguards is often required as some protection equipments depreciate more rapidly than others.

- Cost Benefit Analysis (CBA) This consists of comparing the cost and the benefit of the candidate security countermeasures to select the more appropriate ones. The computation of CBA may experience some difficulties, particularly when the comparison involves dynamic scenarios of discussion.

It should be remarked that asset identification (or analysis) has been considered as a challenging task, which means that it is a focal issue. In the RM terminology, asset identification and valuation corresponds to BIA. Its goal is to estimate the replacement cost of the security critical assets, as well as the values of their security attributes (i.e., confidentiality, integrity, and availability). Depending on whether the risk analysis method is quantitative or qualitative, these values can be expressed in monetary or non-monetary terms.

The purpose of a BIA is to identify the impacts of the unavailability of critical assets. This is indirectly needed to assess the efficiency of the potential countermeasures. Therefore, BIA is strongly linked to CBA as it will appear from the following discussion. Similarly, BIA has an important influence on the Business Continuity Plan (BCP) development, which consists of determining the impact of security incidents on the vital business processes in order to ensure their continuous availability. However, when being directed towards the establishment of a BCP, the BIA should differ a little from the normal case as some parameters, such as the Maximum Tolerated Downtime (MTD), should be considered during the asset valuation.

14.6.2 Classifying risk analysis techniques

A common principle to classify risk analysis approaches consists of reasoning about their continuity. In fact, when looking for locating the risk analysis activity in the RM cycle, it turns out that two possibilities are available: (a) prior to the implementation of the security solutions and (b) during the monitoring (or the incident response) process. The first alternative allows discussing risky events before their occurrence in order to make the infrastructure more convenient with the enterprise environment. Thus, this type of risk analysis approach is called preventive as its main interest is to reduce the risk extent so that the system state becomes as close as possible to the ideal state (where no attack is possible to carry out). The second possibility is to conduct *post facto* risk analysis, meaning that the selected countermeasures should control the effect of malicious acts after they substantially occur. This category is called reactive risk analysis. Although this topic has been abundantly addressed by the security community, it has rarely been put in the frame of RM approaches. It has been rather considered as a functionality of intrusion detection systems.

The major difference between preventive and reactive risk analyses is inherent in the uncertainty formalisms. More concretely, when reasoning about preventive countermeasures, the risk analyst has not a complete picture about the occurrence of the attacks. Henceforth, a probability of occurrence is assigned to every threat event. On the other hand, reactive incident response relies on the actual occurrence of the damaging event. What is unsure, in this situation, is to check whether the attack is effectively related to the detected signs of intrusion. Traditionally, alerts generated by IDSs form the major indicator of the occurrence of an attack. Two factors explain the introduction of uncertainty formalisms at this level: (a) security alerts do not necessarily imply the actual presence of an intrusion as they can correspond to false positives and (b) more than one attack can correspond to a single alert, meaning that a weighted link can be built between the potential attacks and the generated alerts. Consequently, it appears that the attack-alert concept is at the center of the reactive risk analysis. Besides, the security attributes differ from the preventive to the reactive case. For instance, the probability of occurrence, which is used to assess attacks prior to their occurrence, cannot be considered as incident response.

14.7   Risk assessment

For many reasons, the risk assessment component is considered as the core of the computational framework in a RM process. It is usually implemented according to two factors: impact and probability, which represent the damage that would result from the occurrence of a potential threat and the likelihood that the risk becomes real, respectively. As these concepts can be modeled in various ways, a plethora of risk assessment techniques have been developed. This section explains how these techniques can be classified and explores the most relevant models that can be used for computer security risks. Finally, to complete the decision support framework, methods allowing the selection of the best countermeasures are briefly reviewed.

14.7.1   Quantitative vs qualitative approaches

As has been underlined in the above sections, risk assessment can be performed either quantitatively or qualitatively. We briefly highlight the advantages and the drawbacks of each approach.

Advantages of quantitative risk assessment

There are four advantages that can be identified:

- A large spectrum of metrics can be used to represent and evaluate the various risk parameters. This allows a more detailed analysis of the risk events.
- The values of the risk parameters are expressed according to their nature (e.g., monetary units for the asset importance, and threat impact). As no translation to a different scale is necessary, these values are more accurate.
- Complicated decision-making techniques can be used as the quantitative assessment provides a credible set of parameters. This aspect is useful to fulfil the cost-effectiveness requirement.
- The results of the risk analysis process can be expressed in the management's language. This makes it more efficient to help the enterprise in reaching effectively its business objectives.

Limits of quantitative risk assessment

The following drawbacks can be identified:

- The computational methodologies are often complex and undocumented. Managers may face several difficulties to understand some advanced aspects.
- A significant hardware and software infrastructure may be needed to conduct quantitative risk assessment approaches. Automated tools are then required to support the manual effort in order to accelerate the process and make it more precise by avoiding computation errors. Nonetheless, such software is not always available. Developing an appropriate tool for an enterprise may be so expensive that the managers become unenthusiastic to allow a risk management activity.
- Data collection requires a substantial interest and treatment. In fact, the collected information must be accurate enough to make sure that the required analysis is efficient. On the other hand, the data collection mechanisms should be themselves secured to prevent the unauthorized access to sensitive data or the violation of privacy policies.

Advantages of qualitative risk assessment

Similar to quantitative risk assessment, qualitative risk assessment has several advantages:

- The process does not involve complex reasoning and it can be understood by high-level managers. This makes it easier to convince them about the importance of the risk assessment process.
- Qualitative risk assessment can often be performed manually. Computational steps are restricted to simple arithmetic operations on a reduced set of integer numbers.
- The data collection process is less complex than in the quantitative case. Questionnaire-based techniques are often used to this end.
- The use of several parameters that can not effectively be measured, but can be subjectively evaluated gives more freedom to the risk analyst so as to choose the assessment parameter basis.

Limits of qualitative risk assessment

Drawbacks for this method include:

- The lack of theoretical bases increases the uncertainty rates that affect the risk assessment results, which do not give a precise knowledge about the identified risk events.
- The efficiency of the qualitative reasoning relies on the expertise of the risk assessment team (or the population that responded to the questionnaire). This is an important weakness as security specialists are not always available at reasonable costs.
- The results of the qualitative risk assessment process can not be substantially tracked because their subjective nature makes them hard to evaluate.

14.7.2   Risk assessment for preventive risk analysis

Almost all computer security risk assessment techniques rely on a concept called "*Annualized loss exposure computation*" (ALE) which is derived from the traditional concepts called the "Annualized Rate of Occurrence" (ARO) and the "Single Loss Expectancy" (SLE). Formally, let V be the value (the exposure) assigned to the asset under study. E is the percentage of the asset that would be lost should the attack A be realized against it. Then, the Single Loss Expectancy (SLE) is equal to $A \times E$ and we can have:

$$ALE = ARO \times SLE = ARO \times A \times E.$$

In the following we discuss briefly some relevant aspects related to each of these variables:

- The ALE is expressed in annualized terms in order to be easily integrated within the business planning activity.
- The ARO stands for the frequency with which a given threat is expected to occur during a year. This parameter is global in the sense that it does not concern a specific asset.

14.7.3   Risk assessment for reactive risk analysis

A cost model for reactive countermeasures relies on the following prominent decision criteria (Fessi *et al.*, 2004):

Detection cost Prior to the generation of a security alarm by an analyzer, the corresponding sensor should have captured appropriately the required parameters (e.g., metric value, packet header fields) that reveal a related event. Then, the analyzer inspects the collected data to decide whether an intrusion has occurred. Both of these operations have a cost that must be taken into account when estimating the total cost of the incident. Hereinafter, we suppose that this cost is intrinsic to the elementary IDS (consisting of an analyzer and sensors, as described in Chapter 11) that does not depend on the nature of the generated alert.

Cost of reaction Each of the potential reactions, say r_k, $1 \leq k \leq n_r$, where n_r is the total number of security countermeasures, has a cost γ_{r_k} that depends heavily on the response itself. It can be expressed in terms of different attributes such as monetary units or processing resources.

Attack impact Carrying out a malicious act on an information system causes various kinds of undesirable effects. The impact functions t_{a_j}, which are related to effects a_i, can be determined by multiple attributes. Furthermore, another concept, called progression factor and denoted be λ_{a_i}, can be introduced to have a more suitable representation of the benefit of a given reaction (related to effect a_i). In fact, the system can react before or after the attack occurs. Each of these reactions has a different benefit. The former reaction aims at stopping the complete execution of the attack after receiving alarms that reveal it while the objective of the latter is to make the system recover at the right time. To this end, the impact is modeled as a function of the progression factor, as will be illustrated below.

On the other hand, to estimate the impact distribution of security threats, numerous models can be developed. Three examples are given below. They can be obviously enriched by other functions.

1. *Constant impact* The impact function has the following expression in this case:

$$I(t) = I_0, \quad \text{if } 0 \le t \le t_0.$$

2. *Linear impact* In this case, the Maximum Tolerated Downtime (MTD) is integrated in the impact function. Practically, the MTD corresponds to the maximum time interval in which the service of interest can be stopped. In the following expression of the impact, the MTD is denoted by t_0.

$$I(t) = (I_0/t_0) \times t, \quad \text{if } 0 \le t < t_0.$$

3. *Exponential impact* This case corresponds to attacks that have an impact that increases rapidly across time.

$$I(t) = I_0 \times e^{(t/t_0-1)}, \quad \text{if } 0 < t < t_0.$$

The above functions model the variation of the impact according to time. More sophisticated approaches can be conducted to determine more appropriate expressions.

14.8   Monitoring the system state

Up to this point, security policy violations have been shown to be more or less related to the vulnerabilities of the security critical component of an information system. In the previous section, we explained how to react if a security violation is detected. However, we did not address the detection process itself. As a matter of fact, deploying cost-effective security mechanisms to reduce the loss resulting from the occurrence of damaging actions produces a certain result at the business level. Nevertheless, experience has shown that such mechanisms are never sufficient as they can not be asserted to be fully efficient. To this purpose, the security state of the analyzed system should be continuously controlled. At least three reasons confirm this view:

- New vulnerabilities and attack mechanisms can be discovered.
- Violations can be detected at the right time.
- Effective countermeasures can be reviewed to state whether they are sufficient.

Security monitoring constitutes a tricky activity since it aims at controlling the security level in the analyzed system. Security monitoring is customarily performed through the use of IDSs, which use two detection techniques: pattern-based monitoring and behavior-based monitoring.

14.8.1   *Pattern-based monitoring*

This category of monitoring techniques, also referred to as misuse detection, attempts to characterize known patterns of security violations. This basically requires a sensor network, a signature database and an analyzer. Sensors are deployed at the *strategic* segments of the monitored network to collect the needed data. Obviously, an engineering activity should be conducted before setting these sensors to ensure that the target information can be gathered without introducing a considerable overhead at the communication flow level. Signatures consist mainly of descriptions of the known intrusions. An appropriate language is always needed for this representation. Finally, the analyzer matches the collected data with the existing signatures to decide whether an attack did occur.

Another important issue, in addition to the specification language that affects the efficiency of pattern-based monitoring, is identifying the parameters that should be controlled by the sensors. As it is not possible to specify the behavior of the system with full confidence, the security analyst should define exactly what should be monitored. Among the most relevant issues that must be taken into consideration when defining attack patterns, we list the following:

- *Access to the information system objects* The access policy to the system assets should be described at the machine language level. Then, the access events to critical resources should be logged and checked. For example, in UNIX-based operating systems, the execution of the system command *ls* from a user account into /root directory should be detected.
- *Operations sequencing* The order in which some operations are executed is sometimes a good sign of the legitimacy of an access. For example, the execution of some critical commands after being logged in as a normal user is prohibited. This shows how a signature can be based on a sequence of two, or more, events.
- *Protocol violation* Many signatures rely on detecting a communication flow that does not conform to the rules of a given protocol. For example, looking for IP packets where the source and the destination addresses are equal permits us to detect the "*Land attack.*" Similarly, checking the conformance of the fragmentation fields with the corresponding Requests For Comments (RFCs) allows thwarting fragmentation attacks such as Teardrop.

Techniques that belong to this class are characterized by two major shortcomings:

- First, they are not able to detect previously unknown attacks because it is often impossible to map them to any existing signature.
- Second, as different ways can exist to perform a specific attack, it might be difficult to build signatures that cover all the variants of such attacks.

14.8.2 *Behavior-based monitoring*

Anomaly-based intrusion detection is critical as it permits the identification of attacks that do not necessarily have known patterns. Detection results in this mechanism depend on the values of several measurable functions called metrics. Two techniques are used in anomaly detection: activity profiling and thresholding. An important problem that needs to be addressed within the second technique is the threshold setting mechanism. In fact, the rates of false alarms depend highly on the threshold value. Moreover, approaches using anomalies are not built based on a strong relationship between anomalies and attacks. This can lead to a high false alarm rate as is the case with the existing anomaly based intrusion detection systems. In addition, these approaches often utilize more resources than those based on misuse detection.

The first task in the anomaly representation is to define the metrics that have to be monitored. Three types of functions can be considered. They model the following activities:

- *User-level activities* These activities include attributes such as the most used commands, typing frequency, and login/logout period, which help develop the profiles of behavior patterns of users.
- *Host-level activities* These activities include attributes (such as file structure, consumed CPU, and consumed memory) that provide indication of resource usage.
- *Network-level activities* These activities use attributes (e.g., total packet count, packets with specific source or destination ports) to provide information that is gathered on network usage and engineering.

The above functions can represent efficiently the information system state. They should be measured and monitored continuously. Information system administrators have to determine what attributes to record to ensure efficient representations of their systems. Moreover, for each of these metrics, a set of decision rules have to be defined in order to choose whether a value taken by a given function corresponds to an abuse or a legitimate use of the system.

14.9 Conclusion

This chapter has reviewed the major concepts and schemes of Risk Management (RM) in e-based systems and computer networks. A basic definition of security RM includes mainly the ideas, models, abstractions, and schemes that are employed to control security risks. In general, a RM approach is meant to identify the risks threatening the e-system, select the appropriate security measures/solutions to limit this risk to an acceptable level, and go on monitoring the system to maintain its security level and detect deviations. The efficiency of any RM scheme depends mainly on risk analysis, which aims at controlling the state of a system in order to select the appropriate actions that should be taken to enhance its defence means when an attack occurs. This has the potential to increase the efficacy of security solutions as they are selected according to a clear strategy and not on the basis of human intuition.

The main topics covered in the chapter include requirements of RM as well as main techniques of RM. The latter comprise the OCTAVE and CORAS schemes. We also elaborate on the limits of existing schemes, especially the architectural and technical limits. A dedicated section was given to management of risk libraries where we addressed several issues and aspects including the vulnerability of detection and strategies used. The chapter also reviewed the chief concepts of risk analysis and the techniques used: quantitative and qualitative techniques. We also compared these two schemes and showed their advantages and drawbacks. Finally, we discussed the basic concepts related to monitoring the system state.

References

Alberts, C. J. and A. J. Dorofee (2002). *Managing Information Security Risks: the OCTAVE Approach*. Addison-Wesley.

Barber, B. and J. Davey (1992). The use of the CCTA risk analysis and management methodology CRAMM. In *MEDINFO '92*, North Holland, 1589–93.

Campbell, R. P. (1979). A modular approach to computer security risk management. Proceedings of the AFIPS Conference, 1979.

McDermott, J. (2000). Attack net penetration testing. The 2000 New Security Paradigms Workshop, Ballycotton, County Cork, Ireland, September 2000.

Fessi, B. A., M. Hamdi, S. Benabdallah, and N. Boudriga (2004). A decisional framework system for computer network intrusion detection. Conference on Multi-Objective Programming and Goal Programming, Hammamet, Tunisia, 2004.

Hamdi, M., N. Boudriga, J. Krichene, and M. Tounsi (2003). NetRAM: a novel method for network security risk management. Nordic Workshop on Secure IT Systems (NordSec), Gjovik, Norway.

Hamdi, M., J. Krichene, and N. Boudriga (2004). Collective Computer Incident Response using Cognitive Maps, IEEE Conference on Systems, Man, and Cybernetics (IEEE SMC 2004), The Hague, Netherlands, October 10–13, 2004.

Hamdi, M. and N. Boudriga (2005). Computer and network security risk management: theory, challenges and countermeasures. *International Journal of Communication Systems*, to appear, 2006.

Krichene, J., N. Boudriga, and S. Guemara (2003). SECOMO: An estimation cost model for risk management, Proceedings 7th Intern. Conf. Telecom. (ConTel'03), pp. 593–599, Zaghreb, Croatia, June 11–13, 2003.

Moore, A. P., R. J. Ellison, and R. C. Linger (2001). *Attack Modeling for Information Security and Survivability*, CMU/SEI Technical Report, CMU/SEI-2001-TN-01.

Ozier, W. (2000). Risk analysis and assessment. In *Handbook of Information Security*, 4th edn. M. Krause, H. F. Tipton (authors), Auerbach Press, Chapter 15, pp. 247–285.

Peltier, T. R. (2001). *Information Security Risk Analysis*. Auerbach Editions (available at: http://www.sei.cmu.edu/about/website/indexes/siteIndex/siteIndexTR.html).

Schneier, B. (2000). *Secrets and Lies: Digital Security in a Networked World*. Wiley.

Schumacher, M., C. Hall, M. Hurler, and A. Buchmann (2000). Data Mining in Vulnerability Databases, March 2000.

Stolen, K., F. den Braber, T. Dimitrakos, R. Fredriksen, B. A. Gran, S.-H. Houmb, Y. C. Stamatiou, and J. O. Aagedal (2003). Model-based risk assessment in a component-based software engineering process: the CORAS approach to identify security risks. In *Business Component-Based Software Engineering*, F. Barbier (ed.), Kluwer, pp. 189–207.

Index